# Natural Resource Management and Policy

Volume 51

**Series editors**

David Zilberman, California, USA
Renan Goetz, Girona, Spain
Alberto Garrido, Madrid, Spain

There is a growing awareness to the role that natural resources, such as water, land, forests and environmental amenities, play in our lives. There are many competing uses for natural resources, and society is challenged to manage them for improving social well-being. Furthermore, there may be dire consequences to natural resources mismanagement. Renewable resources, such as water, land and the environment are linked, and decisions made with regard to one may affect the others. Policy and management of natural resources now require interdisciplinary approaches including natural and social sciences to correctly address our society preferences.
This series provides a collection of works containing most recent findings on economics, management and policy of renewable biological resources, such as water, land, crop protection, sustainable agriculture, technology, and environmental health. It incorporates modern thinking and techniques of economics and management. Books in this series will incorporate knowledge and models of natural phenomena with economics and managerial decision frameworks to assess alternative options for managing natural resources and environment.

More information about this series at http://www.springer.com/series/6360

Stuart J. Smyth · William A. Kerr
Peter W.B. Phillips

# Biotechnology Regulation and Trade

Springer

Stuart J. Smyth
University of Saskatchewan
Saskatoon, SK
Canada

Peter W.B. Phillips
University of Saskatchewan
Saskatoon, SK
Canada

William A. Kerr
University of Saskatchewan
Saskatoon, SK
Canada

Natural Resource Management and Policy
ISBN 978-3-319-53293-6 ISBN 978-3-319-53295-0 (eBook)
DOI 10.1007/978-3-319-53295-0

Library of Congress Control Number: 2017930949



Printed on acid-free paper

This Springer imprint is published by Springer Nature
The registered company is Springer International Publishing AG
The registered company address is: Gewerbestrasse 11, 6330 Cham, Switzerland

# Preface

The three of us have worked as an informal team for the past decade, individually and collectively teaching, doing research and presenting our findings on trade in GM crops to a mix of academic, industrial and government audiences. Our mix of skills and interests—trade theory, political economy and public policy—has enabled us to challenge some of the underlying assumptions and to contribute in our own small way to the framing of the debate about regulatory and trade regimes for GM crops and to the design of options for more effective systems.

Like at a family wedding, we have dressed up this volume with 'something old, something new, something borrowed and something blue'. The process of knitting together our disparate works from the past has given us new insight into the problems and opportunities. We hope this volume provides insights to you, the reader.

We would like to thank all the anonymous reviewers of the work included here, the global network of scholars who present and debate at the International Consortium of Agricultural Biotechnology Research (ICABR), and our friends, colleagues and family who have put up with our rants, fugues and odd musing over the years on this topic. A work of this type naturally relies on the help of many. We would especially like to thank Kaitlyn Wolfert, Savannah Gleim and Lindsay Pollard for their assistance in consolidating and referencing the works in this volume.

Saskatoon, SK, Canada

Stuart J. Smyth
William A. Kerr
Peter W.B. Phillips

# Contents

# Abbreviations

| | |
|---|---|
| AAFC | Agriculture and Agri-Food Canada |
| AGE | Applied General Equilibrium |
| AoA | Agreement on Agriculture |
| AP | Adventitious Presence |
| BE | Behavioural Economics |
| BSE | Bovine Spongiform Encephalopathy |
| Bt | *Bacillus thuringiensis* |
| CBD | Convention on Biodiversity |
| CCGD | Canadian Council of Grocery Distributors |
| CDs | Consensus Documents |
| CFIA | Canadian Food Inspection Agency |
| CFS | Center for Food Safety |
| CGC | Canadian Grain Commission |
| CGIAR | Consultative Group for International Agricultural Research |
| CGSB | Canadian General Standards Board |
| Codex | *Codex Alimentarius* Commission |
| CPB | Cartagena Protocol on Biosafety |
| DEFRA | Department for Environment, Food and Rural Affairs |
| DFAIT | Foreign Affairs and International Trade Canada |
| DG SANCO | Directorate General for Health and Consumer Affairs |
| DSB | Dispute settlement body |
| EFSA | European Food Safety Authority |
| EIS | Environmental impacts statement |
| ELISAs | Enzyme-linked immunosorbent assays |
| ENGOs | Environmental non-governmental organizations |
| EU | European Union |
| FAO | Food and Agriculture Organization |
| GATS | General Agreement on Trade in Services |
| GATT | General Agreement on Tariffs and Trade |
| GDP | Gross Domestic Product |

| | |
|---|---|
| GM | Genetically modified |
| GMCC | GM Coexistence Conference |
| GMHT | GM Herbicide Tolerance |
| GMO | Genetically modified organism |
| GRAS | Generally regarded as safe |
| GSP-plus | Generalized System of Preference |
| GTAP | Global Trade Analysis Project |
| HACCP | Hazard analysis critical control points |
| HEAR | High erucic acid rapeseed |
| HT | Herbicide tolerant |
| IAD | Institutional analysis and development |
| ICJ | International Court of Justice |
| IMF | International Monetary Fund |
| IP | Intellectual property |
| IPPC | International Plant Protection Convention |
| IPPM | Identity-preserved production and marketing |
| IPRs | Intellectual property rights |
| IRLG | Interagency Regulatory Liaison Group |
| IRR | Internal rate of return |
| ISAAA | International Service for the Acquisition of Agri-biotech Applications |
| ISO | International Organizations for Standardization/International Standards Organization |
| ISPMs | International Standards for Phytosanitary Measures |
| ITO | International Trade Organization |
| ITPGR | International Treaty on Plant Genetic Resources for Food and Agriculture |
| LLP | Low-level presence |
| LMO | Living modified organism |
| MEAs | Multilateral environmental agreements |
| MFN | Most favoured nation |
| MOU | Memorandum of understanding |
| MP | Member of Parliament |
| MRA | Members of regional associations |
| MVR | Mosaic virus resistance |
| NAFTA | North American Free Trade Agreement |
| NAS | National Academy of Sciences |
| NGOs | Non-governmental organizations |
| NIH | National Institute of Health |
| NPV | Net present value |
| OECD | Organization for Economic Cooperation and Development |
| OGTR | Office of the Gene Technology Regulator |
| OIE | International Office of Epizootics/Office International des Epizootics/World Organization for Animal Health |
| OTA | Organic Trade Association |

| | |
|---|---|
| PNTs | Plants with novel traits |
| PPMs | Production and processing methods |
| R&D | Research and development |
| RAF | Risk analysis framework |
| RASFF | Rapid Alert System for Food and Feed |
| SECs | Socio-economic considerations |
| SOD | Saskatchewan Organic Directorate |
| SPS | Sanitary and Phytosanitary |
| TBT | Technical Barriers to Trade |
| TRIPS | Trade-Related Aspects of Intellectual Property Rights |
| TRQs | Tariff rate quotas |
| UNCTAD | United Nations Conference on Trade and Development |
| UNEP | United Nations Environmental Program |
| UPOV | Union for the Protection of New Varieties of Plants |
| USAID | United States Agency for International Development |
| USDA | United States Department of Agriculture |
| WIPO | World Intellectual Property Organization |
| WTO | World Trade Organization |

# List of Figures

# List of Tables

# Introduction

Food policy will remain a global priority in the foreseeable future. While science and technology will help generate new more productive crops and animals, the real challenge will be to ensure that the governance system works efficiently and effectively. Improving global food security in the coming decades is dependent upon ensuring that the food-related products and processes that are created can be distributed to those most in need. This to a significant degree is a two-pronged problem. First, regulatory frameworks are becoming more complex and difficult to navigate for technology developers, increasing the cost and time it takes to receive commercialization approvals. Hence, the transition of technologies and products from developed to developing countries is becoming more difficult and expensive. Second, international commodity trade is in danger of returning to the scenario of 100 years ago where trade barriers were used in a tit-for-tat relationship whereby one country would impose a barrier to a commodity, which would result in the second country responding accordingly, thus ratcheting up trade barriers in commodity trade. If global food security is truly going to be enhanced, regulatory and trade solution are desperately required. One of the major areas where these challenges are manifest is the in the international regulation and trade issues surrounding modern agricultural biotechnology.

It is difficult to get a handle on the scale of the problem, as trade is pursued both out of absolute need and relative benefit. Determining food independence is complex. A country like Canada, for example, could grow many fresh fruits but buys most of the product from abroad, partly because the product is cheaper and partly because of the seasonality of fruit production in Canada due to the climate. This does not mean Canada is dependent on imports. Other countries such as Egypt are, on the other hand, absolutely dependent on wheat produced elsewhere to feed their growing population—they simply do not have sufficient land and other resources to meet their domestic needs. The degree of interdependence in effect comes down to whether a country can feed its population with its own production using its own resources, not whether it actually does so.

The United Nations Food and Agriculture Organization estimates that on that measure very few countries are truly self-sufficient. The only countries to

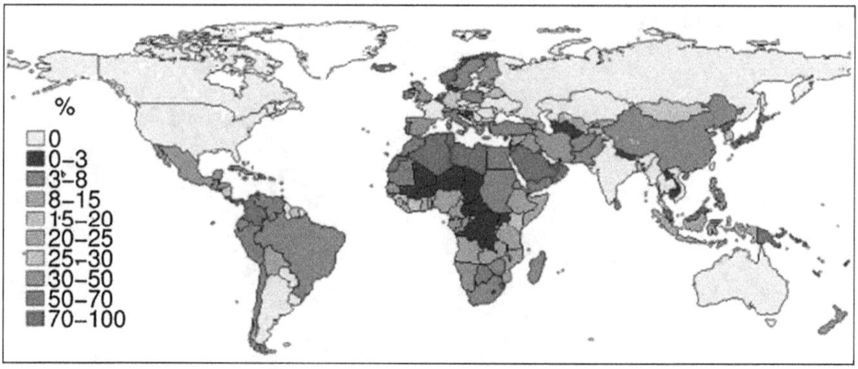


**Fig. 1** Proportion of population dependent on imported food for basic diet. *Source* Fader et al. (2013)

unambiguously have the capacity to produce more food than they consumer are Argentina, Australia, Burma, Canada, France, India, Russia, Thailand and the USA, along with a few other small countries. Approximately 16% of the global population currently relies upon imports of food produced elsewhere. Fader et al. (2013) undertook a trend analysis, concluding that the exhaustion of farmland, climate change and uneven population growth will raise the proportion of the world population that depends on international exchange to secure adequate food to 50% by 2050 (Fig. 1).

This book is explicitly focused on the challenge of facilitating that expansion in trade. It highlights the regulatory and trade challenges that could limit the achievement of global food security and offers a range of strategic responses to overcome these challenges. Scientific advances in the breeding of new crops are proceeding at a rapid pace; however, the benefits of these technologies are only partly reaching food insecure markets due to existing regulatory and trade barriers. The book is organized into three main parts that address these challenges.

Part I of the book delves into the present state of the global market as it pertains to modern agricultural biotechnology. This part opens with a global synopsis of the state of GM crop adoption and the quantified benefits flowing from this adoption. International trade is adversely impacted by two important challenges, coexistence of GM and non-GM products and the low-level presence of unapproved GM products into markets that do not accept GM products. Discussion of these challenges is then followed by the impact on food security by the changing dynamics of risk assessment, particularly within the European Union, where risk is being politicized.

Part II of the book examined the current institutional architecture, discussing how there has been a move away from science-based regulation in the past decade to one that is more inclusive of socio-economic considerations, such as under the Cartagena Protocol on Biosafety. This group of chapters provides an assessment of how products of biotechnology have challenged the ability of international institutions to effectively manage international regulation of, and trade in, these products. The part concludes with a discussion of international conventions and the precedence of these conventions when it comes to resolving disputes between nations regarding GM crops and products.

The final part of the book identifies options for breaking the current gridlock of regulation and trade. Options for both regulatory and trade improvement are presented and discussed.

## Reference

Fader, M., D. Gerten, M. Krause, W. Lucht, and W. Cramer. 2013. Spatial decoupling of agricultural production and consumption: Quantifying dependencies of countries on food imports due to domestic land and water constraints. *Environmental Research Letters* 8(1):1–15.

# Part I
# The Dynamics of Transformative Innovation

# Chapter 1
# GM Crop Development: Solution or Another Problem?

**Abstract** 2015 marked 20 years of commercial production using genetically modified (GM) crops and while numerous non-governmental organizations continue to trumpet that 'there are no benefits from biotechnology or GM crops', the evidence suggests otherwise. This chapter summarizes the economic, environmental and health benefits from GM crop adoptions, concluding with estimates of the distribution of the benefits of GM crops between farmers, consumers and innovators.

## Twenty Years of Biotechnology Benefits

The commercialization of genetically modified (GM) crops in the mid-1990s was the culmination of a long research and development process that heralded an innovative technology that has now delivered 20 years of benefits for adopters and consumers. As discussed and expanded upon below, increased economic, environmental and health benefits are numerous. However, in spite of research dating back 15 years reporting the benefits of GM crops, critics of GM crops and biotechnology generally continue to ignore, dismiss and outright condemn the multitude of benefits resulting from the utilization of GM crops. Critics suggest that the biotechnology industry has co-opted academic researchers and is paying academics to mislead the public in the quantification of the benefits of crops produced using biotechnology (Kloor 2015). These opponents suggest that the distribution of benefits is inequitable—benefit distribution is, however, never equal for new technologies, causes farmers to commit suicide and is polluting the land (Adams 2014). Many of the environmental non-governmental organizations (eNGOs) that actively campaign against GM crops ignore the mounting volume of evidence that quantifies the economic, environmental and health benefits of GM crops. While interesting in and of itself (Clark et al. 2014), this backlash against the use of biotechnology feeds into the domestic regulatory process and spills out, influencing both international trade and global agri-food investment.

 
S.J. Smyth et al., *Biotechnology Regulation and Trade*, Natural Resource Management and Policy 51, DOI 10.1007/978-3-319-53295-0_1

To unpack the challenge of regulating this technology, it is worthwhile examining the case for wider diffusion of GM crops. If the technology had met the Casandra-like expectations of many, the regulatory issue would be moot. The evidence, on the contrary, suggests the technology offers significant and widespread benefits. The gap between the reality and perceptions was captured in a 2013 report released by the United Nations Conference on Trade and Development (UNCTAD) entitled, *Trade and Environmental Review* 2013*: Wake Up Before It Is Too Late*. While containing contributions from more than 60 experts, oddly no single expert in biotechnology or GM crops was listed in the table of contents. Many of the contributors listed have been longstanding critics of biotechnology and GM crops. The essential message of this lengthy report is that for food security to exist over the remainder of this century, agriculture in developing countries needs to return to largely organic, small-scale farming practices.

Africa has become the focal point for the increased commercialization of GM crops to enhance food security, but also the rallying point for critics determined to keep this technology out of Africa. To some extent, Africa has become the final battle ground for those that continue to champion the opposition to a tremendously successful innovative technology in agriculture (Kaldor et al. 2004; Phillips 2007; Smyth et al. 2016). An important message is that biotechnology and GM crops offer a proven and safe option to improve global food security. If food insecure nations were to approve these crops for production and consumption much of the battle to achieve food security would be won. GM crops could provide an immediate food security benefit, if only they were given the opportunity to be adopted globally by farmers.

## Global Distribution of Benefits

From 2000 the debate and research into biotechnology shifted heavily from the potential and hypothetical benefits to the actual impacts, especially in resource-constrained and food-insecure parts of the world. This work contributes to a body of evidence about the ubiquitous and transformative capacity for the technology.

As context, it is useful to remember that although the first commercial GM crop was planted in 1994 (tomatoes), 1996 was the first year in which a significant area of crops containing GM traits were planted (1.66 million ha). Since then there has been a dramatic increase in plantings; in 2015, 17 million farmers in 28 countries planted nearly 180 million ha of GM crops (James 2015). GM traits have largely been incorporated in four main crops—canola, corn, cotton and soybeans—although small areas of GM crops in sugar beet (adopted in the US and Canada since 2008), papaya (in the US since 1999 and China since 2008) and squash (in the US since 2004) have also been planted. Six countries accounted for 92% of total production (in declining order of area): US; Brazil; Argentina; India; Canada; and China. More than half the area planted is in developing nations. Two traits dominate: herbicide tolerant crops account for 65% of the total GM area; while insect resistant crops account for 35% of global plantings. Increasingly these traits are being stacked—including more than one GM trait in a variety—to provide more resilient seeds. GM seeds account for 70% of the global soybean acreage, 52% for cotton, 26% for corn and 20% for canola. In those countries adopting GM varieties, the GM seed market share has risen above 80%. GM crops have also been pro-trade, in that adoption and production is concentrated in leading export nations. Brookes and Barfoot (2014) estimates that biotechnology-based products account for 72% of global trade in cotton and 95% in the case of soybeans. These statistics suggest that there are highly differentiated actors, issues and impacts of the development and commercialization of biotechnology in plant agriculture.

With this as context, the farm-level evidence suggests that benefits have been widely distributed among adopting producers, innovators, consumers, and in some cases, for farmers not adopting the technology.

## *North America*

The US, as the developer of most of the technologies and crop varieties and the single largest adopter, is at the heart of the industry. Combined with Canada and Mexico, which have played small but significant roles in the development and diffusion of GM crops, the North American market cultivated 46% of the global GM acreage in 2016.

The adoption rate of GM soybeans in the US is over 90%, resulting in a yield increase of five bushels per acre compared to non-GM soybean production (Fernandez-Cornejo and Wechsler 2014). The use of conservation tillage has risen substantially with 63% of soybean production employing conservation tillage methods, compared with 30% in 1996. Alston et al. (2014) estimate that the adoption of GM soybeans in the US generated an economic benefit of US$22 billion between 1996 and 2010, with consumers benefiting by over $3 billion.

It is estimated that of the maize production in the US, 90% is done with GM varieties (James 2013). Brookes and Barfoot (2015) undertook an assessment of the economic impacts from the adoption of GM maize in the US from its commercialization in 1996 to 2013. The author's estimate that the net farm economic gain was US$999 million and that the cumulative gain has been US$4.9 billion.

A series of papers (Gusta et al. 2011; Smyth et al. 2011a, b) based on a survey of nearly 600 canola farmers in Western Canada illustrates the nature of the economic and environmental benefits of the technology. Gusta et al. (2011) estimated that GMHT canola generated between $1.063 and $1.192 billion net direct and indirect benefits for producers over 2005–7 period, attributed to both lower input costs and better weed control. In addition to economic benefits the survey identified significant environmental benefits, such as removing summerfallow from crop rotations (Smyth et al. 2011a). Much of the tillage associated with GMHT canola production has been eliminated as 64% of producers were using conservation tillage as their preferred form of weed control, compared to 11% in 1999. 83% of farmers experienced increase levels of moisture conservation and 86% reported reduced soil erosion following the adoption of GMHT canola. Significant changes regarding the use and application of herbicides have occurred in weed management practices used in canola production in Western Canada (Smyth et al. 2011b). When comparing canola production in 1995 and 2006, the toxicity of agro-herbicides applied to canola decreased by 53%, there has been a decrease in producer chemical exposure of 55% and a decrease in chemical active ingredient application of 1.3 million kg. The cumulative environmental impact per hectare (EI/ha) of the top five herbicides applied in 1995 was 46,715, while the figure for the top five herbicides applied in 2006 was 29,458. If GMHT canola had not been developed and Canadian canola farmers continued to use previous production technologies, the amount of active ingredient applied to control weeds in 2007 would have been 38% above that which was actually applied.

One of the few commercialized GM direct to consumer food crops is the production of papaya in Hawaii. Over 90% of papaya production is from GM varieties that are viral resistant (Gonsalves and Gonsalves 2014). In the early to mid-1990s the ringspot virus had destroyed virtually all of the papaya production in Hawaii, which had been worth US$17 million per year. The commercialization of viral resistant papaya has allowed the papaya industry to recover and production is once again returning to the low end of the historic production range of 35–55 million pounds.

## *South America*

Brazil, Argentina, Paraguay, Uruguay, Bolivia, Columbia, Honduras and Chile collectively cultivated more than 75 million ha in 2015, equal to about 42% of global GM acreage. Brazil, Argentina, Paraguay and Uruguay are all among the top ten adopters of GM technology (James 2015).

Brazil produced GM varieties of cotton, maize and soybeans on 44 million ha in 2015, which represents a cumulative adoption rate of 87% (James 2015). With two growing seasons, Brazilian farmers are able to alternately produce GM corn and GM soybeans. James identifies that the most recent estimates for the economic benefits of all three GM crops in Brazil range from US$1.7 to US$6.7 billion for the 2015 production year and a cumulative range of US$8.4–US$18.8 billion over the 16 years of GM crop production.

Argentina is next in the order, as a major adopter with GM soybean production beginning in 1996 and Bt cotton in 1998. Nearly all of the GM soybeans produced in Argentina have been modified to be herbicide tolerant and in 2009, adoption was estimated at 99% with the economic impact of adoption in 2009 estimated at US $302 million, while the cumulative impact was US$3.87 billion (Brookes and Barfoot 2010). Qaim and Traxler (2005) studied the 2001 farm-level and aggregate effects of GM soybean production, finding that producers captured 16% of the benefits, consumers captured 35%, with the technology developers capturing the remainder.

Other Latin American countries are moving closer to the level of GM crop utilization of the leaders in the region. James (2013) reports that Paraguay produces cotton, maize and soybeans, with 85% of this production incorporating GM varieties. The economic benefit of this adoption was estimated to be US$96 million in 2012, with a cumulative benefit of US$830 million over the period 2004–2012. In Uruguay, all soybean production is done using GM varieties, while 70% of the maize production is GM. The 2012 benefits are estimated to be US$20 million (James 2013). Bolivia also produces GM maize and soybean with the cumulative benefit of this estimated at US$430 million between 2008 and 2012, with the 2012 value of US$105 million (James 2013).

## *India*

As the fourth largest adopter of technology, and one of the most populous countries in the world, India is an important case. The 2002 commercialization of Bt cotton in India, with its millions of small landholders, provides an excellent opportunity to assess the impacts of GM crop adoption on developing world farmers. Qaim (2003) assessed the potential impacts of Bt cotton adoption in India (based on 2001 field trials), noting that prior to the commercialization of Bt cotton, farmers were losing an estimated 50–60% of potential output due to insect pests. The analysis found that yields increased by an average of 58% and pesticide budgets dropped by 50%. Subramanian and Qaim (2010) extended this research and reported that after four years of production, Bt cotton yields were 37% higher and pesticide use dropped by 41%. Additional socio-economic benefits were also measured, with the most noticeable impact being increased use of paid female labor. Subramanian and Qaim estimate that Bt cotton-adopting households increased their incomes by 82% and households that were defined by FAO as vulnerable (i.e. income of <$2/day) increased their incomes by 134%.

Following several years of Bt cotton adoption in India, reports from biotechnology critics began to suggest that farmer suicides were increasing in areas of high Bt cotton adoption. Gruère and Sengupta (2011) assess Bt cotton and farmer suicides. Innovation adoption is not homogenous across states and their meta-analysis revealed the commercialization of Bt cotton was plagued by counterfeit seeds, variation in seed quality, lack of agriculture extension and weak knowledge dissemination. In comparing farmer suicides with Bt cotton adoption, Gruère and Sengupta show that when the pre-Bt cotton suicide rate is projected, the actual suicide rate is considerably lower. In two strong GM cotton adopting states suicide rates did not decline to the same level as in other cotton producing states. Analysis of the underlying factors revealed that abnormally low precipitation in some years reduced yields while general household indebtedness was a contributing factor. The authors conclude that Indian cotton farmer suicides are not correlated to Bt cotton adoption.

Further research by Qaim (2014) shows that the application of cotton pesticides has fallen by 0.95–1.3 kg/acre of active ingredient. This results in a cost savings of 879–1284 rupees per acre. In India pesticides are most often applied to cotton by farmers walking through the field using a backpack sprayer, in most cases with little to no protective clothing. Millions of cases of acute pesticide poisonings are reported every year. The adoption of Bt cotton has reduced the number of cases of pesticide poisoning, estimated to range between 2.4 and 9 million annually. This saves the Indian Ministry of Health an estimated US$14–51 million on an annual basis (Kouser and Qaim 2011). Not only has the environment and farmer health benefited, but so too has the yield and profitability of Bt cotton adopters. While Bt cotton adopters do pay a higher price for the seed, this is more than offset by the 24% increase in yield when compared to non-Bt cotton. Profits rose even more dramatically, by an estimated 1877 rupees per acre, or 50%. In 2015, it was estimated that over 28 million acres were planted to Bt cotton, representing 95% of cotton production in India, generating a net gain for farmers of US$1.6 billion. Bt cotton has allowed India to become the largest cotton producing country in the world.

## *China*

China has invested heavily in biotechnology and in 2015 was the sixth largest adopter of GM crops, mostly cotton. Based on a 1999 survey of cotton farmers in northern China, Pray et al. (2001) provided the first insights into the impacts of Bt cotton. Their research measured economic, income distribution, environmental and health effects. While not easy to quantify due to farmer-to-farmer sales and seed saving from year-to-year, the authors estimate that early adoption ranged from 8 to 27%. The authors highlight that while non-Bt farmers saved money on seed costs, they spent considerably more on pesticide purchases and labor. When other production costs were factored in, non-Bt cotton producers experienced negative net income. The authors estimate that approximately 85% of the benefits of adoption accrued to the farmers, with industry receiving 6% and the balance going to government seed agencies. The most substantial impact from Bt cotton adoption may well be the environmental and health benefits resulting from reduced pesticide applications. The adoption of Bt cotton allowed farmers to spray less frequently, in some instances dropping from 30 applications per season to three, but more commonly from 12 to three to four.

Pray et al. (2002) extended their original 1999 survey data with new data from 2000 and 2001. This period was one of rapid change as the Chinese government removed the cotton purchase monopoly of the state-run Cotton and Jute Corporation, allowing the price of cotton to fluctuate with the market. A *New Seed Law* was passed, allowing private seed companies to operate and charge market prices for their seed. The increased Bt cotton adoption reduced market prices for cotton, resulting in some benefits being passed on to consumers through lower prices for cotton and yarn. Even with the lower cotton prices, adopters increased their net income by US$500/ha.

Huang et al. (2010) updated the Chinese Bt cotton story following a decade of commercial production. The authors documented a drop in bollworm infestations, not only in Bt fields, but in all cotton fields in other parts of China. In some non-Bt cotton fields the amount of insecticide in kilograms per hectare dropped from in excess of 40 to less than 10. Across the entire sample region insecticide applications dropped from 14 to 4 kg/ha. These spillover benefits to non-adopters are consistent with the findings of Hutchison et al. (2010) on Bt corn adoption in the US where 75% of the benefits of Bt corn accrue to non-adopters due to lower pest populations.

## *Philippines*

The Philippines has started from a small base but the use of the technology is expanding rapidly, with 0.7 million ha of GM maize under cultivation in 2015. Yorobe and Smale (2012) provide a detailed analysis of the poverty reduction effects from the adoption of GM maize in the Philippines. The mean net income in 2007 for non-GM maize farmers was 16,420 pesos (about US$400), whereas the mean net income for GM maize adopters was 24,700 pesos (US$600), an increase of 50%. While there is a substantial on-farm effect from the adoption of GM maize, there is also a corresponding increase in off-farm income. The authors found that the amount of labor saved from GM maize adoption is so great that it is has allowed many Philippine GM crop adopting farmers to get off-farm jobs, supplementing household incomes.

## *Africa*

Many argue that Africa, which has only three adopting countries and less than 2% of total GM area, has the most to gain from the technology. Where countries allow the technology, the benefits flow. Bennett et al. (2006) identify Bt cotton as the first commercial GM crop produced in Sub-Saharan Africa. It was approved in 1998. One strong adopting region was the Makhathini Flats area of the KwaZulu Natal province in South Africa, where the typical farm size is between 1 and 3 ha. Cotton is a valuable cash crop and typically accounts for most of the land in production. By 2002, an estimated 92% of smallholder cotton producers had adopted Bt cotton. Given male labor migration patterns at this time, it was estimated that 60% of the smallholder producers were female. The authors obtained production records for three growing seasons, showing that Bt cotton yielded substantially higher (89–129%) than conventional cotton, especially under adverse climatic conditions. The authors note that the average daily wage in the area was SAR10-15 making the extra Bt cotton revenue equivalent to two to four months of wages.

Another interesting finding from the South African smallholder GM crop research is the substantial labor savings reported by GM herbicide tolerant (HT) maize adopting farmers. Numerous studies in Africa have shown the devastating effects of uncontrolled and less than optimally controlled weeds can have on crop yields (Shetto and Kwiligwa 1998; Joubert 2000; Chikoye et al. 2004; Atera 2012). Poor weed control is often cited as the single biggest contributor to low maize yields for African smallholder farmers. The number of necessary weedings following planting is the principal limiting factor to African farm sizes (Kent et al. 2001). More efficient use of a post-emergent broad-spectrum herbicide would have a positive impact on weed control effectiveness and therefore on yield, area under crops and thus food security and nutrition. GMHT crop adoption has allowed such usage to occur and in some developing countries has resulted in a less demanding set of managerial skills and increased off-farm income (Fernandez-Cornejo et al. 2005; Gardner et al. 2009; Hurley et al. 2009).

In sub-Saharan Arica's (SSA) labor intensive production systems, a decreased need for manual weed control will result in a reduced need for family and/or hired manual labor. This benefit is realized, in particular, by female farmers and female household members because manual weeding is predominantly a female activity. Based on three seasons of research involving small landholder adopters of GM crops in South Africa, Gouse (2013) found that female GMHT maize adopters spent 10–12 fewer days per season in the field doing arduous weed pulling and hoeing than their conventional maize planting counterparts.

The only other nation in Africa to have sustained GM crop production is Burkina Faso, where Bt cotton was approved in 2009. In comparing yields over the first three production years, Vitale et al. (2014) found that Bt cotton yields were 22% higher than conventional cotton. Input costs were found to be virtually identical, with Bt cotton seed costing $US361/ha while conventional seed cost $US355. The higher seed costs for Bt cotton were offset by the higher insecticide costs for conventional cotton producers. Over the three years of the study, the average economic return per hectare was more than double that of conventional cotton, with returns of $US151/ha for Bt cotton and $US70 for conventional cotton. Health benefits were also found to be substantial with an estimated 30,000 fewer reported cases of pesticide poisoning.

Realizing these types of gains in other African countries, where farmers are differentially less well-off and more restrained by agronomic conditions would go a long way to meeting the United Nation's Millennial Goals. To do so will require governance systems both within Africa and in the international trading system to respond.

## Summary of Economic Benefits

The distribution of GM crop benefits has been subject to rigorous review (Table 1.1). Numerous studies have been undertaken with the results reported. The studies use a variety of models, so the numbers reported below should not be taken

**Table 1.1** Distribution of benefits from GM crops[1]

| Country | Year | Total benefit ($US millions) | Share of total benefits (percentage) | | | |
|---|---|---|---|---|---|---|
| | | | Farmers | Innovators | Consumers | Net ROW[2] |
| *Canola* | | | | | | |
| Canada[a] | 2005–07 | 1192 | – | – | – | – |
| Canada[b] | 2014 | – | 43 | 48 | 5 | 4 |
| *Soybeans* | | | | | | |
| USA[c] | 1997 | 437 | 29 | 18 | 17 | 28 |
| USA and Argentina[d] | 1997 | 206 | 16 | 49 | 35 | – |
| USA[e] | 1997 | 310 | 20 | 68 | 5 | 6 |
| USA[f] | 1999 | 804 | 20 | 45 | 10 | 26 |
| USA and Argentina[d] | 2001 | 1230 | 13 | 34 | 53 | – |
| Argentina[g] | 1996–2010 | 65,153 | 72 | 7 | 21 | – |
| Global[h] | 1996–2010 | 46,000 | 55 | 14 | 31 | – |
| *Maize* | | | | | | |
| USA[i] | 2001 | 334 | 50 | 31 | NA | NA |
| Spain[j] | 2003 | 2 | 60 | 40 | – | – |
| Argentina[g] | 1996–2010 | 5375 | 68 | 20 | 11 | – |
| *Cotton* | | | | | | |
| USA[e] | 1997 | 164 | 37 | 45 | 18 | – |
| China[k] | 1999 | 140 | 98 | 2 | – | – |
| India[l] | 2001 | 315 | 67 | 33 | – | – |
| Argentina[g] | 1996–2010 | 1834 | 96 | 4 | – | – |

*Sources* [a]Gusta et al. (2011)
[b]Smyth and Phillips, Forthcoming
[c]Falck-Zepeda et al. (2000)
[d]Qaim and Traxler (2005)
[e]Price et al. (2003)
[f]Moschini et al. (2000)
[g]Trigo (2011)
[h]Alston et al. (2014)
[i]Wu (2002)
[j]Demont and Tollens (2004)
[k]Pray et al. (2001)
[l]Qaim (2003)
*Notes* [1]Adapted from Fernandez-Cornejo and Wechsler (2014)
[2]Rest of the world (consumers and producers)

as benchmarks, but it is the range of returns and the general sense of distribution that is of interest. Clearly, benefits accrue to more than just the technology developer and the adopting producers; society benefits as well. The social benefits include some that go to the non-adopters. Soybeans have had the greatest number of studies with roughly one-quarter to one-third of the benefits accruing to consumers.

Benefits from technological innovations spill over to other sectors of the economy from that which the innovation occurs. Huang et al. (2010) and Hutchison et al. (2010) document how the benefits from the adoption of GM crops have reduced the pest populations in the adopting region, which benefits all non-adopters of GM crops. The non-adopters incur lower input costs from insecticide applications and higher yields.

In addition to the above commodity specific analysis, several important GM crop meta-analyses have been undertaken. The first, by Carpenter (2010) examined yield comparisons between adopters and non-adopters from 168 studies, finding 124 reporting yield increases, 32 reports of no differences and 13 of lower GM crop yields. Finger et al. (2011) examined 203 peer reviewed publications, finding that yield increases exist due to reduced pest pressures (both insect and weed). Reduced chemical applications contribute to the economic benefits, but are reduced marginally by the higher seed costs for GM seed inputs. Areal et al. (2013) compared 97 observations that compare production between GM and conventional crops, finding that GM crops outperformed conventional crop in both developed and developing countries. In a wider assessment of literature that included journal articles, government reports as well as industry and organization reports, Klümper and Qaim (2014) reported their meta-analysis of 147 studies revealed that chemical use declined by 37%, yields increased by 22% and farmer profits increased by 68%.

## Implications

Clearly there is a wide gap between expectations and realities. UNCTAD's report, which is critical of biotechnology and GM crops, is typical. The report's authors suggest that the solution for food security throughout this century is to return to small-scale organic food production methods. Unfortunately, this report ignores the pressures for food trade examined by Fader et al. (2013) and misinterprets the evidence of the economy and environment impacts of GM crops in both developed and developing countries.

As this chapter identifies, the annual economic benefit of the albeit truncated and limited adoption of the technology is in the billions of dollars and by some estimates has climbed above US$100 billion for the first twenty years of production. Moreover, the environmental benefit from the reduction of millions of kilograms fewer chemical applications and the health benefits from millions of fewer pesticide poisonings suggest the technology has merit, if the regulatory system can overcome a number of hurdles. The next two chapters discuss the regulatory challenge, first from the domestic perspective and then in the international domain.

## References

Adams, M. 2014. Biotech genocide, Monsanto collaborators and the Nazi legacy of 'science' as justification for murder. Natural News, July 21. Available at: http://www.naturalnews.com/046097_biotech_genocide_monsanto_collaborators_media_sellouts.html#ixzz3AxJsnL6j.

Alston, J.M., N. Kalaitzandonakes, and J. Kruse. 2014. The size and distribution from the adoption of biotech soybean varieties. In *Handbook on agriculture, biotechnology and development*, ed. S.J. Smyth, P.W.B. Phillips, and D. Castle, 728–751. Cheltenham, UK: Edward Elgar Publishing Ltd.

Areal, F.J., L. Riesgo, and E. Rodriguez-Cerezo. 2013. Economic and agronomic impact of commercialized GM crops: A meta analysis. *Journal of Agricultural Science* 151 (1): 7–33.

Atera, E. 2012. Farmers' perception and constraints to the adoption of weed control option: The case of Striga Asiatica in Malawi. *Journal of Agricultural Science* 4 (5): 41–50.

Bennett, R., S. Morse, and Y. Ismael. 2006. The economic impact of genetically modified cotton on South African smallholders: Yield, profit and health effects. *Journal of Development Studies* 42 (4): 662–677.

Brookes, G., and P. Barfoot. 2010. GM crops: Global socio-economic and environmental impacts 1996–2008. Dorchester, UK: PG Economics Ltd. http://www.pgeconomics.co.uk/pdf/2010-global-gm-crop-impact-study-final-April-2010.pdf.

Brookes, G., and P. Barfoot. 2015. GM crops: Global socio-economic and environmental impacts 1996–2013. http://www.pgeconomics.co.uk/pdf/2015globalimpactstudyfinalMay2015.pdf.

Brookes, G., and P. Barfoot. 2014. Economic impact of GM crops: The global income and production effects 1996–2012. *GM Crops and Food Biotechnology in Agriculture and the Food Chain* 5 (1): 65–75.

Carpenter, J. 2010. Peer-reviewed surveys indicate positive impact of commercialized GM crops. *Nature Biotechnology* 28 (4): 319–321.

Chikoye, D., S. Schulz, and F. Ekeleme. 2004. Evaluation of integrated weed management practices for maize in the Northern Guinea Savanna of Nigeria. *Crop Protection* 23 (10): 895–900.

Clark, L.F., C.D. Ryan, and W.A. Kerr. 2014. Direct democracy, state governments, and the re-energized GMO debate: Implications of California's proposition 37. *AgBioForum* 16 (3): 177–186.

Demont, M., and E. Tollens. 2004. First impact of biotechnology in the EU: Bt maize adoption in Spain. *Annals of Applied Biology* 145: 197–207.

Fader, M., D. Gerten, M. Krause, W. Lucht, and W. Cramer. 2013. Spatial decoupling of agricultural production and consumption: Quantifying dependences of countries on food imports due to domestic land and water constraints. *Environmental Research Letters* 8 (1): 1–15.

Falck-Zepeda, J.B., G. Traxler, and R.G. Nelson. 2000. Rent creation and distribution from biotechnology innovations: The case of Bt cotton and herbicide-tolerant soybeans in 1997. *Agribusiness* 16 (1): 21–32.

Fernandez-Cornejo, J., and S. Wechsler. 2014. Soybeans. In *Handbook on agriculture, biotechnology and development*, ed. S.J. Smyth, P.W.B. Phillips, and D. Castle, 567–582. Cheltenham, UK: Edward Elgar Publishing Ltd.

Fernandez-Cornejo, J., C. Hendricks, and A. Mishra. 2005. Technology adoption and off-farm household income: The case of herbicide-tolerant soybeans. *Journal of Agriculture and Applied Economics* 37 (3): 549–563.

Finger, R., N. El Benni, T. Kaphengst, C. Evans, S. Herbert, B. Lehmann, S. Morse, and N. Stupak. 2011. A meta analysis on farm-level costs and benefits of GM crops. *Sustainability* 3: 743–762.

Gardner, J.G., R.F. Nehring, and C.H. Nelson. 2009. Genetically modified crops and household labor savings in US crop production. *AgBioForum* 12 (3&4): 303–312.

Gonsalves, C.V., and D. Gonsalves. 2014. The Hawaii papaya story. In *Handbook on agriculture, biotechnology and development*, ed. S.J. Smyth, P.W.B. Phillips, and D. Castle, 728–751. Cheltenham, UK: Edward Elgar Publishing Ltd.

Gouse, M. 2013. An evaluation of the gender differentiated impact of genetically modified crop adoption: A pilot study in South Africa—GM maize and gender: Evidence from smallholder farmers in KwaZulu-Natal, South Africa. Project report to the Program for Biosafety Systems, International Food Policy Research Institute.

Gruère, G., and D. Sengupta. 2011. Bt cotton and farmer suicides in India: An evidence-based assessment. *Journal of Development Studies* 47 (2): 316–337.

Gusta, M., S.J. Smyth, K. Belcher, P.W.B. Phillips, and D. Castle. 2011. Economic benefits of genetically-modified herbicide-tolerant canola for producers. *AgBioForum* 14 (1): 1–13.

Huang, J., J. Mi, H. Lin, Z. Wang, R. Chen, R. Hu, S. Rozelle, and C. Pray. 2010. A decade of Bt cotton in Chinese fields: Assessing the direct effects and indirect externalities of Bt cotton adoption in China. *Science China Life Sciences* 53 (8): 981–991.

Hurley, T.M., P.D. Mitchell, and G.B. Frisvold. 2009. Effects of weed resistance concerns and resistance management practices on the value of roundup ready crops. *AgBioForum* 12 (3&4): 291–302.

Hutchison, W.D., E.C. Burkness, P.D. Mitchell, R.D. Moon, T.W. Leslie, S.J. Fleischer, M. Abrahamson, K.L. Hamilton, K.L. Steffey, M.E. Gray, R.L. Hellmich, L.V. Kaster, T.E. Hunt, R.J. Wright, K. Pecinovsky, T.L. Rabaey, B.R. Flood, and E.S. Raun. 2010. Areawide suppression of European corn borer with Bt maize reaps savings to non-Bt maize growers. *Science* 330 (6001): 222–225.

James, C. 2013. *Global status of commercialized biotech/GM crops: 2013*. ISAAA Brief 44; 2014. Available from: http://www.isaaa.org/resources/publications/briefs/46/executivesummary/default.asp.

James, C. 2015. 20th Anniversary (1996 to 2015) of the Global Commercialization of Biotech Crops and Biotech Crop Highlights in 2015. ISAAA Brief No. 51. ISAAA: Ithaca, NY.

Joubert, A. 2000. Weed control by smallholder farmers in Ciskei, Eastern Cape Province, South Africa. In *Animal power for weed control: A resource book of the animal traction network for Eastern and Southern Africa (ATNESA)*, ed. P. Starkey, and T. Simalenga, 214–217. Wageningen, Netherlands: Technical Centre for Agriculture and Rural Cooperation.

Kaldor, M., H. Anheier, and M. Glasius (eds.). 2004. *Global Civil Society 2004/5*. London: Sage.

Kent, R.J., D.E. Johnson, and M. Becker. 2001. The influences of cropping system on weed communities of rice in Côte d'Ivoire, West Africa. *Agriculture, Ecosystems & Environment* 87 (3): 299–307.

Kloor, K. 2015. GM-crop opponents expand probe into ties between scientists and industry. *Nature* 524 (7564): 145–146.

Klümper, W., and M. Qaim. 2014. A meta-analysis of the impacts of genetically modified crops. *PLoS ONE* 9 (11): 1–7.

Kouser, S., and M. Qaim. 2011. Impact of Bt cotton on pesticide poisoning in smallholder agriculture: A panel data analysis. *Ecological Economics* 70: 2105–2113.

Moschini, G., H. Lapan, and A. Sobolevsky. 2000. Roundup ready soybeans and welfare effects in the soybean complex. *Agribusiness* 16 (1): 33–55.

Phillips, P.W.B. 2007. *Governing transformative technological innovation: Who's in charge?*. Oxford: Edward Elgar.

Pray, C., D. Ma, J. Huang, and F. Qiao. 2001. Impact of Bt cotton in China. *World Development* 29 (5): 813–825.

Pray, C., J. Huang, R. Hu, and S. Rozelle. 2002. Five years of Bt cotton in China—The benefits continue. *The Plant Journal* 31 (4): 423–430.

Price, G.K., W. Lin, J.B. Falck-Zepeda, and J. Fernandez-Cornejo. 2003. Size and distribution of market benefits from adoption biotech crops. Technical Bulletin No. 1906. Washington, DC: US Department of Agriculture—Economic Research Services.

Qaim, M. 2003. Bt cotton in India: Field trial results and economic projections. *World Development* 31 (12): 2115–2127.

Qaim, M. 2014. Agricultural biotechnology in India: Impacts and controversies. In *Handbook on agriculture, biotechnology and development*, ed. S.J. Smyth, P.W.B. Phillips, and D. Castle, 126–137. Cheltenham, UK: Edward Elgar Publishing Ltd.

Qaim, M., and G. Traxler. 2005. Roundup ready soybeans in Argentina: Farm level and aggregate welfare effects. *Agricultural Economics* 32 (1): 73–86.

Shetto, R., and E. Kwiligwa. 1998. Weed control system in maize based on animal drawn cultivator. Rome: FAO Paper. Available at: http://www.fao.org/wairdocs/ilri/x5483b/x5483b0n.htm.

Smyth, S.J., and P.W.B. Phillips. Forthcoming. Thresholds for profitable genomic innovation: University of Saskatchewan, Unpublished study.

Smyth, S.J., M. Gusta, K. Belcher, P.W.B. Phillips, and D. Castle. 2011a. Environmental impacts from herbicide tolerant canola production in Western Canada. *Agricultural Systems* 104 (5): 403–410.

Smyth, S.J., M. Gusta, K. Belcher, P.W.B. Phillips, and D. Castle. 2011b. Changes in herbicide use following the adoption of HR canola in Western Canada. *Weed Technology* 25 (3): 492–500.

Smyth, S.J., P.W.B. Phillips, and W.A. Kerr. 2016. EU failing FAO challenge to improve global food security. *Trends in Biotechnology* 34 (7): 521–523.

Subramanian, A., and M. Qaim. 2010. The impact of Bt cotton on poor households in rural India. *Journal of Development Studies* 46 (2): 295–311.

Trigo, E.J. 2011. Economic impact after 15 years of GM Crops in Argentina. Available at: http://www.argenbio.org/adc/uploads/15_anos_Estudio_de_cultivos_GM_en_Argentina.pdf.

United Nations Conference on Trade and Development. 2013. *Trade and environmental review 2013: Wake up before it is too late*. New York: United Nations.

Vitale, J., G. Vognan, and M. Ouattarra. 2014. Cotton. In *Handbook on agriculture, biotechnology and development*, ed. S.J. Smyth, P.W.B. Phillips, and D. Castle, 604–620. Cheltenham, UK: Edward Elgar Publishing Ltd.

Wu, F. 2002. Bt or not Bt? Tools for regulatory decisions concerning genetically modified corn. PhD thesis, Carnegie Mellon University. Available at: http://sunzi.lib.hku.hk/ER/detail/cof/2707394.

Yorobe, J.M., and M. Smale. 2012. Impacts of Bt maize on smallholder income in the Philippines. *AgBioForum* 15 (2): 152–162.

# Chapter 2
# Winners and Losers from Innovation and Trade

**Abstract** The evidence in support of the various claims of gains and losses following the commercialization of GM crops is examined in this chapter. The gains from adopting GM crops are well reported. The distribution of benefits and costs to non-adopters and others beyond the GM crop space are analyzed, providing an assessment to date of the net gains or losses accruing to all adopters and non-adopters, both in the immediate vicinity and more distanced from the disruptive innovation.

## Introduction

Innovation is pervasive. Just as gravity is a constant in physics, innovation is a constant force in market economies. It would be inaccurate to say that innovation is a sustaining momentum at a constant measured rate over time such as gravity, but innovation is constantly taking place, albeit at a pace that varies by industry, by geographic location and over time. Historically, we witness some periods as being particularly innovative (i.e. the first and second Industrial Revolutions) while others less so (i.e. The Dark Ages). Depending on rulers and societal norms, some societies have been very supportive of innovation, while others have tended to shun it. Russia's Peter the Great comes to mind as a ruler that was supportive of innovation, investing substantial resources to ensure that Russians benefited from technological advances. Generally, rulers view innovation as a beneficial economic activity—the true source of improvements to human wellbeing—and support the development of technologies and encourage their spread throughout society.

While the benefits of innovation are usually loudly trumpeted, it is worthy to note that innovation is not costless. Those that cannot, or will not, adopt, have frequently been tagged as 'losers' in an economic sense, meaning that there are costs incurred within society by those that cannot or do not adopt a technological innovation. Frequently, non-adopters are those without the financial means to take on the costs associated with adopting the new technology. Some non-adopters, however, choose to do so for a variety of other reasons that may be personal,



S.J. Smyth et al., *Biotechnology Regulation and Trade*, Natural Resource Management and Policy 51, DOI 10.1007/978-3-319-53295-0_2

philosophical or religious. It has become accepted economic theory that non-adopters lose from their decision, but there can be important exceptions to this paradigm. For example, an individual who bought a new Corvette in 1966 and did not adopt a newer, safer vehicle, now owns a vehicle that, if it remains in good condition, is worth many thousands more than was paid for it and thousands more than a new vehicle is worth today. Fiscally, it cannot be said that the Corvette owner lost on the decision to not adopt newer, safer vehicles. Had the Corvette owner been in an accident, that might be a different story, but from an economic perspective, the decision not to adopt was sound. This chapter examines one such case in the agri-food space where non-adopters claim to be worse off when they might not be upon closer inspection.

The commercialization of genetically modified (GM) crops in the mid-1990s precipitated the beginning of a suite of innovative technologies and products that has now delivered 20 years of adopter benefits (Smyth et al. 2014). Almost immediately upon the commercialization of GM crops, organic farmers proclaimed that they would be adversely impacted by this technology. In a number of developed market economies, organic farming has expanded, suggesting that the adverse impacts that were predicted at the point of GM crop commercialization have not been universally realized and that possibly the commercialization of GM crops has benefitted organic farming. There is some prima facie evidence:

> "Avoiding GMO's, for themselves or their children, is an increasingly important reason why parents choose organic food," the OTA [Organic Trade Association] said in its U.S. Families Organic Attitudes and Beliefs 2014 Tracking Study, which surveyed more than 1,200 households with at least one child under the age of 18. "Almost 25% of parents buying organic said that wanting to steer clear of genetically modified foods is now one of their top reasons for selecting organic, the most in the four years the survey has been taken, and up significantly from the 16% who said the same in 2013. Of the 15 reasons for buying organic that parents were asked to rate, not buying GMO's showed the biggest jump by far from attitudes a year ago." (Sterk 2014: 62).

## Background

GM varieties of canola, corn and soybean all became commercially available between 1995 and 1997 in both the United States and Canada. Initial concerns from the organic community were somewhat muted given the limited early adoption of GM varieties. At the turn of the millennium adoption rates rose sharply to a level that began to create concern within the organic producing community. Figure 2.1 illustrates the adoption of GM crops in Canada. Until about 2000, organic farmers and the organic industry were able to produce and export organic products without undue concern about provenance and purity being raised. In response to the spread of GM production, and to capitalize on rising concerns of some consumers regarding GM foods (Hobbs and Kerr 2003), the organic industry amended their certification standard to exclude the use of GM seeds by certified organic farmers.

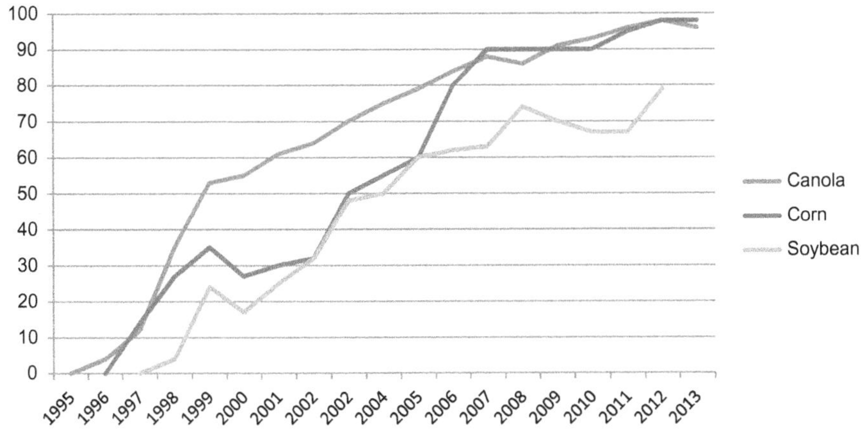


**Fig. 2.1** Adoption of GM Crops in Canada, 1995–2013 (percent). *Source* Canola Council of Canada (2014) for canola 1995–2010; Canadian Soybean Council (2014) for soybean 2010–2012; Hategekimana and Beaulieu (2002) for corn and soybean 2000–2002; ISAAA (James 1996–2013) for canola 2011–2013, corn 1996–1999, 2003–2010, soybean 1997–1999, 2003–2009

The new standards were largely silent on the potential for comingling of GM and organic product. Testing for trace amounts of GM seed was rudimentary at this time, allowing the organic industry the ability to serve foreign exports markets with little or no risk of rejection based on 'impurities'.

As GM seeds became more generally available and the profitability of GM crops became more widely observed, adoption increased. Parallel to this, however, was increasing expressions of concern by the organic industry. While there are no records, to date, of an organic shipment being rejected due to comingling with a GM commodity, the organic industry grew increasingly nervous. Importers began to seek assurances that organic shipments were free of GM commodities (and not simply produced using non-GM seed), something that neither organic producers nor export suppliers could absolutely guarantee. Worried that organic export markets might start to close if guarantees could not be provided, the organic industry turned to the courts.

In Canada in 2002 the Saskatchewan Organic Directorate filed a class-action lawsuit in Saskatchewan's Court of Queen's Bench on behalf of two Saskatchewan organic farmers, Larry Hoffman and Dale Beaudoin, representing all registered organic farmers in the province of Saskatchewan, against the developers of GM canola, Bayer CropScience (the owners of Agrevo's Liberty Link canola) and Monsanto Canada (owner of Roundup Ready canola) (Smyth and Phillips 2016). The putative class-action claimed that commercial production of GM canola was responsible for the loss of organic markets due to the real or potential for comingling of GM canola with organic canola and sought financial compensation for the claimed associated losses. The suit asked for damages for negligence, nuisance, trespass and strict liability. In particular, the plaintiffs argued that the comingling of

GM canola had destroyed the export market for organic canola due to the inability to certify that canola exports, particularly to the European Union, were free of GM canola.

The plaintiffs sought damages on behalf of the 1250 certified organic producers in Saskatchewan, regardless of whether they grew organic canola or not. Indeed, it was revealed in the evidence presented in the hearings that only about 14,074 acres were planted to organic canola in the 1990–2001 period, that only 76 certified organic farmers had ever grown organic canola, and that six or more certified organic farmers still produced organic canola in 2001, after the widespread adoption of GM canola, and that exports of organic canola continued. Based on this evidence, the judge ruled that the evidence did not demonstrate that there was the basis for a class action and ruled against certifying the class action.

The strategy of the organic industry in the US differed slightly in that they sought injunctions against the commercialization of new GM crop varieties. This tactical use of courts was part of the organic industry's strategy for dealing with the commercialization of GM alfalfa and sugar beets. In 2006, a group of conventional and organic farmers (known as the Geertson plaintiffs as it was Geertson Seed Farms that filed the action) sought an injunction against the commercialization of GM alfalfa in the US. They argued that the United States Department of Agriculture (USDA) was negligent in undertaking its regulatory oversight on GM crops as it had not required that an environmental impact statement (EIS) be conducted pertaining to the release of GM alfalfa. The heart of the Geertson plaintiffs argument was that pollen flow from the GM alfalfa would be a 'contaminant' to the organic alfalfa farmers, causing economic and socio-economic harms. The local district court ruled in favor of the plaintiffs, granting the injunction and ordering the USDA to perform an EIS. The USDA and Monsanto (the firm commercializing the technology) appealed, but the district court decision was upheld. In 2010, this was further appealed to the US Supreme Court, which ruled that the district court injunction was improper. After further review by the USDA, Roundup Ready alfalfa was approved without restriction in 2011 and, in spite of further litigation, remains in production in the US.

In similar fashion, in 2008 the Center for Food Safety (CFS), an environmental non-governmental organization, brought a claim against the USDA and Monsanto alleging that commercialization of GM sugar beets should not have been allowed three years earlier, arguing that an EIS was not conducted. The judge allowed planting of GM sugar beets to proceed at the seed multiplication stage. In 2010, the CFS filed for an injunction against any further plantings of GM sugar beets. As GM sugar beets were then commercially cultivated on over 90% of sugar beet acres, the USDA undertook an EIS to ensure that further commercial production was not adversely affected. The USDA completed the EIS in 2012 and deregulated the crop. The CFS further challenged this ruling but the courts rejected the claims and dismissed the case (Bryson 2016; Congressional Research Services 2013).

As indicated above, through the use of court cases, the organic sector claims to be economically disadvantaged as a result of the commercialization of GM crops. However, there has yet to be verifiable evidence from the organic industry that commodity shipments for either domestic use or export are being rejected at point of acceptance due to the low level presence of GM commodities. With no evidence of economic harm due to shipment rejections, one wonders whether, on the contrary, the organic industry is actually benefiting from the commercialization of GM crops. Some evidence exists that supports this conjecture.

Hutchison et al. (2010) examined the environmental impacts of GM corn adoption in the US Midwest. GM corn has been developed to be insect resistance, particularly to the effects of the European corn borer. At the time of their study, GM corn accounted for two-thirds of the corn planted in the US. Their study of European corn borer suppression due to the widespread adoption of GM corn in the states of Illinois, Minnesota and Wisconsin, found that the cumulative economic benefits from 14 years of production was estimated to be US$3.2 billion. Due to the area-wide suppression of corn borers, the authors estimated that 75% (US$2.4 billion) of the economic benefits went to producers not adopting GM varieties (which includes organic corn producers). The high and wide spread use of GM corn has suppressed the presence of the European corn borer to such a degree that those not growing GM corn gain from relatively lower seed costs, lower chemical requirements and generally improved yields due to lower plant damage due to insects.

In China, Huang et al. (2010) updated their Bt cotton study following a decade of commercial production. Based on the authors' analysis, they found a drop in bollworm infestations, not only in Bt cotton fields, but in all cotton fields in parts of China. In some non-Bt cotton fields, the use of insecticide dropped from in excess of 40 kg/ha to less than 10 kg/ha. Across the entire sample region insecticide applications dropped from 14 to 4 kg/ha.

The evidence indicates that with the adoption of GM crop technologies, the benefits spill over to the fields of the non-adopters, providing economic advantages. While the organic industry has been unable to substantiate their claims of economic losses due to shipment rejections, it is possible to demonstrate economic advantages they have gained due to their close proximity to GM crops, achieved through higher yields and reduced pesticide applications (the organic industry allows for the spray application of Bt to control corn borer, the chemical analogue to the Bt gene that gives GM corn its insect resistance). With increased economic benefits from reduced corn borer populations and reduced pesticide applications, organic farmers growing corn have benefited economically by the commercialization of GM corn.

## Methodology

Some civil society groups, and a few regulators, argue for the inclusion of socio-economic considerations (SECs) in assessments of new GM crops. This stems from what Isaac (2007) identified as the 'Social Rationality' perspective on innovations. In this view, a technology that imposes economic or social costs on society, or any of its individual members, is unacceptable and should not be

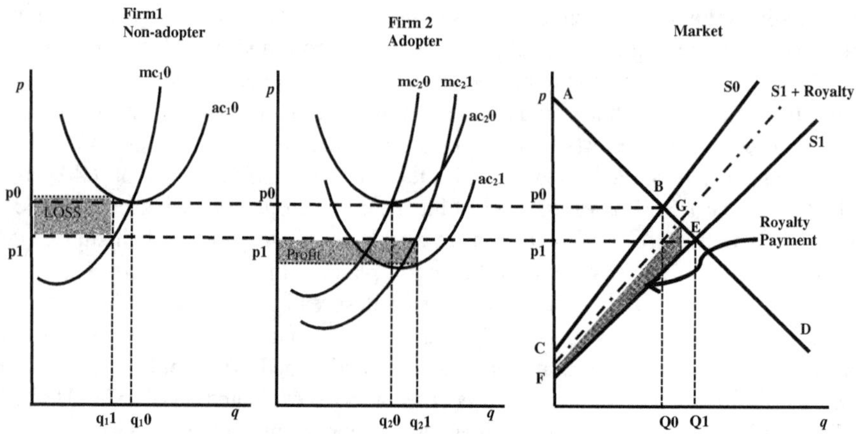


**Fig. 2.2** Technological change

licenced for production or import. In other words, the only technologies that should be accepted are those whose adoption would create only 'winners' (i.e. no 'losers'). The challenge is that all new technologies generate a mix of 'winners' and 'losers'. Implicit in the rejection of any technology that imposes negative economic impacts on some is the rejection of cost-benefits assessments that weigh and make decisions on the aggregate benefits and costs for society. While those who work in the area of technological change understand that it will not be possible to identify 'winner only' technologies, this eventuality is not always intuitive. Given that some who wish to influence technology policy appear to believe that 'winner only' technologies are possible,[1] it is worth laying out the reasons why such a result is not feasible.

Figure 2.2 can be used to illustrate why 'winner only' technologies are not expected to arise. The assessments of technological change are most often based on the 'partial equilibrium, comparative statics' models of economists (Kerr 2010; Gaisford et al. 2001). This approach does not consider cross market effects in its assessments of new technologies—although there is no reason why effects on heavily affected markets cannot be included in any empirical assessment of a new technology. A single market is depicted in the right hand panel of Fig. 2.2.

This discussion of the inclusion SECs in the assessment of a new technology assumes that the technology has been evaluated as being acceptable to society on a scientific basis. In other words, issues such as food safety and environmental risk have been dealt with by competent scientific authorities and the technology is

[1]It is also possible that those putting forth such a policy perspective simply believe that all technologies yield only negative outcomes for all and, hence, wish policy-makers to eschew all new technologies. As it is relatively easy to show that there will be 'winners' from new technologies, the less obvious case of why a 'winner only' technology cannot arise will be explored here.

deemed acceptable for commercialization. If the scientific bar is not cleared then the technology cannot be approved for commercialization and, hence, there is no need for a socio-economic assessment. It should be noted that the scientific-based risk assessment may include economic factors such as the economic cost of human health incidents and thus have been 'dealt with' and should not be included in the subsequent examination of SECs.[2] 'Social welfare' is the normal metric used to evaluate a new technology. The 'comparative statics' methodology compares one static equilibrium (e.g. before the technological change) to a second static equilibrium (e.g. after the technological change). The total (societal) welfare for the pre-technological change equilibrium is defined by the demand curve in the market, D, and the supply curve representing the original technology, S0. In this case the total welfare is area ABC.[3]

A successful new technology represents an increase in technical efficiency—or a lowering of the cost of producing the same output. This is represented by a downward shift in the supply curve—from S0 to S1. The market moves to a new equilibrium at point E where S1 intersects the demand curve. Total welfare is now equal to area AEF which is larger than the original total welfare (area ABC) by area CBEF. Hence, the technological change is 'welfare enhancing'. These increased benefits are divided between consumers—through increased consumer surplus—and producers—through increased producer surplus.

In this example the technological change is assumed to be developed outside the market and provided free of charge to the producers in the market. This would be the case, for example, when new seed varieties developed in government research facilities or publically funded universities are given to the industry at no cost. If, however, the new technology is developed in the private sector that enjoys intellectual property protection through patents[4] and royalties can be charged, then to account for the increased cost represented by those royalty payments the supply curve shifts upward to, for example, S1 + Royalty.[5] The welfare given the technical change when royalties are collected becomes AGC (<AEF). While the technical efficiency associated with the technological change is still represented by the costs underlying supply curve S1, a royalty payment equal to the shaded area in the right hand panel of Fig. 2.2 benefits the holder of the intellectual property rights. Hence,

[2]This does not mean that SECs cannot be part of a larger full assessment that includes both the scientific assessment and the socio-economic assessment. It only means that in terms of undertaking the assessment they should be kept separate.

[3]The metric of social welfare is based on the concepts of consumer surplus and producer surplus. See Gaisford and Kerr (2001) for a discussion of consumer and producer surplus.

[4]Or other forms of intellectual property protection such as plant breeders' rights.

[5]Note, the royalty payment could also be depicted as an increased cost added to the marginal cost of adopting firms thus shifting those costs upward leading, in aggregate, to an upward shift in the supply curve relative to S1. In a similar way, technological improvements developed by firms with a degree of market power would represent a shift upward in the marginal cost curves of firms thus transferring some of the benefits to the firms developing the technology. An example might be a firm that develops a more efficient tractor and is able to charge a higher than competitive price for that new tractor.

the benefits of the technological change are shared between three parties: consumers, producers (farmers) and innovators. The values of these benefits for each party are commonly calculated. Given this particular method of assessing a technological change—partial equilibrium, comparative statics—one might be led to believe that 'winner only' technologies might exist—after all there are no losers noted in the analysis above.

There is, however, a great deal going on 'behind the scenes' when this particular method of assessment is used. This is where the 'losers' are to be found. To discover some of the losers the figures depicting Firms 1 and 2 in Fig. 2.2 are useful. These two firms are representative of all of the firms supplying this competitive market—usually taken to be a 'large' number. The supply curve of an industry is the sum of the marginal costs curves of the individual firms. Let us start with Firm 2—the adopter. In the absence of a technological change the firm is operating on $mc_20$ contributing to S0 in the market. It faces p0 from the market and choses to produce quantity $q_20$. We assume the firm is in equilibrium having an average cost function $ac_20$ and earning normal profit at $q_20$ (i.e. $p0 = ac_20$ at $q_20$). At this point Firm 1 and Firm 2 use the same technology and are assumed to have the same cost structure. Firm 1 is also contributing to supply curve S0 through $mc_10$, producing $q_10$ and earning normal profit.

Consider a technological change that Firm 2 can technically adopt, and is willing to adopt. A technological change lowers the firm's cost (increases its efficiency) moving it from $mc_20$ to $mc_21$. If there are a sufficient number of adopting firms, S0 shifts down to S1. Price declines in the market from p0 to p1. Firm 2 must then re-chose its optimal output based on the new price and its new marginal cost curve (i.e. at $q_21$ where $mc_21 = p1$). If marginal cost shifts down average cost will also shift down so that, as depicted,[6] it now has average cost curve $ac_21$. As $ac_21$ is less than p1 at $q_21$ Firm 2 now makes a 'supernormal' profit equal to the shaded area in the centre panel of Fig. 2.2. Thus all of the firms represented by Firm 2 are 'winners' as a result of the technological change.

In the short run, a new technology seldom has universal adoption. Thus it is important to consider non-adopting firms when evaluating the SECs of a new technology. If there is a new technology that Firms 2 adopts, the effects in the market are as described above. Price falls from p0 to p1. Firm 1 cannot adopt the technology due to technological incompatibility or an unwillingness to adopt it. Thus, it retains the same technologically determined marginal cost curve $mc_10$ (and $ac_10$). Given the decline in price, Firm 1 must re-choose its optimal output. This is now $q_11$ (where $mc_10 = p1$). As $ac_10$ is greater than p1 at $q_11$ the firm makes a loss equal to the shaded area in the left hand panel of Fig. 2.2. Non-adopting firms are 'losers'. The strict model would assume Firm 1 can be expected to leave the industry and their resources are available for deployment elsewhere in the

[6]Note, as drawn $ac_21$ is less that p1 at $q_21$. As there is no direct relationship between the decrease in average cost and the change in price, the new price level, p1, may be greater than, less than or equal to average cost at $q_21$.

economy.[7] While firms leaving the industry shift the supply curve in the market upward, the supernormal profits earned by Firms 2 provide a signal for additional firms that can use the technology to enter shifting the supply curve in the opposite direction. Eventually the market will reach a new equilibrium with full adoption of the most efficient technology.

As the technological change leads to the market price declining—from p0 to p1 there will also be cross market effects. For substitute products the decline in price will lead to a leftward shift in the demand curve of the substitute, meaning less total welfare in the substitutes' market—thus 'losers' are created.[8] Similarly, demand and prices in markets for complements would rise, generating windfall gains. Alston et al. (2014) undertook to put numbers to the cross-market effects for the adoption of GM soybeans, concluding that farmers and countries adopting the technology gained, all consumers (whether they wanted GM crops or not) gained but that producers of competing crops, such as canola, sunflower and palm lost (while consumers of those crops gained).

In a technological change, the quantity of inputs used is unlikely to remain the same. For example, in the centre panel of Fig. 2.2 as the marginal cost curve declines (to $mc_21$) for Firm 2, the total input cost is less than was the case when Firm 2 was operating on $mc_20$. This means that the demand for inputs has declined. The shift to the left in the demand for inputs creates 'losers' in the input market.[9]

Further, and more likely, the mix of inputs is likely to change as the technology changes. For example, if the industry depicted in Fig. 2.2 was the market for personal conveyance vehicles and the technological change was the invention of the automobile, Firm 2 that is producing cars will need a very different mix of inputs (e.g. engine blocks, gas tanks, brake shoes) than Firm 1 that is producing buggies (e.g. leather, hitches for horses).[10] Thus, as the technological change progresses demand for inputs specific to the manufacture of buggies will decline—creating 'losers'.[11] In the case of some GM technologies, their adoption has meant a considerable reduction in the use of pesticides—in effect a reduction in demand for pesticides.

---

[7]Under the assumptions of the strict neoclassical model firms exit freely (i.e. without cost). Of course, in reality firms do not leave easily and resources are not redeployed without cost (Leger et al. 1999).

[8]For example, a technological change in the pork industry leads to a lower price for pork. This will induce consumers to substitute more pork for beef in their diets—a leftward shift of the demand curve for beef. Of course, the demand for complements will increase as, for example, pork consumption increases there will be an increased demand for apple sauce. Thus, if cross market SECs are going to be included in the assessment of a technology it is important that the 'winners' be included along with the 'losers'.

[9]For example, a higher yielding variety of canola will need less fertilizer per unit of output—thus reducing the demand for fertilizer.

[10]At some point the convention is to depict the market for cars as a market for a separate product. Doing this, however, obscures the effect of the technological change.

[11]Of course, demand will also increase for those inputs associated with the production of automobiles—creating 'winners'.

GM technology may also mean a reduction in the quantity or change in quality of labor required in production. For example, in developing countries crops may no longer have to be hand weeded releasing labor from production—often women and girls. This change will represent a reduction in employment. The labor released, however, may be redeployed to other activities such as fetching water or attending school (Gouse 2014). Evaluating secondary market effects of technological changes can be very complex. Of course, changes in secondary markets leads to changes in additional markets. A full assessment of a technology would require a different methodology—a general equilibrium approach that includes an evaluation of all interactive market effects. Unfortunately, disaggregated general equilibrium models require unattainable quantities of data so that any assessments undertaken using computable general equilibrium models is at a level of aggregation that does not yield the degree of detail required for a useful assessment of a technology (Gaisford and Kerr 2001).

The bottom line is that putting policies in place whereby only technologies that have no negative effects—creates only 'winners'—means that all new technologies will be rejected. In effect, this means that the gains in societal welfare observed in Fig. 2.2 will be forgone. Thus, public policy assessments of new technologies need to be conducted in a cost/benefit framework whereby both 'winners' and 'losers' are considered—yet this approach is increasingly under attack by critics of the current regulatory system.

## A GM-Based Technological Change

One area of regulation where the failure to adequately assess the market effects arising from a technological change is the interaction between a market where a GM-based crop is introduced and the market for organic production of the same product. As a result, in some jurisdictions regulatory policy may not produce a socially optimal outcome. These market interactions are explored in Fig. 2.3.

The right hand panel of Fig. 2.3 represents the organic market for a particular crop. In the absence of a GM variety of the crop, organic (O) producers must compete with producers of conventional (C) varieties of the crop. In the organic market, in the absence of GM varieties, the demand for organic product is $D_O0$ and the supply is $S_O0$ yielding price $P_O0$ and quantity $Q_O0$. The left hand panel of Fig. 2.3 depicts the market for the conventional[12] variety of the same crop. In the absence of a GM-based technological change, the demand for conventional varieties is $D_C0$ and conventional supply is $S_C0$ yielding price $P_C0$ and quantity $Q_C0$.

When a GM variety of the crop is introduced into the market, in the absence of labeling and segregation of GM and conventional varieties (such as is the case in Canada), consumers will be unable to discern whether they are consuming GM

[12]And eventually the GM variety.

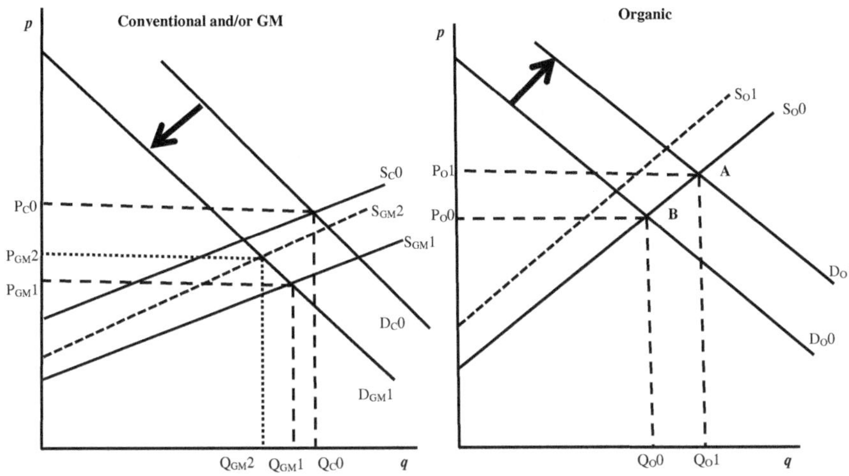


**Fig. 2.3** A GM-based technological change

products. As a result, those consumers that prefer not to consume GM products will begin to purchase organic products—shifting the organic demand curve outward.

At the same time as the demand in the organic market shifts out, the demand in the conventional and/or GM market shifts inward (Gaisford et al. 2001).[13] The greater the proportion of the market that becomes GM the higher the probability that a consumer who does not chose to purchase organic products will be consuming GM and, thus, for those that wish to avoid GM products the larger the incentive to switch to organic products—shifting the demand curve for organic products further outward.[14] If the adoption of GM technology is complete the entire market in the left hand panel of Fig. 2.3 will be GM production meaning consumers are faced with a choice of only GM or organic products.

The new equilibriums in the markets now segmented as 100% GM and organic can be seen in Fig. 2.3. Once all the consumers that wish to avoid GM products have shifted to the organic market, demand in the GM market will be $D_{GM}1$. The post technological change supply curve—the GM supply curve—is $S_{GM}1$. The GM market price is $P_{GM}1$ and the quantity purchased $Q_{GM}1$. The decrease in resource

[13]It is clear that at least a segment of the organic industry anticipated this shifting of demand between markets when it made the decision to declare that organic production would not allow GM technology. This was a propitious marketing decision.

[14]And, of course, shifting the demand curve for conventional and/or GM varieties further inward. If labeling of GM crops is required then consumers will be able to identify conventional products and the demand would segment into three markets. As a result of the transparency of labeling consumers will be able to avoid GM products without switching to organic products and, thus, the market for organic products will not shift outward to the same degree. We abstract from this outcome by assuming that the adoption of GM technology is complete in the non-organic market.

cost per unit (the efficiency gain) is the vertical distance between $S_C0$ and $S_{GM}1$ for any quantity.

In the organic market the additional consumers who have switched their consumption to organic produce move the demand curve from $D_O0$ to $D_O1$. The new price of organic products is $P_O1$ and the quantity $Q_O1$. Thus there is a gain to organic farmers—a gain in producer surplus—of area $P_O0BAP_O1$. This gain to organic farmers is often ignored in assessments of the introduction of GM technology given that policy-makers are focussed on potential 'losers' arising from the advent of a new technology.[15]

While policy-makers and regulators are empowered (and in some cases directed) to use cost-benefit as part of risk analysis, in most cases regulatory decisions are focused on avoiding or mitigating losses rather than realizing those gains that provide adequate returns to compensate losers. In the context of GM regulation some jurisdictions unapologetically concentrate on the potential 'losers' that may arise from the commercialization of GM varieties. One policy focus is on preventing 'losers' in the organic industry. The policy is sometimes identified as ensuring 'coexistence' between organic and GM products but the real intent is to prevent 'losers' arising in the organic industry. As the organic industry has self-declared itself to be producing only non-GM products, any comingling of GM and non-GM varieties in their outputs is posited to threaten producer's ability to sell their products as organic and to receive the price premium. The possibility of genetic drift (or unintentional comingling of organic and GM seed in the supply chain) could jeopardize the certified organic status of some organic farms.[16] In the context of Fig. 2.3, as producers 'leave' the organic industry the supply curve shifts to the left in the right hand panel of the figure—to $S_O1$. As a result there is a loss of producer surplus for the organic industry—the vertical distance between $S_O0$ and $S_O1$. The 'losers' are those organic farms where comingling has been detected.[17]

Coexistence policy aims to prevent these 'losers' from arising. Coexistence policy seeks to create physical or managerial barriers between farms where GM production takes place and organic farms in order to prevent genetic drift from causing organic farmers to lose their organic status. This could result in GM producers being required to create buffers zones around their fields to localize the genetic drift of their GM crops—but as a result reducing their productivity as less

[15]There is also a gain in consumers benefit arising from the shift from the GM market to the organic market by some consumers. This is because consumers must perceive a greater utility from shifting to the new market than the loss of utility from leaving their existing market—otherwise they would not switch from one market to the other. This means that the shift outward in demand in the organic market is larger than the inward shift of demand in the GM market.

[16]While this would not be the case in either Canada or the US where certification only deals with intentional use of GM seeds, in Australia an organic farmer lost his organic certification due to the unintended presence of GM canola in his organic fields.

[17]Of course, organic farms that retain their status—depicted by $S_O1$—receive an increase in producer surplus because the price they receive will rise above $P_O1$ ($S_O1$ intersects $D_O1$ above $P_O1$).

area can be planted to GM crops. In the left hand panel of Fig. 2.3 these reductions in productivity (increased costs per unit of output) mean that the supply curve shifts upward from $S_{GM}1$ to $S_{GM}2$. This decreases producer surplus in the GM industry. The more onerous the coexistence rules for GM farmers the more the supply curve will shift. In the making of coexistence policy this loss is not considered because the focus is on preventing 'losers' arising in the organic industry.

If the costs of coexistence policies are sufficiently high it may 'wipe out' all of the efficiency gains associated with the new technology (i.e. if the cost of coexistence policy shifts the GM supply curve back to $S_C0$ or beyond).[18] From the societal point of view there is a clear trade-off between the loss of surplus in the organic market and the loss of producer surplus in the GM market. The more onerous the coexistence policy is for GM producers the further the GM supply curve shifts and the less the organic supply curve shifts. Thus, there is an optimal coexistence policy whereby the total loss of surplus (from the two markets) is minimized.

Current coexistence policies, as suggested above, are simply focussed on preventing any shift in the organic supply curve. Policies are often constructed with an objective of zero tolerance (Hobbs et al. 2013), which would mean the organic supply curve would not shift. While this could jeopardize important welfare gains in the GM market, these losses are seldom taken into account. The result is sub-optimal from society's perspective, but not of course for the vested interests in the organic industry. In some cases the inability to design a 'viable'[19] coexistence policy has led to 'temporary' bans on licencing GM products[20] with the result that the gains to society from the new technology are forgone in the name of preventing 'losers' arising in the organic industry.

Clearly a serious policy issue arises from the perception that 'winner' only technologies exist and that any technologies that create 'losers' should not be tolerated. All technologies create both 'winners' and 'losers' and policy-makers need to examine both the benefits and costs associated with new technologies and not simply the costs. This is the way SECs are intended to be treated in assessments of new technologies and the most appropriate morally defensible basis for policy engagement.

[18]If the costs of coexistence shift the GM supply curve up to those of $S_C0$ or beyond then producers have no incentive to produce GM varieties and should return to the production of conventional varieties. Society would lose all the benefits of the GM technology. Ironically, if production remains (or returns to) conventional seeds there is no reason for consumers to switch to organic products—because no GM product exists—and the organic industry does not reap the windfall increase in producers' surplus that the existence of GM varieties creates.

[19]Or one that can be agreed upon.

[20]That have cleared the regulatory hurdles for food safety and environmental risk.

## Conclusions

Innovations change societies into which they are commercialized (Phillips 2007). Some in society will benefit more than others (aka the 'winners'), while others that have restricted means or ability to adopt or suffer from negative cross-market effects (aka 'losers') have typically been thought to be disadvantaged. However, this chapter demonstrates that some non-adopters can also benefit from some innovations. In our example, organic farmers have benefited from the commercialization of GM crops as demand for their products has increased due to the desire of some consumers to avoid consuming GM products; all other things being equal, there is no expectation that these consumers would have purchased organic products had GM crops not been commercialized. Organic farmers have also benefitted from lower costs and improved yields due to reduced pest populations arising from widespread adoption of, for example, Bt corn.

In the course of this research, and in the absence of any published data, major downstream firms in the organic industry were contacted and asked to what degree they thought the demand for organic products was affected by consumers seeking to avoid GM foods. One respondent asserted[21]:

> On balance, I believe that organic sales have been boosted perhaps 10% by consumer concerns over GMO. That is belief and not provable fact. Ten percent would be worth perhaps $3 billion. (Clarkson, personal communication).

In direct response to our query,[22] the same firm provided the following information:

> We have some buyers who tell us that they are motivated to buy organic to avoid GMOs. Others indicate that avoiding GMOs is just part of the package they seek in buying certified organic products. Almost all organic buyers place some significant value on avoiding GMOs.
>
> Had the organic rules accepted GMO as an approved practice, organic might have lost 50% or more of its demand to non-GMO verified products raised conventionally with synthetic fertilizers and pesticides.
>
> The premiums for non-GMO trend toward 15% of the underlying commodity price. Current premiums for non-GMO soybeans run about $1.50/bu with a base commodity price of $9.30/bu; for corn, about $0.45 with a base corn price of $3.12/bu. (Clarkson, personal communication).

While not compelling evidence that the organic industry has benefitted from the existence of GM foods, it is consistent with the theoretical framework developed above and suggests that these positive benefits should be included in socio-economic public policy assessments of GM products. It is an area where further applied research is warranted.

[21]While a number of firms were contacted and asked to fill out our brief survey, only Clarkson replied.

[22]A copy of the questions are available from the authors.

The eNGO community, however, have lobbied and pressured many developing countries to implement broad socio-economic considerations as part of their domestic biosafety regulatory framework. In Mali, for instance, eNGOs encouraged the government to enact regulations that will not allow any negative socio-economic impacts to occur from the commercialization of GM crops (CBD 2014). Innovation is not possible within this type of regulatory environment as innovations will inevitably displace older, inefficient technologies, creating losers among those not able or willing to adopt, in cross markets and in the input markets.

The encouragement of such policy development by eNGOs ultimately fosters increased food insecurity. When small landholder farmers are denied access to innovations, they are forced to maintain the existing low-skilled, labor-intensive work that they have undertaken for centuries, while landholders in neighboring countries are able to reap the benefits of GM crops, including greater yields, increased profits, rising household incomes and enhanced food security.

## References


Alston, J.M., N. Kalaitzandonakes, and J. Kruse. 2014. The size and distribution from the adoption of biotech soybean varieties. In *Handbook on agriculture, biotechnology and development*, ed. S.J. Smyth, P.W.B. Phillips, and D. Castle, 728–751. Cheltenham, UK: Edward Elgar Publishing Ltd.

Bryson, N. 2016. Lessons from the legal cases of GM alfalfa and sugar beet deregulation in the United States. In *The Co-existence of genetically modified, organic and conventional foods: Government policies and market practices*, ed. N. Kalaitzandonakes, P.W.B. Phillips, S. J. Smyth, and J. Wesseler, 103–112. New York: Springer.

Canadian Soybean Council. 2014. Industry statistics. Available at: http://www.soybeancouncil.ca/IndustryStatistics/tabid/200/language/en-US/Default.aspx.

Canola Council of Canada. 2014. HT and conventional canola: 1995–2010. Available at: http://www.canolacouncil.org/markets-stats/statistics/estimated-acreage-and-percentage/.

Clarkson, L. 2014. President, Clarkson Grain.

Congressional Research Services. 2013. Deregulating genetically engineered alfalfa and sugar beets: Legal and administrative responses. Available at: https://www.fas.org/sgp/crs/misc/R41395.pdf.

Convention on Biological Diversity. 2014. Global overview of the information on socioeconomic considerations arising from the impact of living modified organisms on the conservation and sustainable use of biological diversity.

Gaisford, J.D., and W.A. Kerr. 2001. *Economic analysis for international trade negotiations*. Cheltenham, UK: Edward Elgar Publishing Ltd.

Gaisford, J.D., J.E. Hobbs, W.A. Kerr, N. Perdikis, and M.D. Plunkett. 2001. *The economics of biotechnology*. Cheltenham, UK: Edward Elgar Publishing Ltd.

Gouse, M. 2014. Labor impacts. In *Socio-economic considerations in biotechnology regulations*, ed. K. Ludlow, S.J. Smyth, and J. Falck-Zepeda, 189–200. New York: Springer.

Hategekimana, B., and M. Beaulieu. 2002. Genetically modified crops: Steady growth in Ontario and Quebec. Catalogue no. 21-004-XIE. Available at: http://www.publications.gc.ca/Collection/Statcan/21-004-X/21-004-XIE2002112.pdf.

Hobbs, J.E., and W.A. Kerr. 2003. Biotechnology and the changing structure of agri-food supply chains. *International Journal of Biotechnology* 5 (3&4): 297–310.

Hobbs, J.E., W.A. Kerr, and S.J. Smyth. 2013. How low can you go? The consequences of zero tolerance. *AgBioForum* 16 (3): 207–221.

Huang, J., J. Mi, H. Lin, Z. Wang, R. Chen, R. Hu, S. Rozelle, and C. Pray. 2010. A decade of Bt cotton in Chinese fields: Assessing the direct effects and indirect externalities of Bt cotton adoption in China. *Science China Life Sciences* 53 (8): 981–991.

Hutchison, W.D., E.C. Burkness, P.D. Mitchell, R.D. Moon, T.W. Leslie, S.J. Fleischer, M. Abrahamson, K.L. Hamilton, K.L. Steffey, M.E. Gray, R.L. Hellmich, L.V. Kaster, T.E. Hunt, R.J. Wright, K. Pecinovsky, T.L. Rabaey, B.R. Flood, and E.S. Raun. 2010. Areawide suppression of European corn borer with Bt maize reaps savings to non-Bt maize growers. *Science* 330 (6001): 222–225.

Isaac, G.E. 2007. Sanitary and phytosanitary issues. In *Handbook on international trade policy*, ed. W.A. Kerr, and J.D. Gaisford, 383–393. Cheltenham: Edward Elgar.

James, C. 1996–2013. ISAAA Briefs. Available at: http://www.isaaa.org/Resources/publications/briefs/.

Kerr, W.A. 2010. *Conflict, chaos and confusion—The crisis in the international trading system*. Cheltenham: Edward Elgar.

Leger, L.A., J.D. Gaisford, and W.A. Kerr. 1999. Labour market adjustments to international trade shocks. In *The current state of economic science*, vol. 4, ed. S. Bhagwan Dahiya, 2011–2034. Rohtak: Spellbound Publications Pvt. Ltd.

Phillips, P.W.B. 2007. *Governing transformative technological innovation: Who's in charge?*. Oxford: Edward Elgar.

Smyth, S.J., and P.W.B. Phillips. 2016. Commercialization strategies and market opportunities for GM canola. In *The co-existence of genetically modified, organic and conventional foods: Government policies and market practices*, ed. N. Kalaitzandonakes, P.W.B. Phillips, S. J. Smyth, and J. Wesseler, 185–196. New York: Springer.

Smyth, S.J., P.W.B. Phillips, and D. Castle (eds.). 2014. *Handbook on agriculture, biotechnology and development*. Cheltenham, UK: Edward Elgar Publishing Ltd.

Sterk, R. 2014. GMO debate grows as crop area rises. World Grain, 61–66.

# Chapter 3
# Domestic Regulatory Approval Costs

**Abstract** Concerns over the effects on human and animal health from consumption of genetically modified food crops, as well as about any potential impacts on the wider environment, have divided public opinion over the merits of using the technology. One consequence is that governments have, for the most part, been very cautious about approving GM crops for cultivation and consumption. A concomitant effect is that the costs involved (and time spent) in achieving regulatory approval for the introduction of a GM crop have risen. This chapter examines the evidence on the absolute and relative costs of gaining regulatory compliance within countries. A range of studies show that there are significant costs of gaining regulatory compliance—product attributes, market context and technical aspects of measurement all have significant effects on the estimated costs of regulatory approval.

## Introduction

The introduction of transgenic technology into agriculture in the 1980s triggered significant regulatory reform. After considerable debate and investigation, the US and eventually most initial developers and adopters of genetically modified (GM) crops put in place new guidelines and procedures for evaluating the safety and efficacy of crops using this new technology. The net effect was that proponents of new crop varieties embodying GM traits were required to undertake a more intensive pre-market assessment which would be adjudicated by national regulators wherever the varieties were intended to be produced or consumed—previously most crop varieties developed through traditional plant breeding methods were initially (and often only) assessed in the country where the variety was developed and first commercialized. By 2013, most countries had enacted some form of regulation or policy directly targeted at GM crops, whether they are intended to be planted, consumed, processed or traded. This policy change imposes both real and opportunity costs on both the technology and the economy.



S.J. Smyth et al., *Biotechnology Regulation and Trade*, Natural Resource Management and Policy 51, DOI 10.1007/978-3-319-53295-0_3

This chapter assesses the scale and scope of the use of GM technologies and the measurable costs of regulating them at the product and market level. In the first instance, the nature of the regulatory system directed at GM crops is examined and then the evidence about costs and impacts presented in a range of studies is discussed. These studies are then assessed using meta-analysis to identify potential causes for differential costs (or at least for differential estimates). Finally, the chapter puts these costs into context of the real and potential impact of the technology on global food security and economic welfare.

## Background

In 1996 the first significant area of crops containing GM traits were planted (1.7 million hectares). As suggested in the introduction to this book, since then there has been a dramatic increase in plantings; by 2015, the global planted area reached almost 180 million hectares (James 2015). As suggested above, GM traits have largely been adopted in four main crops—canola, corn, cotton and soybeans—although small areas of GM crops in sugar beet, papaya and squash (in the US since 2004) have also been planted. In 2015, 28 countries planted GM crops, more than half developing nations. Six countries accounted for 92% of total production: US, Brazil, Argentina, India, Canada and China. Two traits dominate: herbicide tolerant (HT) crops account for 65% of the total area planted to GM crops while insect resistant (IR) crops account for 35% of global plantings; a rising portion of the area involves stacked HT and IR traits. GM seeds account for 70% of the global soybean acreage, 52% for cotton, 26% for corn and 20% for canola. In those countries adopting GM varieties, GM seed market share has risen above 80% in those crops. Brookes (2014) estimates that biotechnology producers account for between 72% of cotton and 95% of global trade in soybeans.

While there has been ongoing examination of the underlying regulatory regimes in the divergent countries (e.g. Isaac 2002; Skogstad 2008; Doern and Prince 2012), that is not the focus of this chapter. The main point to draw from that literature is that there are fundamental differences of approach and impact, which contributes to asynchronous reviews and approvals. In 19 countries that had examined and approved at least one of the 144 events proposed by 2011, the average country had undertaken reviews on only about 5 of the 16 species under investigation and completed environmental reviews on 12% of the possible cases and finished food safety audits on 27% (Phillips 2011b). No single country has reviewed and recorded positive approvals for all of the 144 events proposed as of 2011. The highest rates of positive review were in Canada and the US, two of the largest producers and exporters of GM crops, and Japan, a key importer of foodstuffs with GM traits. The low level of successful review for environmental release reflects the politics of the adopting countries while the somewhat higher percentage of completed food reviews reflects the reality that much of the international trade now includes GM elements.

One aspect to keep in mind is that companies make their own decisions about where and when to seek approval for commercialization. Soybeans and maize/corn have the highest penetration rates, averaging around 10% of the total countries producing those crops. Nevertheless, those countries adopting GM varieties account for an estimated 73% of global soybean area and 30% of global maize/corn area. While one might interpret incomplete adoption as reflecting the unwillingness of countries to accept GM technologies, in many cases it is simply a business-based decision.

Truncated adoption can be the result of two separate decisions. Many interpret the limited regulatory acceptance of GM crops as a judgment of regulators—some assume that stalled introductions are because companies have failed to satisfy the regulators of the health and environmental safety of their products. In fact, in many cases proponents are simply waiting for regulators to make their judgments. Few if any products have been explicitly rejected for health or environmental safety reasons by any regulator anywhere. What appears to be transpiring is that many proponents have simply not applied for regulatory assessment in small markets. In many cases it is simply a business decision based on the expectation that there would not be adequate revenues for the technology owners to justify the investments in acquiring regulatory compliance and developing a local supply chain. A number of tentative estimates of the costs of regulatory compliance in developing countries suggests the upfront cost per country have ranged from US$500,000 to US$5 million for the first GM event in a species and approvals have taken from 2 to 7 years to complete, with subsequent GM traits in the same crop being less expensive and somewhat more timely (Kalaitzandonakes et al. 2007; Pray et al. 2005, 2006; Bayer et al. 2010). To illustrate the scale of the problem, assume a biotechnology company could generate a free cash flow of US$10 per hectare planted (which would depend critically on the structure of local intellectual property laws and the structure of the local seed industry) and they attained an above average rate of farmer adoption, there are at least 40 countries producing maize/corn where it is unlikely for a biotechnology company to recoup even the lowest likely regulatory costs within 10 years of starting the process. Given there would also be extra development costs to backcross their proprietary traits into cultivars appropriate for those markets, the number of unprofitable markets is probably even higher. Few companies are willing to take commercial risks of this magnitude.

This chapter assesses the primary and secondary evidence that delimits the scale and scope of regulatory costs and uses meta-analysis to untangle the drivers of differential estimates by product and market.

## Meta-analysis

The practice of meta-analysis has been developed and applied to a number of disciplines in recent years. Meta-analysis is essentially an analysis of analyses (Hedges and Olkin 1985). The basic approach is to aggregate research findings statistically and calculate a set of explicit or implicit weights for the key factors underlying the findings. A more traditional approach to surveying the evidence would involve narrative reviews and tables of a selection of studies (as will be presented below). The terminology of meta-analysis is often credited to the Cochrane Collaboration (http://www.cochrane.org/) which in 1993 introduced multivariate regression to examine the impact of moderator variables on study effect size using regression-based techniques. The goal is to combine the results of two or more separate studies to generate weighted averages of moderator effects on the results in order to increase explanatory power, answer questions not posed by individual studies and to settle controversies arising from conflicting claims.

In the economics, meta-analysis has been applied to a range of questions through the use of 'instrumental' variables in regression, including returns to agricultural research, environmental evaluation, firm structure and strategy and their impact on innovation and wealth creation, consumer valuation, industrial structure and risk tolerance and the statistical value of a life (Table 3.1). In each case, analysts identify an area of substantial research that has generated a diversity of empirical results for an instrumental variable, agglomerate a set of representative estimates (often the number of estimates is greater than the number of studies as authors investigate different specifications and simulations) and identify a set of 'moderators' that may help to explain the diversity of estimates from the instrumental variables. All of these studies use multiple regression techniques to meta-analyze the effect of key factors on the estimate of the instrumental variable.

While meta-analysis has been adapted and adopted by regulators for use inside the system (e.g. to assess and examine a body of empirical studies of epidemiological or environmental impacts), so far no one has used it to assess the system

**Table 3.1** Representative meta-analyses

| Study | Topic | Studies | Observations | Moderators |
|---|---|---|---|---|
| Alston et al. (2000) | Returns to agricultural R&D | 294 | 1858 | 46 |
| Brouwer et al. (1999) | Wetland contingent evaluation | 30 | 103 | 30 |
| Camisón-Zornoza et al. (2004) | Organizational size and impact on innovation | 53 | 57 | 25 |
| Datta et al. (1992) | Factors influencing wealth creating from mergers and acquisitions | 41 | 409 | 13 |
| Lusk et al. (2005) | GM food valuations | 25 | 57 | 16 |
| Marra and Schurle (1994) | Farm level risk and farm size | 731 | 2193 | 6 |
| Mulatu et al. (2003) | Impact of environmental regulation on international trade | 13 | 691 | 22 |
| Viscusi and Aldy (2002) | Value of a statistical life | 49 | 49 | 27 |

itself. Executive Order 12866 on Economic Analysis of Federal Regulations (OMB 1996) offers a cautious endorsement to meta-analysis, noting that combining data or results from a number of different studies can allow one to re-estimate key model parameters, thereby improving confidence in the parameter estimates or alternatively use parameter estimates (e.g. elasticities of supply and demand or implicit values of mortality risk reduction) from different studies as data points to analyze variations as functions of potential causal factors. The Order cautions that meta-analysis must be used carefully, ensuring that the data used are comparable, that appropriate statistical methods are used and that the analysis is undertaken only on studies that measure comparable independent and dependent variables. This signpost is used to guide our choice of studies to include in our analysis.

## The Data

The basic challenge is that the costs of full regulatory approval of a biotech crop are not publicly available (and may not actually exist anywhere). At the root of the problem is that full information regarding private approval costs have been closely guarded by biotechnology developers (Kalaitzandonakes et al. 2007). Firms appear to believe that the evidence of their capacity to efficiently and effectively complete the regulatory process is a core competency and, hence, a proprietary asset that, if more fully disclosed, might be reverse-engineered and increase competition in a market often characterized as being an oligopoly.

Three types of cost analyses have been undertaken thus far that have produced a very wide range of estimates of the costs of regulation. A number of economists have worked to construct estimates of the direct cash costs for proponents and regulators of completing the regulatory review of new products, largely based on data gathered from regulators, a few firms and a number of public sector scientists. A subset of economists have looked at regulation as a structural barrier, largely drawing on assumptions about the direct cash costs of regulatory compliance to calculate the opportunity costs of the current regulatory structure. Meanwhile, the big six agricultural biotechnology companies commissioned a study of their all-in costs of bringing a product to market. Given the diversity of approaches, it should not be any surprise that the estimates range widely. Studies put the cash cost of compliance as low as US$73,000 for a single event in a single market to as high as US$35 million for cultivation approval in two countries and import approvals in five markets (what the industry asserts is the norm for full introduction of a major crop trait). Meanwhile, the all-in opportunity cost of developing and introducing a trait is estimated to be more than US$130 million, with the dead-weight opportunity cost of regulatory delay adding at least as much to the regulatory burden for firms and the market.

As with many public policy issues, the devil is in the details. The main difference in the estimates can be found by examining what is included and excluded. Phillips (2011a) offered a timeline for the development of a new GM variety (see Fig. 3.1),

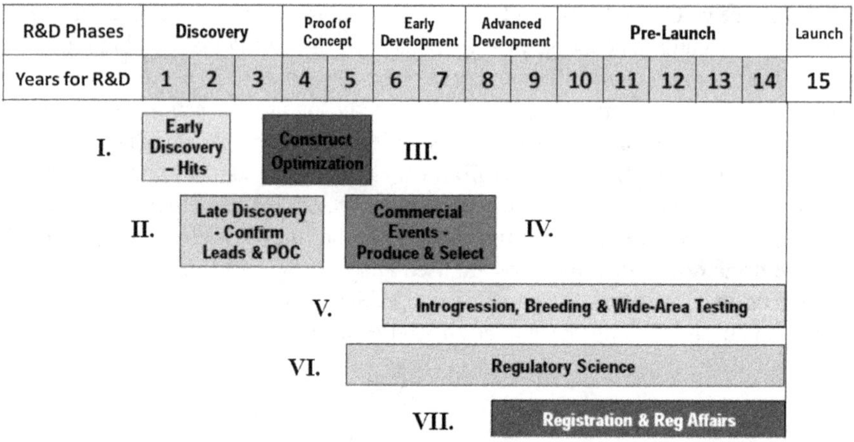


**Fig. 3.1** The research, development, regulatory and commercial pathway. *Source* Phillips (2011a)

positing that the approximate 15 year research, development and commercialization process involves up to seven interconnected stages, only two of which many would explicitly call regulatory compliance: regulatory science (VI); and registration and regulatory affairs (VII). Nevertheless, the earlier stages of discovery, construct optimization, commercial event selection, introgression and wide-area testing each provides some of the scientific evidence that are explicit inputs to the regulatory process. Comparing and contrasting the different estimates requires knowing what is included and how it is calculated.

In an effort to be comprehensive, a literature survey was conducted to collect available studies on regulatory approval costs. The collection of literature involved a keyword search of 'GMO' and 'regulatory costs' in the databases of Thomson Reuters' Web of Science, Research in Agricultural and Applied Economics (AgEcon-Search), AgBioForum, Google Scholar, BEcon and Repec. Other literature (often termed the 'grey market') included conference presentations, policy briefs and a consultancy report on costs in the EU and the US. The literature search was limited to English language publications.

The survey identified more than a dozen articles and reports that offer more than 70 estimates of cost of regulatory approval in 16 countries (Argentina, Australia, Brazil, Canada, China, Costa Rica, the European Union, India, Indonesia, Japan, Kenya, South Korea, the Philippines, South Africa, Taiwan and the United States) involving a dozen crops and more than 10 different genetic constructs. These studies offer three complementary but substantially different estimates of the cost of regulation. Firstly, a core group of economists has constructed direct cost estimates of regulatory compliance incurred by both proponents and regulators for single events in single markets. Secondly, a number of welfare economists have estimated the opportunity costs of the regulatory process, which includes both the out-of-pocket expenses for both parties and the imputed expense of delays in commercialization, both for the proponent and for society (using measures of social welfare). Thirdly, private sector firms have constructed a set of private costs incurred throughout the discovery and commercialization process, attributing some but not all to the burden of regulatory compliance. As one would expect, the estimates overlap but the means estimation differ by at least one order of magnitude due to the underlying assumptions and breadth of the methods (Table 3.2).

To put this into context, Fig. 3.2 shows how these different estimates correlate and overlap. The constructed cost approach generally limits the analysis to areas A + B of Fig. 3.2, while welfare economists are concerned about A + B plus the opportunity costs (C). In contrast, the biotechnology firms are exclusively interested in their cost and benefits, focusing on A + D; while they are concerned about their share of the imputed opportunity costs (part of C), they seldom try to make their case based on this type of analysis. This disconnect is somewhat puzzling, as the opportunity costs of delay (C) have the largest impact on the long-term returns of any new technology. One possible explanation is that while opportunity costs

**Table 3.2** A concordance of included and excluded variables in cost estimates

| | Costs constructed by economists | Economic welfare analysis | Costs constructed by industry |
|---|---|---|---|
| Example | Kalaitzandonakes et al. (2007) | Bayer et al. (2010) | Phillips (2011a) |
| Total discovery cost | Not included | Not included | Private direct costs calculated; some undetermined portion attributed to regulatory requirements |
| Construct optimization | | | |
| Commercial event selection | | | |
| Breeding and wide-area testing | | | |
| Regulatory science | Public and private direct costs constructed for single events in single markets | Public and private direct costs assumed or based on exogenous estimates | Private direct costs calculated for global introduction of event |
| Registration and regulatory affairs | | | |
| Opportunity cost | Not included | Indirect costs calculated | Not included |
| Relative areas in Fig. 3.2 | A + B | A + B + C | A + D |

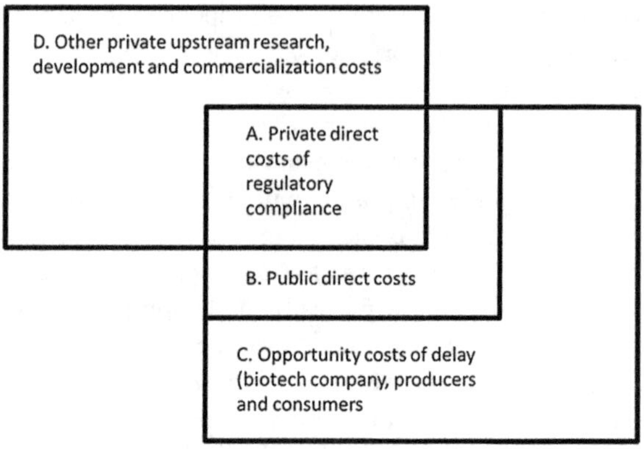


**Fig. 3.2** The relationship between different estimates of regulatory costs

represent real value in the context of new technology development, most of it does not directly accrue to the biotechnology firms themselves, but instead flows to producers and consumers.

## *Constructed Cost Analyses*

More than 65 cost estimates were developed by three groups of economists (which were organized into 10 different teams to calculate different types of estimates for various products and markets): Carl Pray from Rutgers led a series of cost estimates for products in Asia; Nicholas Kalaitzandonakes from University of Missouri/Columbia worked with a team to estimate costs for regulatory compliance for HT and Bt maize/corn in the US, EU, Canada, Argentina, Australia, China, Japan, Korea and the Philippines; and Jose Falck-Zepeda from IFPRI has worked with a range of economists to develop estimates across a range of food and non-food crops in developing countries.

One challenge with much of the general literature is that the explicit methodologies for constructing the cost estimates are often not fully disclosed. In this case, however, the three main teams have offered detailed disclosures of what is included and excluded, and in many instances offer illustrations or details on how the costs were imputed or derived. Table 3.2 shows the array of explicit costs identified in each of those studies. While the terminology and descriptions vary, for the most part they consistently include most of the same activities generally required for regulatory compliance. All biotechnology crops are subjected to a battery of tests and significant regulatory scrutiny before commercialization. To identify these general categories of compliance costs, lead scientists and regulatory affairs practitioners (identified through a review of regulatory submissions) were interviewed in Kalaitzandonakes et al. (2007). In addition, data were collected through assessment of representative dossiers for various novel maize traits submitted over the past 10 years. Using these representative dossiers and the relevant data provided by biotech developers, costs were structured by identifying aggregate categories that were characteristic across all types of dossier submissions. Finally, the authors undertook an evaluation of the degree of overlap among multiple submissions for the same technology across various national regulatory systems and the incremental compliance costs associated with each additional international market where regulatory approval was sought. Pray et al. (2005) report data were also gathered from a number of private companies and from national regulators (Table 3.3).

While there are more than 65 cost estimates, the constructed cost studies have been divided into two categories. Forty nine cost estimates developed by the three leading research teams for three key crops (maize/corn, cotton and soybean) and two key technologies (HT and Bt) that had well disclosed methodologies were able to be isolated. Each of these products achieved regulatory compliance and commercial entry into at least one market in the period under review. Those studies are presented in Table 3.4 and are the focus for the meta-analysis that follows.

**Table 3.3** Methodologies and categories of costs considered

| | Kalaitzandonakes et al. (2007) | Pray et al. (2005) | Falck-Zepeda et al. (2006) |
|---|---|---|---|
| Pre-submission costs | Yes | Yes | Yes |
| Molecular work | Composition assessment; development of ELISA test; production of tissues | Expression in oils and lint; gene stability | Molecular characterization and testing; ELISA development |
| Animal safety | Performance and safety studies | Performance and safety studies | Poultry, goat, cow and fish feeding studies |
| Food safety | Protein testing; 90 day rat toxicology trial | Brown Norway rat allergenicity trial | 90 day rat toxicology and allergenicity trials |
| Environmental safety | Non-target organisms studies; agronomic and phenotypic assessment; environmental fate studies; stewardship | Pollen flow; soil microflora; greenhouse trials, limited field trials, multi-location field trials, large-scale trials, large-scale farm trials | Gene flow; impact on non-target organisms; agronomic and phenotypic assessment baseline and follow-up resistance studies; non-target organism studies |
| Regulatory fees | Yes | Yes | Yes |
| Regulator expenses | Yes | Yes | Not clear |
| In-house costs | Facility and management overhead | Salaries and overhead expenses | Salaries and overhead costs |
| Socio-economic studies | Na | Yes | Yes |
| Post approval costs | Na | IPM package | Na |

**Table 3.4** Estimates for regulatory approval costs for cotton, maize and soybeans

| Product | Trait | Crop type | Country | Approval cost $US 000 | Approach | Author/year |
|---|---|---|---|---|---|---|
| Cotton | Bt | Non food | India | $1999 | Ex post | Pray et al. (2005) |
| | Bt | Non food | China | Private: $89.50 Public: $53.00–$61.00 | Ex post | Pray et al. (2006) |
| | Bt | Non food | Indonesia | $93.50 | Ex post | Falck-Zepeda et al. (2006) |
| | Ht | Non food | Indonesia | $112.48 | Ex ante | |
| Maize | Bt | Food | China | $500 (private) | Ex ante | Pray et al. (2006) |
| | Bt | Food | Philippines | $1700 | Ex post | Falck-Zepeda et al. (2006) |
| | Bt | Food | India | $500-$1500 | Ex ante | |
| | Bt | Food | Kenya | $980 | Ex ante | |
| | Ht | Food | Indonesia | $106.85 | Ex ante | |
| | Bt | food | US | $6790–$14,840 | Ex post | Kalaitzandonakes et al. (2007) |
| | Ht | Food | US | $5910–$13,910 | | |
| | Bt | Food | EU | $6870–$14,530 | | |
| | Ht | Food | EU | $6140–$14,315 | | |
| | Bt | Food | Argentina | $6640–$14,575 | | |
| | Ht | Food | Argentina | $5910–$13,910 | | |
| | Bt | Food | Canada | $6680–$14,320 | | |
| | Ht | Food | Canada | $5950–$14,105 | | |
| | Bt | Food | Australia | $6640–$14,515 | | |
| | Ht | Food | Australia | $5910–$13,910 | | |
| | Bt | Food | China | $6640–$14,515 | | |
| | Ht | Food | China | $5910–$13,910 | | |
| | Bt | Food | Japan | $6640–$14,515 | | |
| | Ht | Food | Japan | $5910–$13,910 | | |
| | Bt | Food | Korea | $6,64–$14,515 | | |
| | Ht | Food | Korea | $5910–$13,910 | | |
| | Bt | Food | Philippines | $6640–$14,515 | | |
| | Ht | Food | Philippines | $5910–$13,910 | | |
| Soybeans | Ht | Food | Brazil | $4000 | Ex ante | Falck-Zepeda et al. (2006) |

Another 17 studies offered estimates of costs to gain regulatory compliance for a range of other crops, but all were ex ante assessments of products that had not completed regulatory review during the period under review, and in many cases the methodology and data sources were less than complete (Table 3.5). While these

**Table 3.5** Estimates of regulatory approval costs for other crops

| Product | Trait | Crop type | Country | Approval cost $US 000 | Approach | author/year |
|---|---|---|---|---|---|---|
| Rice | Bt | Food | Indonesia | $73 | Ex ante | Falck-Zepeda et al. (2006) |
| | Bt | Food | India | $1500–$2000 | Ex ante | Falck-Zepeda and Cohen (2006) |
| | Bt | Food | Costa Rica | $2800 | Ex ante | Falck-Zepeda and Cohen (2006) |
| | Beta-carotene SGR1 | Food | Philippines | $134.5 | Ex ante/Ex post | Falck-Zepeda et al. (2006) |
| | Xa21 Bac. Blight R. | Food | Philippines | $127.6 | Ex ante/Ex post | Falck-Zepeda et al. (2006) |
| Rice | Bt | Food | Philippines | $690.7 | Ex ante | Bayer et al. (2010) |
| Beans | VR | Food | Brazil | $700 | Ex ante | Falck-Zepeda and Cohen (2006) |
| Vegetable | Na | Food | India | $4000 | Ex ante | Falck-Zepeda and Cohen (2006) |
| Potatoes | Bt* | Food | South Africa | $980 | Ex ante | Falck-Zepeda and Cohen (2006) |
| | Na | Food | Brazil | $980 | Ex ante | Falck-Zepeda and Cohen (2006) |
| Jute | Na | Non Food | India | $1000–$1500 | Ex ante | Falck-Zepeda and Cohen (2006) |
| Mustard | Hybrid | Food | India | $4103–$5103 | Ex post/ex ante | Pray et al. (2005) |
| Eggplant | Bt | Food | India | $53.6 | Ex post/ex ante | Pray et al. (2005) |
| | Bt | Food | Philippines | $475 | Ex ante | Bayer et al. (2010) |
| Tomato | MVR | Food | Philippines | $475 | Ex ante | Bayer et al. (2010) |
| Papaya | RSV | Food | Philippines | $248.5 | Ex ante | Bayer et al. (2010) |
| | Delayed Ripening | Food | Philippines | $349.2 | Ex post | Falck-Zepeda et al. (2006) |

offer a tantalizing glimpse into the dispersion of costs over a wider range of countries, crops and traits, they are qualitatively different than the studies presented in Table 3.4 and mindful of the directions in Executive Order 12866, they have not been formally incorporated into our meta-analysis.

**Table 3.6** Average costs categorized by moderator

| | N | Mean | Std dev | Min | Max |
|---|---|---|---|---|---|
| All studies | 49 | 7,808,272 | 5,384,964 | 53,000 | 14,840,000 |
| Bt trait | 28 | 7,112,000 | 5,670,146 | 53,000 | 14,840,000 |
| HT trait | 21 | 8,736,635 | 4,825,553 | 106,850 | 14,315,000 |
| Food uses | 43 | 8,841,787 | 4,924,380 | 106,850 | 14,840,000 |
| Non-food uses | 6 | 401,413 | 714,742 | 53,000 | 1,999,000 |
| Net exporter | 16 | 8,217,438 | 5,025,254 | 500,000 | 14,840,000 |
| Net importer | 33 | 7,609,889 | 5,540,118 | 53,000 | 14,530,000 |
| Ex ante study | 7 | 1,099,904 | 1,268,629 | 106,850 | 4,000,000 |
| Ex post study | 42 | 8,926,333 | 4,981,180 | 53,000 | 14,840,000 |

The 49 estimates presented in Table 3.3 suggest that the average cost of regulatory compliance for a single new trait in a single market is US$7.8 million, with a minimum cost of US$53,000 and maximum of US$14.8 million (Table 3.6). For the purposes of this study, the interest was in determining whether the costs differ significantly based on four key sets of moderators. First, about half the reviews examined candidate crops with Bt and half with HT traits; given their different agronomic and environmental effects in different markets, that could affect the cost of compliance. Second, not all of the crops were targeted for acceptance for human consumption. Most Bt cotton was not intended for human consumption (even though a small amount of cottonseed oil is consumed) but most corn and soybeans were assumed to be available for direct human consumption (even though most early innovations were directed to varieties that were predominantly used as animal feed). Third, the nature of the market where approval was being sought could conceivably affect the nature of the review, about one third of the estimates looked at reviews by importing countries and the rest by exporting countries. Finally, it is possible that the timing of the evaluation may affect the nature of the estimate, most of the evaluations involved drawing explicit ex post evidence from regulators or proponents on the costs of the process; a few were ex ante constructions based on theoretical assays or expert judgment.

On the basis of a cursory examination, while there is no statistical evidence to suggest that Bt or HT crops are dealt with differently or that exporters and importers impose different regulatory burdens on GM crops, there is some support for the notion that crops intended for use as human food are more costly to regulate than non-food uses (Table 3.6). Finally, the data suggests that ex post studies generate cost estimates on average eight times larger than ex ante projections—which is statistically significant.

One aspect not considered is whether the nationality or corporate status of the developer affects regulatory costs. There is some anecdotal evidence that some developing countries (esp. China, India and Brazil) may impose higher assessment costs on events proposed by multinational corporations (perhaps because they are first-movers or perhaps as an anti-competitive action by governments); in contrast, local public sector institutes producing competitive traits appear to have lower costs and speedier reviews (Pray 2013). Both the Chinese and Indian governments have programs to subsidize the cost of regulatory approvals for government research institutes and local private companies, but not for private companies (this could affect the constructed costs but would not lower the overall cost of regulation). While it would be worth testing further for this, it is not always clear in the 49 identified studies how the regulators or subsidy programs judge the provenance of the events considered—quite often candidate GMOs are the product of a collection of partnerships involving public, private and collective actors, making it difficult to clearly assign provenance. As a result, we have not included this as a moderator.

## *Economic Welfare Analysis*

A second approach is to use the direct costs of regulatory compliance to generate a range of indirect costs that result from those outlays. Direct costs include all those items in Table 3.2, such as research, laboratory and field trials, preparation of dockets, fees and charges paid by proponents and the corresponding expenditures by the regulators to assess and adjudicate the evidence. Indirect costs include the opportunity costs faced by biotechnology companies (as they incur debt servicing charges while revenues are delayed), farmers (who forgo income that could be generated by adopting the new crop) and consumers (who under normal circumstances derive larger consumer surpluses from more productive and lower priced or higher quality products).

A number of economists have attempted to calculate the absolute and relative costs of the regulatory process. Most of the economists conclude that the time it takes to complete regulatory oversight is more important than the actual out-of-pocket direct cash costs. Smyth and Phillips (2002) estimated that if Monsanto and AgrEvo had waited to commercialize their HT canola until they gained regulatory acceptance in Japan, they would have incurred a two-year delay and a major opportunity cost. This ex post analysis calculated that, by adopting an identity-preserved production and marketing regime to channel the asynchronously approved varieties to the accepting North American market, the companies accelerated adoption by two years, which was estimated to have generated a net present value in 1995 $ of more than C$100 million. Pray et al. (2005) also discovered that the two-year delay (pause) in the approval of Bt cotton in India led to aggregate losses to farmers alone of more than US$100 million.

Bayer et al. (2010) undertook a sensitivity analysis in their regulatory study to evaluate the effects of increasing regulatory costs and altering the time required for regulatory approval and, hence, adoption of the technologies by farmers. They found the effects on total net benefits of increasing the direct cash costs of regulatory compliance in each case were small, even with four-fold increases. They estimated less than a US$1 million change in NPV in most cases, with the impacts varying from a 1% decrease for rice and papaya traits to a 7% decrease in the case of MVR tomato. These losses were small compared to the opportunity costs that were incurred when commercialization was delayed from one to three years in the regulatory system. A one-year delay in adoption resulted in a 12% decrease in the projected NPV for Bt rice and up to a 36% decrease for MVR tomato while a 3-year regulatory delay was estimated to cause a 34% drop in the NPV of Bt rice compared to the baseline and a 93% decline for MVR tomato.

There is some supporting evidence that these costs may actually be incurred, as the Phillips (2011a) study discussed below reports that firms in 2011 faced a regulatory process that was on average 21 months longer than before 2002. This undoubtedly generates significant opportunity costs to developers and users.

## *The Industry Estimates*

In 2011, Crop Life International, a global federation of the top eight agricultural biotechnology companies and 16 national industry associations, commissioned Phillips McDougall, a consultancy based in England, to survey member companies to determine the cost and duration associated with the discovery, development and authorization of a new biotechnology-derived plant trait that had received cultivation approval in two countries and import approvals from at least five countries. Six multinationals which have undertaken this task were surveyed (i.e. BASF Corporation, Bayer CropScience, Dow AgroSciences, DuPont /Pioneer, Monsanto Company and Syngenta AG); five provided cost estimates for the discovery and construct optimization phases and all offered advice on the costs they incurred downstream, including their regulatory costs. The study reported that the mean cost associated with the discovery, development and authorization of a new biotechnology derived crop trait introduced in the 2008–2012 timeframe was US$136 million (see Table 3.7). While the study reported that the costs of meeting regulatory requirements amounted to US$35.1 million (only 25.8% of total costs), that may only be part of the story. The evidence presented regarding the mean number of units evaluated at each stage of the development process suggests that firms are now doing more upstream research on discovery opportunities and constructs before they converge on a narrower set of commercial events that they target for introgression, selection and assessment. While the study does not explicitly relate this change in research design to the regulatory costs, it is likely that some of those changes are in response to the evolving regulatory processes.

Interestingly, the industry estimates are in the same range as the constructed cost analyses reported above. The average cost of the 49 evaluations reported in Table 3.6 was about US$7.8 million, with exporting countries being about 5% more costly than the average and compliance in importing nations costing about 3% less than the average. If those were scaled up to that used by Phillips McDougall involving two producing countries and five importers, the total cost would be about US$55 million, which while US$20 million higher than in the industry study is still

**Table 3.7** Industry estimates of the discovery, development and authorization costs of a plant biotechnology trait

| Category | Cost $US million | Mean number of units evaluated | | Duration of each stage in process (months) | | |
|---|---|---|---|---|---|---|
| | | For event before 2002 | For event 2008–12 | For event before 2002 | For event 2008–12 | To complete in 2011 |
| Total discovery cost | 31.0 | | | 55.3 | 53.9 | 46.3 |
| *Early discovery* | *17.6* | *1638* | *6204* | *38.0* | *33.9* | *25.4* |
| *Late discovery* | *13.4* | *302* | *4005* | *17.3* | *20.0* | *20.9* |
| Construct optimization | 28.3 | 135 | 511 | 18.0 | 27.0 | 32.8 |
| Commercial event production and selection | 13.6 | 2853 | 1302 | 24.0 | 30.0 | 34.0 |
| Introgression breeding and wide-area testing | 28.0 | 4 | 2 | 40.0 | 37.2 | 42.0 |
| Regulatory science | 17.9 | 2 | 1 | 50.5 | 37.2 | 47.0 |
| Registration and regulatory affairs | 17.2 | 1 | 1 | 44.5 | 48.8 | 65.5 |
| Total cost and time for an event | 136.0 | | | 232.3 | 234.1 | 268.0 |

*Source* Phillips (2011a)

within one standard deviation of the constructed cost estimates. Moreover, one might expect some economies of scale of in undertaking the regulatory science, registration and regulatory affairs in seven markets simultaneously. So at one level, there is some convergence between the scholarly and practitioner views of the direct costs of regulatory compliance.

## Empirical Analysis

Given the above analysis, we selected the 49 estimates that examined the direct costs of regulatory compliance (areas A + B in Fig. 3.2) as the dependent variables and identified four moderators along with the intercept as the appropriate variables to test using regression analysis.

The key moderators that one might anticipate would change the scale and scope of the regulatory cost estimates include the GM trait involved, the intended use of the crop, the market context for the regulatory action and the nature of the analysis undertaken. The 49 cost estimates address one of either herbicide tolerance (HT) or insect resistance (Bt); one might anticipate that the costs of compliance might vary given that Bt crops have tended to raise a host of environmental and pest management questions that HT crops have not; based on reviews to date, both crops have raised similar considerations about human and animal safety. Similarly, the intended use of the novel trait under consideration could change the nature of the review. Cotton is generally used as an industrial input to textiles and clothing (although small amounts of cottonseed oils are consumed as human food in some niche markets). Meanwhile, maize/corn and soybeans are both major components of processed foods and significant animal feeds. Market context could similarly matter. Many of those countries that examined these technologies are major producers and exporters while others are largely importers and consumers. A priori it is not clear whether exporters would undertake more extensive reviews than importers, but there is potential that they might differ. Finally, most meta-analyses try to test to see if the conceptual and practical approach to the studies could influence the divergence of results (e.g. Alston et al. 2000, show that the structure of the lags chosen has a major impact on the estimates of the return on investments in gains to research studies). In our case, we were able to determine whether the cost estimates were constructed before the regulatory process began (e.g. an ex ante analysis) or whether they were undertaken after the process had begun or been completed. One might expect the two approaches to deliver different results, depending on the prescience of the analysts and the degree of optimism underlying the assumptions about the scale and scope of what would need to be done to satisfy the regulators.

Table 3.8 shows the correlation matrix for the dependent variable, regulatory approval costs, and the eight possible moderators. The coefficients of correlation for the trait, use, context or method indicate a variation in the relationship between these independent variables and the cost of approval. Traits do not seem to have a major effect on costs, the HT variable has a low positive correlation with overall cost while Bt crops appear to, ceteris paribus, incur lower regulatory costs; neither is statistically significant in this analysis. The sign and coefficient for non-food use suggests a negative correlation with cost of approval; in other words, the cost of

**Table 3.8** Correlation matrix of the instrumental variables

| | Total cost | Bt | HT | Non-food | Food | Net exporter | Net importer | Ex ante |
|---|---|---|---|---|---|---|---|---|
| Total cost | 1 | | | | | | | |
| Bt | −0.19 | 1 | | | | | | |
| HT | 0.19 | −1.00 | 1 | | | | | |
| Non-food | −0.76* | 0.20 | −0.20 | 1 | | | | |
| Food | 0.76** | −0.20 | 0.20 | −1.00 | 1 | | | |
| Net exporter | 0.17 | −0.01 | 0.01 | −0.13 | 0.13 | 1 | | |
| Net importer | −0.17 | 0.01 | −0.01 | 0.13 | −0.13 | −1.00 | 1 | |
| Ex ante | −0.48 | 0.00 | 0.00 | 0.03 | −0.03 | 0.09 | −0.09 | 1 |
| Ex post | 0.48** | 0.00 | 0.00 | −0.03 | 0.03 | −0.09 | 0.09 | −1.00 |

Statistically significant at 95% (*) and 99% (**)

approval is likely to be lower if the trait is intended for a non-food use. The food-use variable shows the obverse effect. Both are statistically significant, suggesting they may be major drivers of the cost structure. The net-exporter variable, which represents those countries producing exportable surpluses of the specific crop under review, shows a small positive correlation that is not statistically significant; the net importer variable is the obverse. Finally, the type of analysis for measuring costs shows some evidence of a relationship. Those studies that used an ex post approach are positively correlated with higher costs; this is statistically significant. Given that the moderators are paired and have perfect negative correlation, only one of each of the pairs will be used (that is, Bt/HT, non-food/food, net exporter/net importer and ex ante/ex post). The rest of the correlation matrix shows that the correlation coefficients between independent variables are not significantly different than zero, which provides prima facie support that ordinary least squares (OLS) will generate unbiased and efficient estimates.

We undertook a basic regression where:

$$Ln(y_t) = a + b'X + \epsilon$$

and

Yt is the cost of regulatory approval (put into log linear form),
a is the intercept,
$b$ is the vector of slope coefficients,
$X$ is the matrix of independent variables and
$\epsilon$ is the error term.

The regression model consists of four independent variables that are dichotomous dummy variables indicating the presence or absence of a particular characteristic for the key moderators. The specific moderators chosen were HT, non-food use, net exporter and ex-ante analysis. The OLS model uses the natural log of the dependent variable (the total cash costs of regulatory compliance in US$ millions), which allows us to test the effect of the explanatory variables in percentage terms instead of units.

As a general rule, regression models include a constant or intercept term (Greene 2002). The constant term explains the variation in the registration costs that are independent of the explanatory variables. For example, a consumption regression model includes a constant term to show that individuals consume regardless of income. In the context of this chapter, omitting the intercept term would suggest that variation in registration costs are solely determined by the independent variables under study, which would not likely be realistic. Hence, we have included the intercept, which can be interpreted as the benchmark cost of regulating a new GM trait in a single undifferentiated market. To test this, we removed the intercept to see if it might change the nature of the analysis.

Two models (with and without the intercept) were tested (Table 3.9). Model 1 results suggest that the independent variables account for approximately 81% of the variation in costs of approval. The coefficient for HT is modestly positive but

**Table 3.9** Meta-analysis regression results

| Default category | Independent variable | Model 1 | Model 2 |
|---|---|---|---|
| | Intercept term | 15.77 | – |
| Bt | Ht | 0.15 | 11.6 |
| Food | Non Food | −3.68* | 7.79 |
| Net importer | Net exporter | 0.43* | 2.76 |
| Ex post | Ex ante | −2.25** | 3.1 |
| | Model $R^2$ | 0.81 | −0.31 |
| | No. of observations | 49 | 49 |

**Statistical significant at 99%; *Statistically significant at 90%

statistically insignificant, so we are unable to draw any inference of whether the trait has any material effect on costs of regulatory compliance. In other words, the trait of the GM crop has little or no impact on costs of approval *based on the model presented and the data examined.* All of the other variables are statistically significant. The coefficient for non-food use suggests that regulatory costs are lower for GM crops that are not destined for human consumption. The coefficient of −3.68% for non-food is statistically significant at the 10% level of significance, which implies that the intended use for a GM crop (i.e. as a food, feed or industrial application) is a significant factor in the total regulatory approval cost. This intuitively fits with observations from regulators and product proponents—human safety is paramount in regulatory systems. The trade context also has a major impact on costs—exporting countries impose more expensive regulatory processes than importing nations. This is open to a number of different interpretations. One possibility is that exporters are also going to be major producers, so they would need to do both extensive food and feed trials *and* environmental assessments; importers might be able to simply focus on food and feed impacts. A second interpretation is that some of the incremental costs of gaining compliance in net exporting countries may be skewed by the inclusion of the mega-adopters (US, Argentina, China, Brazil, Canada and India) with some countries less enthusiastic about the technology (especially the European Union and some African nations), which subject foreign GM crops to 'less favorable' conditions (especially delays in decisions, which do not impose extra cash costs) to protect domestic crops from competition from imported GM products. A third possibility is that the lower cost of approval in net importing countries may be due to the nature of the specific approvals that have been costed in our dataset. Many developing countries are quite open to GM technology and see it as a way of enhancing food security; in that context, some countries accept foreign data in their regulatory system. Bayer et al. (2010) argue that the acceptance of foreign data resulted in lower costs of approval in the Philippines. Finally, the negative sign for the ex ante assessments suggests that on average those analysts conducting their analysis prior to submission to the regulatory approval body are more optimistic (or less grounded in the nature of the process). The large difference in ex ante and ex post analyses may arise from unexpected multiple trials and requests for evidence that emerge in the approval process which would clearly make the costs 'after' submission much higher than estimates before the GM crop enters the approval process. In essence, optimism appears to dominate and underestimates of the costs are made.

For completeness, we present model two that tests the hypothesis that trait, product use, trade status and measurement are the only factors impacting costs of regulatory approval. We removed the intercept and found that the overall quality of the regression and the role of specific variables deteriorated—the $R^2$ dropped and none of the coefficients were significant or had the expected signs in this specification. Therefore, model 1, with the intercept, is preferred empirically.

## Conclusion and Implications

The review of studies done to date shows some degree of convergence, with the constructed cost estimates higher but still not statistically different than the industry views. The meta-analysis further showed that the context for regulatory reviews—particularly the trait and the market orientation of the respective country—have significant effects on the cost of compliance. Moreover, the nature of the review—whether ex ante or ex post—fundamentally alters the estimates.

One might think that over time one should see greater clarity in the estimates. While that would be helpful, the current estimates are deficient in a few ways. First, there are important indirect costs of staggered approvals, with different traits often being approved at different times in different countries; this is not captured in any of the extant studies. Moreover, none of the studies to date have explicitly examined the costs of gaining approval for stacked traits, which now represent about 25% of global GM area. Regulators presented with stack traits seem to have a practice of not simply assessing the added traits—they also revisit and require full assessment of those previously approved traits that are part of the stack, which could add significantly to costs.

In other technology areas, familiarity and experience has tended to lead to convergence of methods and costs of assessing applications. While in a perfect world this might occur, the international conflict over the role of GM technologies in the global food system, combined with the competitive practices of individual firms seeking to develop and introduce new traits, suggests that convergence may be a ways off. This chapter offers some evidence that the conflict between the mega-adopters and a few key importing and competing markets (especially the EU) has a real and measurable effect on the costs of regulatory compliance and the use-benefits that would be expected to flow to producers, innovators and consumers around the world. Other studies, including Haggui et al. (2006) and Paarlberg (2008) offer complementary evidence of this multiplier effect. Meanwhile, there is ample evidence that firms regard their capacity to satisfy regulators as a major competency and competitive advantage, so that they have limited incentive to share their tacit knowledge of the regulatory process and to develop best practices that could lower the barriers to market entry.

## References


Alston, J.M., M. Marra, P. Pardey, and T. Wyatt. 2000. Research returns redux: a meta-analysis of the returns to agricultural R&D. *Australian Journal of Agricultural and Resource Economics* 44 (2): 185–215.

Bayer, J., G. Norton, and J. Falck-Zepeda. 2010. Cost of compliance with biotechnology regulation in the Philippines: Implications for developing countries. *AgBioForum* 13 (1): 53–62.

Brookes, G. 2014. Global adoption of GM Crops, 1995–2012. In *Handbook on agriculture, biotechnology and development*, ed. S.J. Smyth, P.W.B. Phillips, and D. Castle, 25–35. Cheltenham, UK: Edward Elgar Publishing Ltd.

Brouwer, R., I. Langford, I. Bateman, and R. Turner. 1999. A meta-analysis of wetland contingent valuation studies. *Regional Environmental Change* 1 (1): 47–57.

Camisón-Zornoza, C., R. Lapiedra-Alcamí, M. Segarra-Ciprés, and M. Boronat-Navarro. 2004. A meta-analysis of innovation and organizational size. *Organization Studies* 25 (3): 331–361.

Datta, D.K., G.E. Pinches, and V.K. Narayanan. 1992. Factors influencing wealth creation from mergers and acquisitions: A meta-analysis. *Strategic Management Journal* 13: 67–84.

Doern, B., and M. Prince. 2012. *Three bio-realms: Biotechnology and the governance of food, health, and life in Canada*. Toronto: University of Toronto Press.

Falck-Zepeda, J., J. Yorobe, A. Manalo, G. Ramon, A. Bahagiawati, E.M. Lokollo, S. Supriyati, and P. Zambrano. 2006. The cost of compliance with biosafety regulations in Indonesia and The Philippines. July–August: Paper presented at the American Agricultural Economics Association Annual Meeting.

Falck-Zepeda, J., and J. Cohen. 2006. Biosafety regulations of genetically modified orphan crops. In *Regulating agricultural biotechnology: Economics and policy*, ed. R.E. Just, J.M. Alston, and D. Zilberman, 509–533. New York: Springer.

Greene, H.W. 2002. *Econometric analysis*, 5th ed. New York: New York University Press.

Haggui, F., P.W.B. Phillips, and R. Gray. 2006. Opposition to genetically modified wheat and global food security. In *International trade and policies for GM products*, ed. R. Evenson, and V. Santaniello, 175–190. Wallingoford, UK: CABI Publishing.

Hedges, L.V., and I. Olkin. 1985. *Statistical methods for meta-analysis*. Orlando, FL: Academic Press.

Isaac, G. 2002. *Agricultural biotechnology and trans-atlantic trade*. Wallingford: CABI Publishing.

James, C. 2015. 20th Anniversary (1996 to 2015) of the Global Commercialization of Biotech Crops and Biotech Crop Highlights in 2015. ISAAA Brief No. 51. ISAAA: Ithaca, NY.

Kalaitzandonakes, N., J.M. Alston, and K.J. Bradford. 2007. Compliance costs for regulatory approval of newbiotech crops. *Nature Biotechnology* 25: 509–511.

Lusk, J.L., M. Jamal, L. Kurlander, M. Roucan, and L. Taulman. 2005. A meta-analysis of genetically modified food valuation studies. *Journal of Agricultural and Resource Economics* 30(1): 28–44.

Marra, M., and B. Schurle. 1994. Wheat yield risk and aggregation: A meta-analysis approach. *Journal of Agricultural and Resource Economics* 19 (1): 69–77.

Mulatu, A., R.J.G.M. Florax, and C. Withagen. 2003. Environmental regulation and competitiveness: An exploratory meta-analysis. In *Empirical modeling of the economy and the environment*, ed. C. Böhringer, and A. Löschel, 23–54. Heidelberg: Physica-Verlag HD.

Office of Management and Budget. 1996. Economic Analysis of Federal Regulations Under Executive Order 12866, January 11, 1996. Available at: www.whitehouse.gov/omb/inforeg_riaguide.

Paarlberg, R. 2008. *Starved for science: How biotechnology is being kept out of Africa*. Boston: Harvard University Press.

Phillips McDougall. 2011a. The cost and time involved in the discovery, development and authorization of a new plant biotechnology derived trait: A consultancy study for CropLife International. Midlothian, UK: Phillips McDougall.

Phillips, P.W.B. 2011b. LLP impacts on global trade, agricultural innovation and commodity prices. Presentation to GM Co-existence Conference, Vancouver, Canada, October 27.

Pray, C.E. 2013. Personal communication.

Pray, C., P. Bengali, and B. Ramaswami. 2005. The cost of regulation: The India experience. *Quarterly Journal of International Agriculture* 44 (3): 267–289.

Pray, C.E., B. Ramaswami, J. Huang, R. Hu, P. Bengali, and H. Zhang. 2006. Benefits and costs of biosafety regulations in India and China. In *Regulating agricultural biotechnology: Economics and policy*, ed. R.E. Just, D. Zilberman, and J.M. Alston, 481–508. New York: Springer.

Skogstad, G. 2008. *Internationalization and Canadian agriculture: Policy and governing paradigms*. Toronto: University of Toronto Press.

Smyth, S.J., and P.W.B. Phillips. 2002. Competitors co-operating: Establishing a supply chain to manage genetically modified canola. *International Food and Agribusiness Management Review* 4 (1): 51–66.

Viscusi, W.K., and J.E. Aldy. 2002. The value of a statistical life: A critical review of market estimates throughout the world. Harvard Law School John M. Olin Center for Law, Economics and Business Discussion Paper Series. Paper 392. Available at: http://lsr.nellco.org/harvard_olin/392.

# Chapter 4
# Global Trade Impacts from Low Level Presence

**Abstract** Asynchronous regulatory approval and commercialization of new GM traits in a range of extensively traded food crops can cause significant trade disruptions. The adventitious presence of a number of unapproved or incompletely approved traits in canola, corn, rice and flax has been associated with economic harm. Market losses from diversion to second-best markets, increasing marketing costs for securing quality assured supplies, added regulatory compliance time and costs as well as costly and often destructive litigation have been linked to such events. Fears of similar events in wheat, soybeans and other staple crops have generated significant concern. This chapter examines the root of the regulatory challenge, evaluates the immediate impacts on the market, assesses the long-term implications on research and investment in advanced crop varieties and considers the global impacts on consumers and food security.

## Introduction

This chapter addresses the conditional question of what happens if efficient and effective market structures do not emerge at the right time in the right places. In that context, asynchronous regulatory approval and commercialization of new GM traits in a range of extensively-traded, food crops will exacerbate the problem. Since 1998, markets have diverged. The US, Brazil, Argentina, India, Canada, and China and a number of other major food producers and exporters expeditiously developed, adapted and adopted genetically modified GM traits in a range of large area crops while the European Union and a number of significant food importing countries have signalled to suppliers their unwillingness to accept either unregulated varieties or undifferentiated supplies. Consumers, regulations, technology and markets have all worked to adapt to serve those increasingly parallel or divergent markets. The divergence, however, inhibits the benefits that can accrue from international trade, reduces incentives to make further investments in technological advances and ultimately reduces the ability to address the future food security challenges that have been identified.



S.J. Smyth et al., *Biotechnology Regulation and Trade*, Natural Resource Management and Policy 51, DOI 10.1007/978-3-319-53295-0_4

## Background

There is now considerable evidence that market requirements regarding the products of modern agricultural biotechnology have diverged. While there is little rigorous market evidence that consumers both want differentiated products and are willing to pay for them, citizens and consumers have disclosed their preferences through a range of surveys and studies, for the most part signalling that 'all other things being equal' they wanted quality-assured supplies that would allow them to exercise their consumer sovereignty, with access to differentiated qualities at differentiated prices. Surveys almost unanimously show that if asked directly, individuals will indicate a preference for labeling. A large number of surveys show more than 85% of those surveyed in most countries would choose proactive labeling of the presence of GM elements as a policy option. One key ancillary question is whether they would pay for that policy. Lusk et al. (2005) undertook a meta-analysis of 25 studies that collectively report 57 valuations for GM food. Across all studies, consumers on average placed a higher value on non-GM relative to GM foods, ranging anywhere from 42% (unweighted average using all data) to 23% (weighted average excluding extreme outlier). They then undertook a regression analysis and concluded that a number of factors significantly affect the size of the estimated values. In particular, consumers in the EU placed a higher value (29%) on non-GM food than consumers from North America. They also showed that the method of value elicitation had a significant effect: when the actual shopper in a family was surveyed, they required a 72% lower premium than non-shoppers; premiums for real products were 40% lower than for hypothetical products; premiums estimated using the willingness-to-avoid method exceeded by 58% premiums estimated by the willingness-to-pay value measure; and conducting the valuation in-person bumped the premiums by 62% as compared to when the valuation was solicited over the phone or by mail. The authors attributed this bias to the Hawthorne effect, where respondents try to answer the question in ways they perceive the interviewer is seeking. Alternatively, this may simply be because an in-person process forces people to think more about their answers, which amplifies their concerns. The product being analyzed was also found to significantly affect valuations—GM meat premiums would need to be 49% higher than GM oil premiums to make them acceptable, while GM products that were characterized as providing tangible personal benefits, such as increased freshness or nutrition, decreased premiums by between 28% and 49%. While the markets have expanded in recent years, there is no a priori reason to assume that these relationships have changed demonstrably. If true, then the significant gulf between consumers in US and EU is evident and the signals they are sending are likely having an impact on policy makers, regulators, investors and markets.

This divergence of consumer perceptions is reflected in an array of processes that have set in motion regulations or standards as a basis for differentiation. There has been significant study of the underlying regulatory regimes in the divergent countries (e.g. Isaac 2002; Skogstad 2008; Doern and Reed 2000; Perdikis 2000).

The main point to draw from the literature for the purposes of this chapter is that there are fundamental differences of approach and impact. At the macro level, the four major crops have had differing levels of innovation (from 30–138 events) that have been differentially approved for food, feed and cultivation (Table 4.1).

Breaking that down by country shows there is a significant degree of regulatory divergence among countries. Phillips (2009) showed that in the first 15 years of adoption (1995–2011) on average only about 12% of the potential environmental approvals and 24% of the food approvals were actually achieved over the 18 most significant producers and importers. The highest rates of positive review were in Canada and US, two of the largest producers and exporters, and Japan, a key importer. In the 89 countries examined, the average country undertook reviews on only about 5 species, of the 16 under investigation. The low level of successful review for environmental release reflects the politics of the adopting countries while the somewhat higher percentage of completed food reviews reflects the reality that much of the international trade now inevitably include GM elements; prospective review and approval is less disruptive than simply waiting for comingling and detection of adventitious presence. Table 4.2 updates that data for the 1995–2016 period, focusing in on the six mega adopters and the two major importers. The picture hasn't changed much. In this analysis, we can see that in no case did even two thirds of the events get approved. Food and feed approvals tended to be above approvals for cultivation.

Another way to view the challenge is to look at the array of approvals for different events within a specific crop type, Fig. 4.1 shows the date of first approval for each of 90 different events, and then the follow-on approvals over the succeeding years. One can see that some traits take more than 15 years to gain substantial regulatory approval across the major producing and trading countries while some never get universally adopted. This uneven uptake and use of the technology compounds the risks for investors and producers alike.

**Table 4.1** Approval of GM events

| Crop | Number of events approved | Events approved for | | |
|---|---|---|---|---|
| | | Food | Feed | Cultivation |
| Canola | 32 | 32 | 30 | 22 |
| Cotton | 54 | 44 | 42 | 45 |
| Corn | 138 | 133 | 123 | 90 |
| Soybean | 30 | 28 | 26 | 28 |
| Total | 254 | 237 | 221 | 185 |

*Source* Calculated by Gleim et al. (2016) using data from ISAAA

**Table 4.2** Regulatory decisions related to GM events in top 8 ad adopters and importers, 1995–2016 (Argentina, Australia, Brazil, Canada, China, EU, Japan and US)

| | Maize/corn | Sugar beet | Canola | Soybeans | Flax, linseed | Cotton | Potato | Rice | Alfalfa |
|---|---|---|---|---|---|---|---|---|---|
| Maximum number of events | 114 | 3 | 22 | 24 | 1 | 34 | 43 | 3 | 4 |
| *Average number of approvals for* | | | | | | | | | |
| Food | 44 | 2 | 13 | 15 | 0 | 16 | 11 | 1 | 2 |
| Feed | 40 | 1 | 12 | 13 | 0 | 16 | 8 | 1 | 1 |
| Cultivation | 30 | 1 | 9 | 13 | 0 | 13 | 7 | 1 | 1 |
| *Average percentage of total approvals possible for* | | | | | | | | | |
| Food | 38% | 50% | 58% | 63% | 25% | 46% | 25% | 25% | 41% |
| Feed | 35% | 42% | 52% | 53% | 25% | 46% | 19% | 17% | 31% |
| Cultivation | 26% | 25% | 41% | 53% | 25% | 38% | 17% | 29% | 31% |

*Source* Authors' calculation using ISAAA GM Approval Database. (http://www.isaaa.org/gmapprovaldatabase/default.asp)

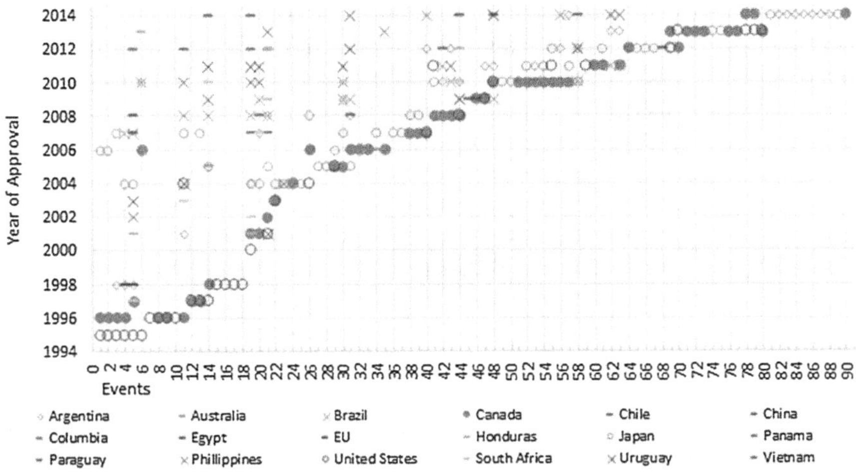


**Fig. 4.1** Country approvals by event. *Source* Gleim et al. (2016)

## Theoretical Considerations

One can assess the impact of these divergent forces on the market and on social welfare in a number of ways. Some analysts examine specific events or specific actors while others focus on the systemic effects.

There are three key decisions actors make which are critical for understanding the underlying pathways of technological diffusion and value generation. While the decisions regarding GM foods are linked, one could look at them as the decisions to invest in discovery, research and development by multinational firms, governments and seed consortia, the decision to adapt that technology to the marketplace (which involves both the commercial supply chain and the regulators), the decision of individual farmers to adopt the technology in their operations and finally the decision of a consumer to purchase and consume products arising from the use of the technology. Generally economists tend to treat those as separate decision points, with 'upstream' decisions creating down-stream opportunities. In this context, profit optimizing firms are assumed to seek to realize some economic profits (sometimes called 'rents') from developing and commercializing new technologies. They are generally assumed to make investments in opportunities offering risk adjusted, internal rates of return above their cost of capital, which means they tend to ignore value generated by others but not appropriated by them. The other supply chain actors are assumed to undertake similar analyses—it is often assumed, however, that they operate under perfect competition, so that they are effectively docile 'takers' of new technologies. Farmers, meanwhile, are assumed to undertake implicit cost-benefit calculations and adopt what are expected to be profitable opportunities. Finally, consumers are assumed to be rational, all knowing beings who purchase without emotion.

None of these assumptions fit the circumstances of GM foods. Firms are continually testing their choices, recalibrating their net present value calculations in response to changes in market circumstances, changes in expected costs and benefits, alterations in their discount rate and the timing of their activities. This can, at times, lead to significant mid-course corrections in their R&D pipelines and budgets as their expectations change. While one tends to assume highly rational decision-making, most market commentators are used to observing, on the contrary, emotional responses that lead to irrational exuberance. In short, small changes in expectations can lead to large shifts in performance. There is also some evidence that competing and complementary supply chain participants have market power and are willing to exploit it. They no longer automatically simply accept what flows downstream—wholesalers want assurances of the quality of products they handle, branded food processors have very explicit moral or corporate strategies that dictate what they will accept and retailers are increasingly moving upstream to exploit their market power to develop differentiated products (often under own-label brands) to enhance their profits.

Moving downstream, farmers are similarly canny when it comes to new technologies. They are looking at how a technology may fit in with their farms operation (e.g. fit with complementary technologies or competing crops in rotations) and what risks they might incur in terms of market access. Recent market disruptions related to adventitious presence of unapproved GM elements—e.g. Star Link Corn, LL 601 rice and CDC Triffid Flax—have heightened awareness, while signals coming from local and international consumers have influenced farmer attitudes to adopting new crops, such as Monsanto's shelving of round-up ready (RR) wheat. One way of handling this conceptually is to add supply chain and consumer attitudes as variables in the decision matrix, which may work to either block options or alternatively to alter a farmer's specific cost-benefit calculations (Haggui 2004).

Finally, as noted above, consumers are far from certain about how to assess and value these items, and their opinions have the possibility of feeding back into the biotechnology research cost-benefit analyses, into considerations of firms in the supply chain and, as just noted, into farmer decisions regarding adoption. In short, it is increasingly risky to treat each decision point as an isolated choice that can be made independently of others, or that one can ignore the role of expectations and preferences.

Beyond the pathways of influence and their effect on the micro-level decision-making, there are a range of methods of aggregating the impacts and estimating the effects on groups, the market and the economy. Welfare analysis is the usual starting point for assessing the impacts of changes in the production function or consumer preferences. Alston et al. (1995) wrote what has become a seminal work on welfare analysis in agriculture and a large number of scholars have applied it across a range of applications of GM technology in the agri-food sector (see Brookes and Barfoot 2010 for a survey of studies). Essentially this approach relies on estimating the shape and slope of the appropriate supply and demand curves in a specific market, shocking one or both of the curves in response to a new technology and then estimating the changes in welfare—with consumers and producers gaining or losing 'surplus' measured as the value of the space between the demand curve and supply curves, as transected by the price line. As set out in Chap. 2, increasing capacity to produce is assumed to shift the supply curve out which, all other things being equal, will raise welfare. The result of most assessments is that net incremental welfare gains are generated by most changes brought by biotechnology, with developers gaining (eventually revenues repay their investments) and (especially early) adopters generally gaining (as yield gains in cases analyzed more than offset any price declines), consumers gaining due to lower prices (especially if they are indifferent to the provenance of the technology), non-adopters losing (due to market lower prices), consumers with concerns possibly losing (depending on their tastes) and firms with technology that is no longer competitive losing (as their profits or rents are bid away by the owners of the new technology). Depending on whether it is a single country case or an open market, multi-country model with trade, one may also be able to assess the effect on consumers and producers in other markets (e.g. Moschini et al. 1999; Phillips 2003).

A second approach is to undertake a 'meta-analysis' of the empirical studies done on specific technological introductions in specific markets and either aggregate the results or undertake regression analysis using the results of the studies. Brookes and Barfoot (2010) undertook the first approach, providing a detailed assessment of the economic and commercial effect of GM crops around the world. While no one has yet engaged in the regression approach, the method has been laid out in the seminal study by Alston et al. (1998) of more than 1800 impact assessments of new agri-food innovations.

A third empirical approach is to develop a more extensive Applied General Equilibrium (AGE) model that can provide an elaborate and realistic representation of the economy including the linkages between multiple agents, sectors and other economies. One commonly used model is the GTAP (Global Trade Analysis Project) multi-regional AGE model which incorporates economic activity in 57 industries of 66 regions. The model includes accounting equations which ensure receipts and expenditures of each agent in the economy are balanced and behavioral equations based upon microeconomic theory. This model has been used by Anderson (2010) and his results are discussed below.

In essence, all of these methods offer estimates of the impacts of new technology and the potential to extrapolate to infer the foregone benefits, what economists call 'opportunity cost', of delayed, truncated or cancelled adoptions. Opportunity costs can show how large the gap exists between the actual and the optimum and can also offer insights into where unrealized benefits may be located. The latter could offer options for finding win-win solutions to hold-ups in the market.

## Practical Effects

While more nimble, efficient, synchronous regulatory action—by proponents, national regulators and the supply chain—may be the welfare maximizing ideal, in the interim uncertainties and risks have acted to inhibit trade in GM commodities, delayed and truncated uptake by farmers and the supply chain, reduced investment in new GM crops and ultimately led to higher prices and lower benefits than would prevail otherwise. Ultimately, food security for the world's poorest inhabitants will be negatively impacted.

The first evidence of this mismatch of supply and demand has been observed in the trade system. Compounding this disconnect in regulatory approaches is the fact that biotechnology is pro-trade. Almost every product category where biotechnology has been adopted has enhanced production and trade. Oehmke et al. (2001) argue the most important influence on agricultural trade patterns in the past 15 years was the advent of biotechnology. They use Argentina to illustrate the case. In 1996 Argentine soybean production was 11.2 million tonnes, of which 0.75 million tonnes were exported. Transgenic soybeans helped boost production to 19.5 million tonnes and exports to 3.2 million tonnes in 1997, making Argentina the third largest exporter of soybeans in the world; exports increased further to 5.8 million tonnes in 2000. Argentina and Brazil in 2013 produce about 47% of the world supply of soybeans, up from only about 22% in the early 1980s. The big jump in market share coincided with the rapid adoption of herbicide tolerant soybeans. Smyth et al. (2011) show a similar result for canola in Canada, where the adoption of herbicide tolerant varieties precipitated a sharp rise in average acreage cultivated to canola. As adoption of HT varieties reached greater than 90%, the canola area rose from 9 million acres in 2002 to more than 20 million acres each year since 2012, precipitating greater trade.

The effect can be estimated more formally through the regulatory and trade data. If one takes the number of countries actually producing GM crops (using James 2015), the market shares of GM producing nations can be determined. The data shows that GM producers account for the lion's share of the three largest GM food crops, the share of production ranging from 32% for Canola to 88% for soybeans, and export market share ranging from 51% for canola to 96% for soybeans. Key producers, moreover, export to between 60 and 193 countries, many of which have specific rules about labeling and marketing GM foods. In this context, firms are challenged to seek ways to sustain both production and trade. In absence of domestic differentiation strategies, any produce coming from an exporting country has the risk of being co-mingled and triggering domestic regulatory oversight (Table 4.3).

Against this backdrop, new, specific, timely and cost effective scientific tests have revealed the adventitious presence of a number of unapproved or incompletely approved traits in canola, corn, rice and flax and raised fears of similar events in wheat, soybeans and other staple crops. This has generated significant concern (Hobbs et al. 2014). In the first instance, these events have destroyed economic value —generating market losses as unapproved crops need to be tracked down and removed from commercial channels and then diverted to second-best markets. Beyond the events themselves, exporters report increasing transaction costs and increasing regulatory compliance time and costs as well as costly and often reputation damaging litigation. In most cases, the end result is lower uptake and use of what are considered by regulators to be safe and efficacious technologies and depressed prices in commodity markets.

**Table 4.3** Global GM production and trade, 2013

| Crop | Number of states approved and producing | Percentage of global production from GM states (%) | Percentage of exports from producing states (%) | Total number of importers |
|---|---|---|---|---|
| Maize/corn | 13 | 57 | 78 | 189 |
| Soybeans | 11 | 88 | 96 | 153 |
| Canola | 4 | 32 | 51 | 95 |
| Potato | 1 | 5 | 4 | 193 |
| Papaya | 1 | 0.1 | 5 | 104 |
| Sugar beet | 2 | 12 | 11 | 60 |

*Sources* Authors calculations using data from James (2015) and FAO. Stat. (http://www.fao.org/faostat/en/#data)

There is also some evidence that the supply chains that receive and handle the key GM commodity crops are struggling to adapt. Adapting is not a simple matter as sophisticated structures and arrangements are involved—missteps by any of the actors in the supply chain can raise costs and dissipate benefits. In the broadest view, annually there are 340 million metric tonnes of grains and oilseeds moving through the international trading system involving 195 countries. Each of the major crops has upwards of 20 different grades or differentiations (in some cases more than 100), each with their own specifications. Underlying those shipments is a research and development system which turns over the seed stock in some crops every 3–5 years, with new varieties, new traits and new attributes—some which enhance value and some which simply create differentiable attributes. In Canada, for example, there are more than a dozen centres of public-private research effort involving hundreds of actors, more than 3000 seed growers who multiply the foundation seed for commercial sale and about 250,000 farmers who plant an array of crops and varieties in sophisticated rotations on more than 50 million acres. The grains and oilseeds are then harvested and largely stored on-farm and called forward by shipment aggregators through about 1000 delivery points. The average commodity shipments then involve a minimum of three domestic transfers, in unit sizes ranging from 300 bushels to 6000 tonnes, ultimately being aggregated in unit trains for shipment to the US or at export position at ports. Beyond Canadian frontiers, shipments usually involve at least a further two transfers before they reach the specific buyer who contracted for that product.

In response to these uncertainties and challenges, there are an increasing number of examples where supply chain actors have interceded in the market and refused to handle crops that might be difficult to trade. In the late 1990s many food processors worked to remove or differentiate GM crops from their supply chains in order to meet the pending rules of EC Reg 258/97 of the European Union which set out new labeling rules (Phillips and McNeill 2000). Seven large supermarket chains in the EU joined forces to eliminate GM ingredients in their own-label products. These included Carrefour (France), Delaize (Belgium), Marks and Spenser (UK), Sainsbury's (UK), SuperQuinn (Ireland) and Tesco (UK). A range of other retailers in Europe (e.g. Effelunga, Edeka, Somerfields, Waitrose, Iceland Foods and Northern Foods) and elsewhere (e.g. Woolworth's in South Africa and Park N'Shop in China and Hong Kong) responded by claiming their own-label products were free of GM materials (Phillips and McNeil 2000). In response, aggregators offered to meet these new requirements by either replacing inescapably comingled ingredients with alternate non-GM ingredients or developed parallel supply chains to serve this differentiated market. The leadership was largely in the hands of the retailer or branded foods processer. In the intervening period, a number of wholesale aggregators have also threatened to refuse to trade in GM crops (e.g. when round-up ready wheat was proposed). One of the few documented cases where a firm rejected legally approved GM crops was litigated in the US. A US Federal District Court ruled in 2011 that St. Louis-based grain handler, Bunge North America, did not have to accept a GM corn produced by Syngenta AG. Syngenta sued Bunge after Bunge rejected Syngenta's Agrisure Viptera corn. Bunge, which runs grain elevators and receiving stations, said it was 'unable to accept' delivery of the corn because the bio-engineered corn was not cleared for approval in export markets. The Federal Court ruled that, "Bunge's decision to reject Viptera corn at all of its locations was a legitimate and reasonable business decision. The injunction would impost prodigious costs on Bunge for a situation that Bunge did not create" (http://www.soyatech.com/news_story.php?id=25399). There are no estimates of the costs of this decision, but there surely are both real economic costs as producers have to seek other venues for sale and some negative signalling to producers, seed suppliers and biotechnology companies that business is getting more uncertain.

Adoption can be assessed at two levels. In the first instance, the number of countries that produce a crop represents one type of maximum market penetration—soybeans and maize/corn have the highest penetration rates, averaging around 10% of the total countries producing those crops. While one might interpret this low penetration as reflective of the unwillingness of countries to accept GM technologies, in many cases it is simply a business decision based on the expectation that there would not be adequate revenues for the technology owners to justify the investments in acquiring regulatory compliance and developing a supply chain. As reported in Chap. 3 a number of tentative estimates of the costs of regulatory compliance in developing countries suggests the upfront cost per country have ranged from US$500,000 to $5 million for the first GM event in a species and taken from 2 to 7 years, with subsequent GM traits in the same crop being less expensive and more timely (Kalaitzandonakes et al. 2007; Pray et al., 2005, 2006; Bayer et al. 2010).

As second way to look at adoption is the share of the global area planted to GM crops. In an analysis done using 2010 data, Phillips (2009) found that while only about 12% of the total countries producing soybeans had adopted GM varieties, the GM area in those countries accounted for an estimated 73% of global soybean area. GM maize/corn accounted for about 30% of global area while GM canola represented about 22%. In countries that have approved and adopted the technology, adoption rates are much higher. Herbicide tolerant canola accounted more than 90% of Canada's canola area and GM soybeans accounted for a similar share in the US

**Table 4.4** Global adoption of GM technologies by crop

| | Number of countries producing crop 2009 | Number of countries producing GM crops | 2010 GM area (million hectares) | 2009 global area planted (million hectares) | GM area as % world crop area (%) |
|---|---|---|---|---|---|
| Soybean | 91 | 11 | 73.3 | 99.5 | 73.7 |
| Maize | 163 | 16 | 46.8 | 158.6 | 29.5 |
| Rapeseed/canola | 61 | 4 | 7.0 | 31.1 | 22.5 |
| Potato | 157 | 3 | Trace | 18.7 | ~0 |
| Papaya | 62 | 2 | Trace | 0.4 | ~0 |
| Sugar beet | 53 | 2 | Trace | 4.3 | ~0 |
| Squash | Na | 1 | Trace | Na | Na |
| Tomato | 172 | 1 | Trace | 4.4 | ~0 |

*Source* Author's calculations using data from ISAAA (2011) and FAO. Stat. (http://www.fao.org/faostat/en/#data)

market. Given only limited expansion of the market since then, the following results are representative of the current context (Table 4.4).

Brookes and Barfoot (2010) have used an array of economic studies of GM crops and an economic model to calculate the economic effects of GM technologies both on the crops that have been targeted for introduction of GM traits and for competing or complementary crops. Given the high adoption of the three key GM food crops (with adoption rates ranging from 22.5 to 73.7%), there has been a significant impact on global prices of those crops. They estimate that soybean prices are 9.6% lower than they would otherwise be, while maize/corn prices are 5.8% lower and canola prices are 3.8% lower. Given that these crops are also substitutes for other food stuffs, such as wheat, barley, sunflower seeds and sorghum, the greater competition from these three core foodstuffs has causes price pressures in those markets, ranging from a decline in wheat prices of 2.7% from what otherwise might have prevailed, to as large as a 4.2% drop for sorghum (Table 4.5).

**Table 4.5** The impact of GM crops on global prices for key commodities

| | 2010 GM production | 2009 area planted | GM as % world | Estimated price impact |
|---|---|---|---|---|
| Soybean | 73.3 | 99.5 | 73.7 | −9.6 |
| Maize | 46.8 | 158.6 | 29.5 | −5.8 |
| Canola | 7.0 | 31.1 | 22.5 | −3.8 |
| Wheat | 0 | 225.6 | 0 | −2.7 |
| Barley | 0 | 54.1 | 0 | −3.3 |
| Sunflower | 0 | 23.7 | 0 | −3.1 |
| Sorghum | 0 | 40.0 | 0 | −4.2 |

*Source* ISAAA (2011) and Brookes and Barfoot (2010)

These price effects translate into significant welfare effects. Anderson (2010) used the GTAP AGE model to first estimate the global impact of adoption of GM coarse grains and oilseeds. He estimated that the introduction of the crops to that point in time raised global GDP by US$2.3 billion (1997$), with about 41% of the benefit accruing in the adopting countries and the rest in non-adopting but importing and consuming countries. With the EU moratorium on approval of new GM events, in contrast, consumers and producers suffered a major loss (approximately US$3.1 billion in 1997$). Meanwhile the benefits to adopters were reduced and other major consumers gained a bit more due to price differentials that emerged. He then went on to estimate the opportunity cost of the truncated diffusion of GM soybeans, corn and canola and the delayed adoption of GM rice and wheat. If adoption was limited to USA, Canada and Argentina, annual global welfare estimates could still rise by $4.3 billion (1997$), with about 25% of the benefits captured in the three leading adopting countries and the rest in consuming countries (mostly in the developing world, but some in the EU). If all countries (except the EU) adopt the technology, it was estimated that annual global welfare could increase by more than $7.5 billion (1997$), with almost all the incremental benefits flowing to producers and consumers in developing nations. An extended EU moratorium on both production and consumption would inflict a range of economic costs on both EU producers and consumers and on the rest of the world. Anderson (2010) estimated that the longer the EU sustained the de facto moratorium, the more likely that the aggregate global impact would decline, ultimately to a global net loss of $0.8 billion if the moratorium became permanent and discouraged developing nations to delay or forgo this technology in these key food crops. While the moratorium has been removed, approvals of GM crops in the EU remain very slow, in effect mirroring the era of the moratorium (Viju et al. 2012). While the EU bore the lion's share of the losses due to their moratorium, the annual gains in the three key adopting nations are cut to $0.75 billion from $1.0 billion (for a loss of $250 million), clearly reducing the incentives for biotechnology companies to sustain their investments in new traits and varieties and for farmers to adapt and adopt the technology.

Using Brookes and Barfoot's (2010) data as a base, the effect of limited adoption can be put into context in two important ways. First, if the technology diffused fully, there would be both greater economic benefits for consumers, with prices down by at least 13% for the three crops, compared with actual drops of only 3.8–9.6%. More importantly from the perspective of sustained innovation in the agri-food sector, greater adoption would generate greater technology revenues which would provide more of an incentive for the leading biotechnology companies to sustain or expand their research efforts. With adoption rates in 2010, technology revenues for the three key GM food crops were estimated to be approximately $2.8 billion—Table 4.6. As adoption rises, those revenues increase. Scaling-up Brookes and Barfoot's (2010) estimates, one could anticipate that at maximum adoption gross annual technology revenues on these crops could rise to as high as $7.2 billion, more than $4.4 billion higher than currently realized.

**Table 4.6** Estimated impact of truncated diffusion on GM technology revenues

| | 2010 GM area (millions ha)[a] | 2009 area planted (millions ha)[b] | GM as % world (%)[c] | Estimated price impact[d] | Extrapolated price with full adoption[e] | Annual technology revenue (US$ billion) | |
|---|---|---|---|---|---|---|---|
| | | | | | | Actual[f] | Full Adoption[g] |
| Soybean | 73.3 | 99.5 | 73.7 | −9.6% | −13.0% | $1.1 | $1.4 |
| Maize | 46.8 | 158.6 | 29.5 | −5.8% | −19.7% | $1.6 | $5.4 |
| Canola | 7.0 | 31.1 | 22.5 | −3.8% | −16.7% | $0.1 | $0.4 |
| Total | 127.1 | 289.2 | 43.9 | – | – | $2.8 | $7.2 |

*Sources*
[a] ISAAA (2011)
[b] FAOStat. (http://www.fao.org/faostat/en/#data)
[c] Author calculation
[d] Brookes and Barfoot (2010)
[e] Author calculation
[f] Brookes and Barfoot (2010)
[g] Author calculation

One important question is what the agro-biotechnology industry would do with \$4.4 billion in additional annual revenues, and possibly more if the technology was able to be adapted and adopted in rice, wheat and other major food crops. James (2003) evaluated the research effort in 2001, concluding that there was about \$4.4 billion in crop biotechnology investment, 70% by private firms and 30% in the public domain. He further concluded that 96% of the investment was in industrial countries. Put in context, the private biotechnology sector was probably investing about \$3 billion in anticipation of generating gross annual revenues somewhere in the \$3–7 billion range, relative to a value for annual global seed trade estimated to be approximately \$25 billion at that time.

A review of the financial statements and publicly revealed research plans of the five largest private investors as of 2011 shows that they currently invested approximately six percent of gross revenues in research and development. Monsanto was at the high end, investing about 11.5%; at the other end, Dow invested about 3.1% of gross revenues (possibly due to Dow's more diversified portfolio of activities). If we take the average ratio of R&D to sales for the sector, the absence of up to \$4.4 billion in revenues in 2011 would translate into a drop of \$264 million in annual investments. Given the rule of thumb that it takes about \$100 million to bring a major new trait/crop combination to market, this translates into 2–3 fewer major innovations per year. If we used Monsanto's R&D rate, which is primarily in agro-biotechnology, then 11.5% of the missing \$4.4 billion would translate into more than \$530 million of lost investment, or about an average of five new traits per year. There is some evidence of the narrowing research effort. A 2011 review of the publicly stated research pipelines of the five key multinationals (Monsanto, Bayer, Pioneer Hi-Bred, Syngenta and Dow Agrosciences) showed that the bulk of their effort remains focused on corn, soybeans and canola as well as non-food crop cotton), with what appear to be only tentative efforts related to other large area field crops or vegetables—Table 4.7. In Canada, for example, in 2010 all but 15 of the 979 field trials were conducted on canola, corn and soybeans—one single alfalfa trial (by Monsanto) was the lone exception.

Since then there has been significant actual and planned consolidation in the industry. Monsanto has accepted an offer to merge with Bayer, DuPont Pioneer agreed to merge with Dow Agrosciences and ChemChina has made offers to merge with both SinoChem and Syngenta AG. How this level of concentration and consolidation will affect research and new product development is far from certain, but generally industry concentration reduces rather than increases research and competition.

Beyond the immediate opportunity costs of the truncated market, there is a real concern that impaired trade and suppressed investment in research into new crops will have long term effects. Oehmke et al. (2001) undertook an analysis that gives a sense of what this might mean in the long-term. They modified a Hecksher-Ohlin trade model to include endogenously determined levels of R&D activity, which then induces economic growth in a Schumpeterian fashion, through a quality-ladder approach (see Grossman and Helpman 1991 for pioneering work on quality ladders). They extended the Hecksher-Ohlin framework to include developing

**Table 4.7** GM crops in research pipelines for five key multinationals, 2011

| | Syngenta | Monsanto | Bayer | DuPont | Dow | Total |
|---|---|---|---|---|---|---|
| Sales ($ millions) | 11,641 | 10,502 | 35,088 | 32,733 | 53,674 | 143,638 |
| R&D ($ millions) | 1032 | 1205 | 3053 | 1652 | 1660 | 8602 |
| R&D (as % revenues) | 8.9% | 11.5% | 8.7% | 5.0% | 3.1% | 6.0% |
| R&D (% ag focused) | NA | ~100% | 24% | NA | NA | NA |
| *Crops targets* | | | | | | |
| Corn | Yes | 11 | 1 | 14 | Yes | ~14 |
| Soy | 1 | 12 | Up to 7 | 11 | Yes | ~12 |
| Canola | 1 | 4 | Up to 7 | 2 | Na | ~7 |
| Cotton | Na | 6 | 3 | Na | Yes | ~6 |
| Alfalfa | Na | 2 | Na | Na | Na | ~2 |
| Wheat | Na | 2 | Na | Na | Na | ~2 |
| Sugarcane | Na | 1 | Na | Na | Na | ~1 |
| Rice | Na | Na | 2 | 3 | Na | ~3 |
| Other | Sunflower, vegetables NES | Broccoli, cucumber, lettuce, pepper, sweet corn, tomato | Cucumber, melons, onions, tomato | | 7 traits | |

*Sources*

- Financial data from http://moneycentral.msn.com/investor/invsub/results/statemnt.aspx?symbol={companysymbol}
- Syngenta: http://www.syngenta.com/GLOBAL/CORPORATE/EN/PRODUCTS-AND-INNOVATION/RESEARCH-ANDDEVELOPMENT/BIOTECHNOLOGY/Pages/biotechnology-pipeline.aspx
- Monsanto: http://www.monsanto.com/products/Documents/pipeline-flash/pdfs/TechPipelineALL.pdf
- Bayer: http://www.bayer.com/en/Bayer-CropScience.aspx
- Pioneer Hi-Bred: http://www.pioneer.com/home/site/about/research/pipeline/specification-sheets/
- Dow: http://www.dowagro.com/innovation/pipeline/index.htm

countries and biotechnology. They identified a number of 'important and robust' results, including that the EU policy on prohibiting the production and consumption of agricultural biotechnology products slows agricultural and economic growth. Meanwhile North America and a number of developing countries continually adopt the latest biotechnology innovations, increasing their effective factor endowments and become more capital intensive, which leads them to reallocate land, labor and capital to the production of R&D and biotechnology goods (i.e. the more capital-intensive goods). The EU correspondingly becomes relatively more labor intensive and consumes more labor-intensive goods, which has a negative effect on EU production and trade with the rest of the world. The non-adoption of biotechnology innovations and the diminished trade amplifies the relative stagnation of agricultural growth in the EU.

Given that the EU is the largest single producer of agricultural and food products, this raises serious questions about where new supplies of food will come from to feed as many as two and a half billion additional people by 2050.

## Conclusions

Asynchronous approvals for GM crops—whether due to different regulatory decisions or varying effort by proponents to gain approvals—are important in the key product categories of GM grains and oilseeds. This divergence is creating a number of immediate challenges for the global agri-food system that, if unresolved, could have long-term effects on prices, trade, investment, innovation and ultimately the global capacity to produce adequate food to feed the growing world population.

Asynchronous approvals are a systemic concern for the global agri-food system. They are caused by disconnects between consumers (with varying expectations and preferences), between regulators (due to their inability and unwillingness to seek harmonization of regulations or the synchronous assessments of specific products) and between proponents and regulators (as proponents selectively prosecute their products rather than seeking universal market approval). Asynchronous approvals would be a minor issue if the price premiums were large enough to encourage global supply chains to adequately differentiate those systems to deliver quality assured products with different provenance. However, they are not.

Identifying the sources of asynchronicity is only half the challenge. The solution will only come if the opportunity costs identified in this chapter can be monetized and mobilized to induce one or more of the actors to close the gaps. This is not as easy as it might seem. While all of the economic studies suggest that the opportunity costs of the current situation are large and have real and significant effects on trade, prices and investments, mobilizing those interests is not straightforward.

Options include mobilizing consumers, regulators, the research community or one of the actors in the supply chain. First, this could involve one or more actors finding the right package of product attributes and supply chain mechanisms to leverage a large enough price premium to trigger a quality assured product category that bridges the gaps and satisfies both producers and consumers. Second, the regulators with most at stake (e.g. US, Argentina, China and Canada, the leading adopters, and the EU and other leading importers) either alone or in partnerships with the biotechnology companies and global supply chain actors could create more synchronicity in the regulatory system. This will probably take both more money and more willingness to adapt than we have seen so far—which may require a new way of regulating, which is discussed in the next section. Third, the research community could try to fill the investment gap created by sub-optimal revenues. Governments, because they are the only actors able to tap into and monetize the global welfare gains, could possibly fill the gap—the difficulty will be to find a way to translate any resulting research efforts to market uses. Finally, one could look to the supply chain itself for leadership. Given that much of the value left on the table right now would accrue to those in the supply chain, it is possible that one of actors could take the lead and differentiate part of the system to close the gap.

The one thing that is clear from the existing research is that this is not simply an academic concern, or a matter for dialogue among interest groups. It has real impacts on global welfare today and has the potential to have a long-term, adverse effect on global food security. The welfare gains not being realized now are the immediate concern but the potential that this lost revenue will lock-in and depress global welfare over the longer term is a bigger concern. Truncated markets and depressed technology revenues have the potential to lower innovation in the agri-food sector and, perhaps of a higher long-term concern, lower productive capacity in the EU, ultimately lowering the global capacity to meet the inexorably rising demand from a world population forecast to rise to over nine billion by 2050.

The next question is how well the existing international trade and regulatory architecture is able to respond to the issues raised by the arrival of modern agricultural biotechnology—clearly a transformative technology? Transformative technologies often present major challenges for existing institutional structures. One only has to think of the arrival of the automobile in the era of institutions designed for the 'horse and buggy' era (Kerr 2014). There were no institutions to deal with the constructions of highway networks, the safety of pipelines and gasoline refineries, the transport of gasoline, parking, tire safety, speed limits, licencing of drivers, emission standards, disposal of cars at the end of their useful life, rules of the road such as traffic signs and lights—the list of gaps is almost endless. There were also trade-inhibiting regulatory failures. One example will suffice; half of the world drives on the right and half on the left. This means that to be fully tradeable cars must be engineered to allow for controls on either side of vehicle—a non-trivial cost. For years US car makers, for example, complained that Japan had non-tariff barriers to trade in automobiles when the reality was that US-made cars were only engineered for the domestic, left hand drive market while Japanese motorists needed right hand drive cars (Kerr and Perdikis 2014). Of course, when Japanese car makers wanted to enter the US car market they had to make the expenditures required to engineer their cars to be flexible in the positioning of controls.

Biotechnology is a similar transformative technology and one should expect it to strain the existing international architecture, and require institutional adaptation or, in some cases, entirely new institutions. The next section examines the resilience of the existing institutional environment.

## References


Alston, J., M. Mara, P. Pardey, and T. Wyatt. 1998. Research returns redux: a meta-analysis of the returns to agricultural R&D. Environment and Production Technology Division Discussion Paper No. 38. Washington: International Food Policy Research Institute.

Alston, J., G. Norton, and P. Pardey. 1995. *Science under scarcity: principles and practice of agricultural research evaluation and priority setting*. Ithaca NY: Cornell University Press.

Anderson, K. 2010. Economic impacts of policies affecting crop biotechnology and trade. *New Biotechnology* 27 (5): 558–564.

Bayer, J.C., G.W. Norton, and J. Falck-Zepeda. 2010. Cost of compliance with biotechnology regulation in the Philippines: Implications for developing countries. *AgBioForum* 13 (1): 53–62.

Brookes, G., and P. Barfoot. 2010. GM crops: global socio-economic and environmental impacts 1996–2008. Dorchester, UK: PG Economics Ltd. http://www.pgeconomics.co.uk/pdf/2010-global-gm-crop-impact-study-final-April-2010.pdf.

Doern, B., and T. Reed. 2000. *Risky Business*. Toronto: University of Toronto Press.

Gleim, S., S.J. Smyth and P.W.B. Phillips. 2016. Regulatory system impacts on global GM crop adoption patterns. *Journal of International Law and Trade Policy* 17(20): 96–116.

Grossman, G., and E. Helpman. 1991. *Innovation and Growth in the Global Economy*. London: The MIT Press.

Haggui, F. 2004. EU opposition to GM wheat in terms of global food security. PhD Thesis, University of Saskatchewan. http://library2.usask.ca/theses/available/etd-09212004-133502/unrestricted/thesis.pdf.

Hobbs, J.E., W.A. Kerr, and S.J. Smyth. 2014. How low can you go? The consequences of zero tolerance. *AgBioForum* 16 (3): 207–221.

ISAAA. 2011. Global Status of Commercialized Biotech/GM Crops. ISAAA Brief 42-2010. www.isaaa.org.

Isaac, G. 2002. *Agricultural Biotechnology and Transatlantic Trade: Regulatory Barriers to GM Crops*. Wallingford, UK: CAB International.

James, C. 2003. Preview: Global Status of Commercialized Transgenic Crops: 2003. ISAAA Briefs No. 30. ISAAA: Ithaca, NY.

James, C. 2015. isaaa.org/resources/publications/briefs/51/default.asp 20th Anniversary (1996 to 2015) of the Global Commercialization of Biotech Crops and Biotech Crop Highlights in 2015. ISAAA Brief 51. ISAAA: Ithaca, NY.

Kalaitzandonakes, N., J.M. Alston, and K.J. Bradford. 2007. Compliance costs for regulatory approval of new biotech crops. *Nature Biotechnology* 25: 509–511.

Kerr, W.A. 2014. The Trade System and Biotechnology. In *Handbook on Agriculture, Biotechnology and Development*, ed. S.J. Smyth, P.W.B. Phillips, and D. Castle, 217–229. Cheltenham: Edward Elgar.

Kerr, W.A., and N. Perdikis. 2014. *A Guide to the Global Business Environment*. Cheltenham: Edward Elgar.

Lusk, J., M. Jamal, L. Kurlander, M. Roucan, and L. Taulman. 2005. A meta-analysis of genetically modified food valuation studies. *Journal of Agricultural and Resource Economics* 30 (1): 28–44.

Moschini, G., H. Lapan, and A. Sobolevsky. 1999. Roundup ready soybeans and welfare effects in the soybean complex. *Agribusiness* 16: 33–55.

Oehmke, J., M. Maredia, and D. Weatherspoon. 2001. The effects of biotechnology policy on trade and growth. *The Estey Centre Journal of International Law and Trade Policy* 2(2):283–296.

Perdikis, N. 2000. A conflict of legitimate concerns or pandering to vested interests? Conflicting attitudes towards the regulation of trade in genetically modified goods - The EU and the US. *Journal of International Law and Trade Policy* 1 (1): 51–65.

Phillips, P.W.B., and H. McNeill. 2000. Labeling for GM foods: Theory and practice. *AgBioForum* 3 (4): 219–224.

Phillips, P.W.B. 2003. The economic impact of herbicide tolerant canola in Canada. In *The Economic and Environmental Impacts of Agbiotech: A Global Perspective*, ed. N. Kalaitzanondakes, 119–140. Dordrecht, the Netherlands: Kluwer.

Phillips, P.W.B. 2009. Coexistence as a Problem of Market-making. 4th International Conference on Coexistence between Genetically Modified and non-GM based Agricultural Supply Chains. Melbourne, Australia, November 10–12. http://www.gmcc-09.com/wp-content/uploads/phillips.pdf.

Pray, C.E., P. Bengali, and B. Ramaswami. 2005. The cost of regulation: The India experience. *Quarterly Journal of International Agriculture* 44 (3): 267–289.

Pray, C.E., B. Ramaswami, J. Huang, R. Hu, P. Bengali, and H. Zhang. 2006. Costs and enforcement of biosafety regulations in India and China. *International Journal of Technology and Globalization* 2 (1&2): 137–157.

Skogstad, G. 2008. *Supranational Regulation and Contested Accountability: The Case of GMO Risk Regulation in the European Union*. Florence: EUI.

Smyth, S.J., W.A. Kerr, and P.W.B. Phillips. 2011. Recent trends in the scientific basis of sanitary and phytosanitary trade rules and their potential impact on investment. *The Journal of World Investment and Trade* 12 (1): 5–26.

Viju, C., M.T. Yeung, and W.A. Kerr. 2012. The trade implications of the post-moratorium European Union approval system for genetically modified organisms. *Journal of World Trade* 46 (5): 1207–1238.

# Part II
# The Dynamics of the Institutional Regulatory Architecture

# Chapter 5
# International Governance in the 21st Century

**Abstract** Governments have spent most of the post-Second World War period building an open and liberal trading regime. Since the creation of the World Trade Organization (WTO) in 1995, some governments have backslid, especially in the context of trade in food. Adjudication by dispute panels at the WTO has largely dismissed these justifications as illegitimate, raising concerns that consumers (and others) are disenfranchised. The result has been a loss of credibility for the WTO and sometimes its demonization.

## Introduction

As outlined in the previous chapters, since the advent of the use of biotechnology in agricultural production in 1995 there has been a rapid and widespread adoption of the technology by farmers in all markets where the technology has been approved for production. As a result, the world is now divided into four groupings of countries: (1) those that have allowed the use of the technology in agricultural production; (2) those that have eschewed the use of the technology and ban or otherwise heavily restrict imports; (3) those that have not approved biotechnology in production but allow imports under a variety of regulatory restrictions or impediments; and (4) those that have yet to develop domestic or import policies related to genetically modified (GM) products, often because there are no GM products that are suitable for their farming systems. These divergent domestic regulatory and trade contexts for GM products lead to trade frictions and sometimes formal trade disputes. As described above, only four crops make up the bulk of global GM production—soybeans, corn, cotton and canola. A few other crops have entered into commercial production (e.g. papaya) but are not yet important enough raise significant international trade issues. As more and more GM crops are approved in some jurisdictions, but not others, however, the potential for comingling of unapproved GM traits in shipments of conventionally produced crops increases (Kerr 2016).



S.J. Smyth et al., *Biotechnology Regulation and Trade*, Natural Resource Management and Policy 51, DOI 10.1007/978-3-319-53295-0_5

Trade restrictions or fear of losing foreign markets has inhibited the development of additional GM crops such as wheat and flax (Furtan et al. 2005; Viju et al. 2014). A large number of new GM crops are, however, in the pipeline. With new crops and new countries approving the use of biotechnology in agriculture, unintended comingling will increase considerably, leading to more refusals of shipments at ports of entry leading to further trade issues (Hobbs et al. 2014). Traditional economic protectionism has played little part in these trade frictions. The economic model that underpins the rules of trade embodied in the WTO and most preferential trade agreements assumes only producer-vested interests will seek to disrupt trade. They seek to enact trade restrictions to raise the price that they must compete with, thus improving their economic prospects. Consumers are never expected to ask for trade restrictions because they would work against their interest by raising prices and lowering net consumer benefits. Latterly, and in particular in the case of requests for restrictions on trade in the products of agricultural biotechnology, the requests for protection come from a range of groups in civil society that claim to speak for consumers, the environment or some other social group. A closer look at the rules of trade and their underlying premises is warranted given the rising importance of trade in the products of biotechnology and the vociferousness with which the arguments over trade in GM products are undertaken.

While these new protectionist pressures are concentrated in the food sector, there are many recent examples where requests for trade protection in other sectors have come from environmentalists, consumers and other interested parties. Some consumers in the European Union (EU) advocated for an import ban on seal pelts from Canada and have persuaded the European Parliament to legislate limits on imports. Environmentalists in the United States (US) asked for a ban on both imports of tuna caught in dolphin unfriendly ways and shrimp caught in turtle unfriendly ways. Consumers in the EU were successful in banning imports of beef produced using growth hormones—and in having the EU Commission accept retaliation rather than comply with a WTO disputes panel ruling. The latter, while certainly within the EU's rights under the WTO, was an unprecedented action. Consumers in a range of developed countries demanded broad-based restrictions on imports from China in the wake of media reports of a spate of product safety problems in 2007 (Liu et al. 2009). Child labor, leg-hold traps, organic standards, animal welfare, green labeling and a wide range of other issues have also led to calls for restrictions on imports, including restrictions on imports of the products of agricultural biotechnology. International trade institutions, however, do not make provision for any group other than producers asking their governments for protection. As a result, while some of these justifications for imposing trade barriers have been adjudicated through the WTO's dispute settlement body, these issues have not been satisfactorily resolved. The groups involved feel marginalized and frustrated, and the reputation of international trade institutions has plummeted within civil society. Ultimately, this decline in credibility threatens the architecture and management of the global economy.

To explain the inability of international trade institutions such as the World Trade Organization to respond to the new civil society groups requesting protection, the remainder of this chapter outlines the economic theory that underpins the WTO and examines some examples of how governments faced with demands for protection have dealt with the problem; it also looks at the WTO's response. The effects of the WTO's actions are then examined. In the end, likely little can be expected from international trade institutions until economists provide a theoretical model that directly acknowledges and incorporates the broader interests and concerns of the twenty-first century. The major challenges facing the development of an appropriate international trade model are then set out. It provides the building blocks for what follows in the remainder of the book.

## The International Trade Model, Circa 1947

To understand the logic that is embedded in the institutional architecture for international trade, one has to first examine what economists teach about international trade and in particular what they teach in basic university courses. This is because negotiating delegations are largely comprised of individuals whose academic training is not centred on international trade, although individuals may have taken a course or two in international economics (Kerr 2007a).[1] In the mid-1940s when the Havana Charter[2]—including the General Agreement on Tariffs and Trade (GATT) —was negotiated, this was certainly the case. Thus, it is important to understand what those charged with negotiating the international trade architecture might have understood regarding the economics of international trade.

In the academic study of international trade by economists there are three major divisions—trade theory, empirical analysis of trade relations, and trade policy (Kerr 2007b). Trade theory and trade policy are relevant to the structuring of the institutional architecture for international trade that currently exists. Trade theory seeks fundamental insights through rigorous modeling and tightly specified assumptions.[3] Trade policy deals with the economic effects of government intervention in markets where international transactions take place.[4] Both of these feature heavily in introductory and intermediate courses in international economics.

[1]Of course, they often have considerable on the job experience in trade policy.

[2]In the wake of the Great Depression and the Second World War, the US and the United Kingdom worked hard to put a new set of international institutions in place that would reduce the likelihood of conflict in international relations: the United Nations (UN) for dealing with political conflict; the International Monetary Fund (IMF) for limiting the strategic use of currency devaluations; and the World Bank to deal with differences in levels of development. One other international institution was envisioned—an International Trade Organization (ITO) (Kerr 2000). The Havana Charter—negotiated in 1947—was the agreed architecture for the ITO. While the US Administration had agreed to the Havana Charter, it seemed clear that the US Congress would not ratify it. It was never submitted to the US Congress and the ITO was stillborn. One of the sub-agreements of the Havana charter—the GATT—became the de facto multilateral international trade institution (Miner 2007).

[3]General equilibrium is the predominant analytical approach.

[4]Using a partial equilibrium, comparative statics approach for the most part.

In the first instance, simple general equilibrium trade theory is used to prove the counter-intuitive result of comparative advantage—that both countries gain from an opening of trade even if one country is more efficient at producing both (all) goods. Magee (1979, p. 91) reports, for example, that:

> Harry Johnson [a famous international trade economist] once offered the offhand suggestion that an important job of trade theorists was to develop results that would be paradoxical to non-specialists.

Nobel Laureate Paul Samuelson reported that he was once challenged by mathematician Stanislaw Ulam to "name me one proposition in all of the social sciences which is both true and non-trivial." After some thought, Samuelson's response was *comparative advantage* because:

> That it is logically true need not be argued before a mathematician; that it is not trivial is attested by the thousands of important and intelligent men who have never been able to grasp the doctrine for themselves or to believe it after it was explained to them (Samuelson 1969, p. 9).

While general equilibrium is used to shed light on other issues in introductory courses in international trade, comparative advantage is the center piece for illustrating the ability of trade theory to shed light on economic phenomena. While more recently there has been some re-thinking of unqualified support for the theory of comparative advantage (Samuelson 2004), this was not the case in the 1940s. The first edition of Samuelson's extremely successful introductory level textbook was published in 1948, the year after conclusion of the Havana Charter-GATT negotiations. First-year texts largely present generally accepted paradigms, otherwise they are unlikely to garner wide classroom adoption. Samuelson's book provides a careful, but simple, proof of the mutual benefits of trade liberalization, as well as some discussion generalizing the theory's results and some criticisms. In the first Canadian edition, the discussion of comparative advantage ends as follows:

> Yet for all its oversimplification, the theory of comparative advantage provides a most important glimpse of the truth. Political economy has found few more pregnant principles. A nation that neglects comparative advantage may pay a heavy price in terms of living standards and potential growth (Samuelson and Scott 1966, p. 720).

How does the teaching of international trade translate into the international trade architecture put in place in 1947 with the GATT? The prologue to the GATT-1947 reads as follows:

> Recognizing that their relations in the field of trade and economic endeavour should be conducted with a view to raising standards of living, ensuring full employment and a large and steadily growing volume of real income and effective demand, developing the full use of the resources of the world and expanding the production and exchange of goods,
>
> Being desirous of contributing to these objectives by entering into reciprocal and mutually advantageous arrangements directed to the substantial reduction of tariffs and other barriers to trade and to the elimination of discriminatory treatment in international commerce (GATT 1947).

Hence, as trade liberalization was expected (from theory) to be welfare enhancing, it was enshrined as the sole and unqualified objective of the GATT. Of course, member states would have all sorts of political reasons for keeping existing trade barriers or wanting to put new ones in place. The *rules* negotiated in the rest of the GATT-1947 agreement simply set agreed limits on how governments of member states are allowed to respond to protectionists. Thus, there is a general acceptance of the proposition that trade liberalization is the economic goal, but also a recognition that governments may, at times, need to respond to requests for protection for domestic political reasons—to act in ways that are not welfare enhancing. There is no questioning of the basic general equilibrium result arising from comparative advantage. This is important to note, as it will inform the discussion of consumer and other groups in civil society's requests for protection later in the chapter and in other chapters in the book.

Beyond the important commitment to trade liberalization, general equilibrium trade theory does not seem to have influenced the development of the GATT's institutional architecture to any significant degree. The actual rules of trade that were negotiated are based on the partial equilibrium model that has long been the workhorse of trade policy. Partial equilibrium was, and to a considerable extent still is, used to put undergraduate and graduate students through their paces when learning applied welfare analysis (Gaisford and Kerr 2001; Kerr 2007b).

According to Schmitz (1979, pp. 273–274):

> One branch of international trade economics, which has a long history, is partial analysis. This approach to the study of trade problems, which dates back to at least Barone [1908] is widely used in both theoretical work and as a basis for quantitative analysis. …
>
> Early economists used this approach (i.e., partial analysis, along with the concept of economic surplus as a measure of welfare) to determine whether or not countries gain from engaging in international trade and the extent to which countries are hurt or made better off by tariffs. Interestingly, Barone very early concluded that consumers' and producers' surplus are at a maximum under free trade – a proposition which still holds true in theory today. …
>
> A less sophisticated but widely used version of Barone's basic model assumes that there are no effects on the terms of trade as a result of the imposition of tariff (sic.) by the country under consideration … This is referred to as the "small country" assumption model. It appears that Cordon [1957] was among the first to use this surplus model.[5]

While trade policy was formulated on a market-by-market basis long before there were trade agreements—tariffs are, after all, imposed on an individual product basis—the GATT-1947 is clearly informed by the partial equilibrium model, the ruling economic paradigm of the time.

[5]According to Schmitz (1979), Haberler (1936) was the first explicitly to discuss the small country case. Of course, as Schmitz notes, the strong assumptions of the simple model limit its general applicability. One suspects, however, that these limitations do not receive much emphasis in introductory and intermediate courses in international economics.

When applying the normal supply and demand framework of the partial equilibrium model, economists either assume that the cross market effects are sufficiently small that they can be safely ignored or that any cross market effects are of no interest to the question being analyzed. Thus, the analysis is confined to a single market. Of course, a *market* is loosely defined so that it can encompass what the economist might find useful—the market for cream linen shirts in Lubbock, the Texas market for beef, the US market for cotton, or the North American market for barley. In the GATT/WTO a single market has a specialized meaning, in that it is the basis upon which the principles are applied. One of the major principles of the GATT/WTO is non-discrimination (Gaisford and Kerr 2001). Non-discrimination has two manifestations: (1) national treatment, whereby foreign firms can be treated no less favourably than domestic firms; and (2) the most-favoured-nation principle, whereby foreign firms from all member states must be treated equally. For example, the same tariff rate should be applied to goods from all member states. This means that there cannot be different trade policies applied to the products of individual countries arriving in the same market, as this would constitute discrimination.

This raises the question as to when a country can apply different trade policies? The answer, given the partial equilibrium framework, is when there is a different market. This raises a second question: when is there a separate market? The GATT-1947 deals with this question using the concept of *like products*. Having different trade policies applied to *like products* would amount to discrimination. If products are not alike, then different trade policies could be applied. This leads, of course, to the question of what constitutes a *like product*, a question that was the subject of difficult negotiations (Isaac et al. 2002). The result was an arbitrary dividing line that has been the root of much of the subsequent difficulties that international trade institutions have had in dealing with sources of requests for protection emanating from groups in civil society (e.g. consumer groups or environmental non-governmental organizations).

A major concern with the definition came from developing countries which feared that differences in technology could be used to make goods *not-like*. For example, cotton trousers produced using a labor-intensive technology such as hand looms would be considered *not-like* cotton trousers produced using capital-intensive mechanized technology, and hence potentially subject to trade barriers. Developing countries perceived that considering technology as a reason for goods to be considered not alike would allow developed countries to put barriers in place against their goods. As a result, the agreement provides that goods can only be considered not alike if they could be physically identified as being different—e.g. a railway locomotive powered by steam could be considered unlike a locomotive powered using an electric engine. If *production and processing methods* (PPMs) differed then a product would still be considered a *like product* and the imposition of trade barriers would be considered discriminatory and, thus, in violation of GATT obligations. Of course, this is one of the central problems in the trade of products arising from the use of agricultural biotechnology where the requests for protection are often based on the use of the technology in production and not on specific identifiable product attributes (Isaac et al. 2002),

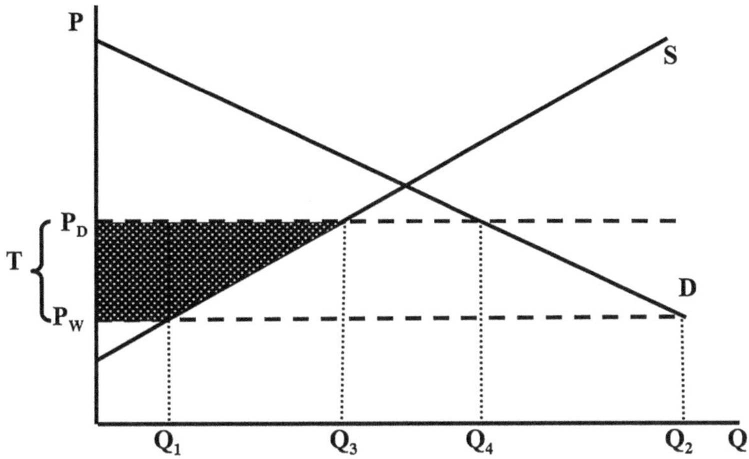


**Fig. 5.1** Partial equilibrium model with trade barrier showing effect on producer's surplus

The partial equilibrium trade model leads to predicted outcomes that further define the international trade architecture. Figure 5.1 illustrates the simple partial equilibrium model developed in Chap. 2 when imports are priced at less than the domestic equilibrium price. With no trade barrier, imports are available at $P_W$ (meaning this is the price with which domestic firms must compete), domestic producers would chose to produce $Q_1$ but domestic consumers would wish to consume $Q_2$ and imports would equal $Q_2$ minus $Q_1$. If a trade barrier—say a tariff equal to T—is imposed then a wedge is imposed between the import price, $P_{W,}$ and the domestic price, $P_D$. Domestic producers now compete against imports priced at $P_{D,}$ and will alter their output to $Q_3$. Domestic producers clearly benefit from the trade barrier, receiving a higher price and a corresponding increase in producer surplus as illustrated by the shaded area.

The model clearly predicts that producers have an incentive to ask governments for protection. Governments, at times, may feel the need to provide the requested protection. While trade liberalization may be welfare enhancing,[6] governments have been unwilling to move to free trade either unilaterally or collectively. Beggar-thy-neighbour trade wars, however, are recognized by governments as being detrimental—as a non-cooperative outcome—and acknowledge benefits may arise from cooperation through trade agreements (Gaisford and Kerr 2003; Gaisford and Hester 2007). The entire international trade architecture is exclusively focussed on collectively agreed limits on how governments are allowed to respond to pressure from domestic producers to provide protection. For example, up until the 1995 Uruguay Round Agreement on Agriculture (AoA) countries were allowed to

[6]In the sense of consumer and producer surplus commonly used in the context of the partial equilibrium model.

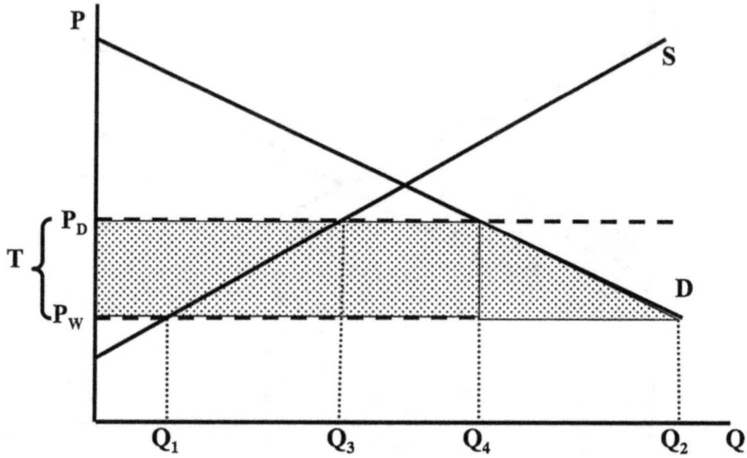


**Fig. 5.2** Partial equilibrium model with trade barrier showing effect on consumers' surplus

use quantitative limits on imports and variable levies to control the import of agricultural products. In the AoA, it was collectively agreed that these forms of protection would be replaced by tariffs, or in some cases by tariff rate quotas (TRQs) (Skully 2007), thereby limiting and many cases reducing the options available to governments for responding to requests for protection from agricultural producers.

Figure 5.2 shows the effect on consumers of imposing a tariff. Similar to Fig. 5.1, the tariff drives up the price paid by consumers from $P_W$ to $P_D$. Hence, when a trade barrier is imposed, consumers pay a higher price and suffer a decline in consumer surplus as shown by the shaded area. The unambiguous prediction from the standard partial equilibrium model is that consumers will lose from protection. A lower quantity will be consumed at a higher price. As a result, the framers of the GATT-1947 never expected consumers to ask their governments for protection and no provision was made in the agreements to allow governments to respond to consumers' requests for protection. Thus, while it was recognized that governments have a right to respond to producers' requests for protection, this is not the case if consumers seek protection.

The GATT-1947 also makes no allowance for economic actors who are not direct participants in markets that might ask their governments for protection. For example, if environmentalists perceive a large negative externality in a market, one the exporting country cannot be convinced to address, then they may ask that the product be excluded from their domestic market, as in the case of tuna caught using methods considered unfriendly to dolphins—clearly a second best solution (Isaac et al. 2002).

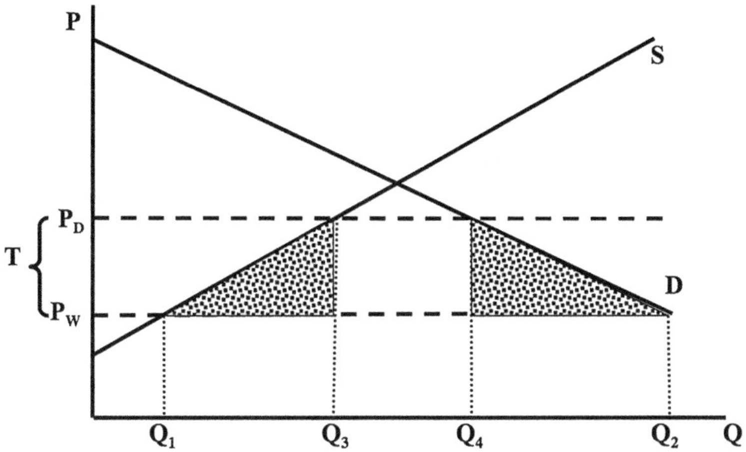


**Fig. 5.3** Partial equilibrium model with trade barrier showing effect on welfare

There is one other unambiguous comparative static result that arises from the use of the partial equilibrium trade model. Any deviation from the unrestricted optimum will generate dead weight losses. In Fig. 5.3, if the trade barrier, T, were to be removed total welfare would increase by the trapezoid framed by the two price lines and the supply and demand curves—while the clear rectangle is simply a transfer of welfare from producers to consumers, the shaded triangles are dead weight losses (reductions in total welfare arising from the distortion in the market due to inefficient domestic production and artificially restricted consumption). This powerful result from the partial equilibrium model supports the trade liberalization goal of the GATT-1947.

The economic model that underlies the GATT-1947 provides other clear and unambiguous results. It is entirely possible that consumers might suffer a loss of utility from having a foreign product in their market. Figure 5.4 provides the framework for examining a real-world example. Suppose that imports of a product are made using child labor which some consumers prefer not to purchase. As whether or not a product is made with child labor is not discernible by examining the product, it would be considered under GATT-1947 to be a *like product* to that produced domestically without the use of child labor. Without any government intervention, the product could be imported at $P_W$. The domestic price would equal the world price and domestic producers (using adult labor) would produce only to $Q_1$. While consumers cannot discern whether the goods they are purchasing are made with child labor—a pooled market result (Gaisford and Kerr 2001)—they do know that there is some probability that the goods they would purchase in this context were made using child labor. To avoid purchasing goods produced using

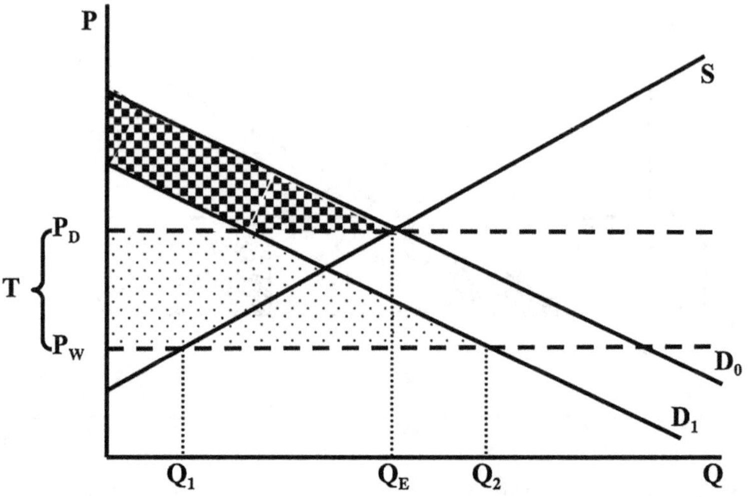


**Fig. 5.4** Removal of trade barrier when consumers lose as well as gain

child labor, some consumers may simply choose not to purchase the product at all. Remember $D_0$ is an aggregate demand curve so some consumers could simply leave the market leading to a leftward shift in demand—$D_1$.[7] The market moves to a new equilibrium at $P_W$ with $Q_2$ consumed and $Q_1$ produced domestically. Consumers still benefit from the positive *price effect* (the lightly shaded area between $P_D$ and $P_W$ and $D_1$). However, there is a loss of consumer surplus due to the leftward shift in demand—a negative *quality effect*.[8] In this context, trade liberalization may no longer be unambiguously beneficial for consumers. Rather, it becomes an empirical question regarding the relative size of the beneficial *price effect* and the detrimental *quality effect*. In this simple case, those who would choose to leave the market have an incentive to ask for protection. Governments may wish to respond to these consumer demands for protection—and to enhance their consumer surplus—just as they may want to respond to producer requests for protection, but there is no mechanism in the international trade architecture to allow such a response. Of course, devising empirical estimates of the size of surpluses is fraught with difficulties and heavily influenced by the assumptions selected. As a result, estimates will be controversial and contested. As a result, disputes will fester.

In a similar vein, trade liberalization is no longer unambiguously welfare enhancing as consumers suffer a negative quality effect. In Fig. 5.5 the removal of

[7]The case of a subset of consumers choosing to leave the market is only one possibility. Similar results have been found when consumers generally suffer a loss of utility due to the entry of foreign products with undesirable, but physically undifferentiable, attributes into the market (Hobbs and Kerr 2006). Similar results have also been found when a trade barrier that does not totally close a market is removed (Gaisford et al. 2001).

[8]Denoted by the cross-hatched area between $D_1$ and $D_0$ above the original price, $P_D$.

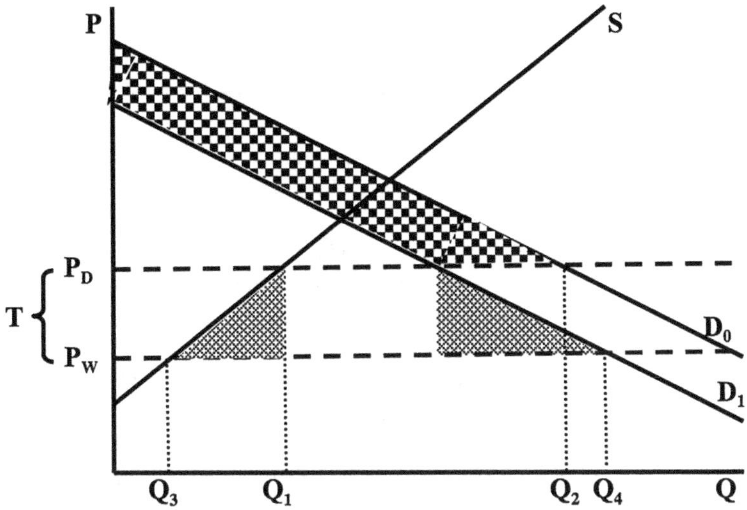


**Fig. 5.5** Partial equilibrium model showing welfare trade-offs for consumers given a negative quality effect

the trade barrier T leads to a decline in price from $P_D$ to $P_W$. As a result of this *positive price effect* there is a reduction in *dead weight loss* (equal to the shaded areas between $P_D$ and $P_W$). However, similar to Fig. 5.4, there is a decline in consumer surplus (the cross-hatched area) due to the *negative quality effect*. Hence, whether trade liberalization is welfare enhancing becomes an empirical question that depends on the relative sizes of the two effects. Obviously, this is a major departure from the unambiguous support for trade liberalization inherent in the standard model and calls into question the rationale for the commitment to trade liberalization that all GATT/WTO members have made.

One other important result arises from the potential for consumers to request protection. If the government decides to acquiesce to consumer requests for protection, any trade barriers put in place will also benefit producers—they will enjoy a rise in price and an increase in surplus. At a minimum, this may cause confusion regarding who is asking for protection. The importance of this result will be returned to later in this chapter.

## Exceptions: Discrimination

The GATT/WTO provides for exceptions to the Principle of Non-Discrimination. Five of these exceptions are briefly reviewed here because they have been (or have the potential to be) used by governments to justify their response to consumer requests for protection.

## *Protection of Public Morals*

GATT Article XX provides a number of General Exceptions to the Principle of Non-Discrimination. Article XX (a) allows trade barriers to be imposed when it is necessary to *protect the public morals*. This allows governments to restrict the import of pornographic materials or, in states with official religions, to ban the import of food products whose consumption is prohibited on religious grounds. Until recently it had not been the subject of a major disputes panel ruling (Isaac 2007).[9] In 2009, however, it was the reason put forward by the European Union to justify its import ban on seal products which was disputed by Canada and Norway. The Panel agreed with the EU in this case. Given this judgement, however, it is not clear what the evidence requirements will be. What determines when public morals are breeched? Given the ruling in the seal products case, the EU might be tempted to justify its import regime for products of biotechnology on Article XX (a) rather than Sanitary and Phytosanitary grounds as has been the case thus far. Without clear criterion as to when public morals are to be considered violated, this exception could provide an avenue for widespread justifications of protectionist measures.

## *Conservation of Exhaustible Natural Resources*

GATT Article XX (g) allows trade barriers to be put in place if they are necessary for the *conservation of exhaustible natural resources*. A number of requirements are built into Article XX (g). Environmentalists have looked to this exception as a means to enlist the WTO in furthering environmental protection. Thus far this exception has been interpreted narrowly, such that trade restrictions are only allowed when all other avenues for resolving the environmental dispute are exhausted (Isaac et al. 2002). However, "there is significant confusion concerning the interpretation and uncertainty in the application of its specific terms" (Belcher 2007, p. 434), so the opportunity to use Article XX (g) has not been entirely closed off. If this opportunity does exist, it has not yet been successfully exploited by those who might seek protection.

## *Marks of Origin*

GATT Article IX allows importing countries to require an indication of the country of origin to be applied to imported products. Again, this might be construed as the

[9]The US successfully invoked protection of *public morals* in its dispute with Antigua and Barbuda over online gambling. See http://docsonline.wto.org/imrd/gen_searchResult.asp?RN=0&searchtype=browse&q1=%28+%40meta%5FSymbol+WT%FCDS285%FC%2A%29&language=1.

framers of the GATT-1947 being cognizant of, for example, *consumers' right to know* concerns[10] but a close reading suggests that the primary concern of the GATT rules is to prevent the abuse of these regulations in aid of providing protection for producers.[11] The only explicit reference to consumers relates to them being protected from fraud. Thus, trade barriers could be put in place to keep products labeled as being from country X when they were actually from country Y—for example, cigars made in Haiti being labeled as Cuban.

## Sanitary and Phytosanitary Measures

Sanitary barriers dealing with human health benefit consumers but the rules pertaining to sanitary and phytosanitary (SPS) trade barriers are written to prevent their use as disguised economic protection for producers. Discrimination among countries is allowed because disease profiles vary—e.g. a disease may exist in a tropical country so barriers to imports from that country may be justified, whereas the disease may not occur in a temperate country so barriers against the same product could not be justified. Scientific evidence is to be the basis for the imposition of such barriers and a risk assessment must be carried out (Isaac et al. 2002). Barriers can be put in place if there is insufficient scientific information—but the imposing country is obligated to actively seek the required information and the barriers are expected to be both proportionate to the risk and temporary.

## Technical Barriers to Trade

Technical barriers to trade relate to questions such as standards and labeling which can be put in place to protect or inform consumers (Hobbs and Kerr 2006). Non-discrimination is the dominant principle. There is also a provision for *novel* products. If a product is *novel* it can be considered *unlike* existing products, and trade barriers to a new product can be put in place. A product cannot be designated as novel based on *production or processing methods* (PPM) alone (Isaac et al. 2002). Hence, for example, a country could not require imports of a new strain of genetically modified corn to be labeled 'GM corn' because genetic modification is a PPM.

---

[10]See Hobbs (2001) for a discussion of how marks of origin may mislead consumers seeking a *positive quality effect* from protection.

[11]The recent experience with US country-of-origin labeling initiatives for certain food products is an example of protectionism arising from producers (Kerr and Hall 2003).

## Marginalized Members of Civil Society

According to the Webster's Dictionary (1985, p. 303), a crank is "marked by eccentricity … crazy, imbecile". Given that the international trade architecture does not provide for countries wishing to respond to consumers (and others) asking for protection, their concerns are effectively ignored and the individuals who express protectionist preferences—preferences that are often strongly held—feel marginalized. From their perspective, the international trade institutions treat them as if they are *cranks*. Five examples illustrate the challenge.

First, a variety of growth hormones are used to enhance beef cattle productivity across North America and a range of other major beef producing countries. Due to strong concerns expressed by some consumers in the European Union regarding the effect on children of the residual presence of these hormones in beef, the use of these hormones was virtually banned within the European Union (Roberts 1998). Subsequently, an import ban on beef produced using hormones was also put in place. The US and Canada brought a case forward to the disputes system of the WTO. The WTO agreements did not allow the EU to justify the barrier on the basis that its consumers did not want beef produced using hormones in their market. It could not claim that beef produced using growth hormones was an *unlike* product because there is no observable difference in the resulting beef and the use of growth hormones is a PPM. With no direct way to justify their import ban to the WTO Panel, the EU attempted to defend the ban based on there being a risk to human health. Insufficient scientific evidence was the rationale put forward. This was done despite a strong case being made that a *scientific consensus* existed that growth hormones do not pose a risk to human health. The evidence supporting this position included reports by the EU's own scientists. Nevertheless, the EU claimed that insufficient testing had been done on sub-populations who were taking particular drugs—raising complex questions regarding what would constitute sufficient scientific evidence. It was also clear that some EU consumers were not willing to accept the *scientific consensus,* which called into question the use of science as a decision criterion (Kerr 2003). The Panel eventually came to the conclusion that the EU had not conducted a viable risk assessment and did not provide sufficient scientific justification for a ban. It ruled that the ban should be lifted. The EU, as is within its rights under the WTO, refused to comply with the Panel's ruling and chose, instead, to accept retaliation (Kerr and Hobbs 2005).

While permitted, choosing retaliation was unprecedented and indicates the magnitude of the political problem with consumers that was perceived by the EU Commission. From the perspective of those consumers, the WTO gave no credence to their concerns, in effect saying to them: Why don't you accept that beef produced using growth hormones is a *like product* to beef produced without hormones?; or Why don't you accept the scientific consensus? In effect, you are cranks. Certainly EU consumers who feel ignored are not likely to be supporters of multilateral international trade institutions. On a final note, there is a common perception in North America that EU consumer concerns are a myth and that the import ban was simply disguised economic protection for the EU beef industry. It has also been argued that it was a measure that reduced beef subsidy costs for the EU. It is true that the ban did provide protection for the EU beef industry, but there is compelling evidence that the ban was due to consumer pressures and not industry rent seeking (Lusk et al. 2003; Alfnes 2004).

Second, environmental activists pressured the US federal government to impose import bans on tuna caught in ways that are harmful to dolphins, as well as import prohibitions on shrimp caught in ways that were detrimental to turtles—for many this was an unacceptable negative externality associated with imports of these products. Complaints were brought to the GATT and subsequently the WTO that these US domestic laws violated the non-discrimination principle of the WTO. The US could not justify its barriers on *unlike product* grounds because fishing methods are a PPM. The US attempted to use the exemption provided in Article XX (g) pertaining to *conservation of exhaustible natural resources*. The Panel, however, interpreted XX (g) narrowly and ruled that the ban should be removed, and the US complied (Isaac et al. 2002).[12] The narrow basis of the ruling was consistent with a system designed to control the abuse of this exemption in the economic interests of producers. From the perspective of US environmental groups, however, the message from the WTO was that those who asked for trade barriers simply did not understand that there was no difference between tuna/shrimp fished in different ways. Their concerns were, in effect, ignored and they were treated as cranks. Needless to say, US environmental interests felt marginalized and became dismissive of the WTO (Fox 2002). It was after these rulings that a common sight at WTO gatherings were protestors dressed as turtles and dolphins

Third, in response to rising concerns among some consumers, a number of countries, including the EU, have been raising domestic animal welfare standards. Animal welfare non-government organizations have been lobbying fiercely to allow only products that meet EU animal welfare standards to be sold in the EU. In other words, imports would have to comply with EU criterion. The EU has taken a proposal to the WTO to have the issue put on the negotiating agenda in the Doha Round. Developing countries reacted to attempts to discuss trade restrictions based on animal welfare standards by suggesting this was thinly veiled economic protection for developed country livestock and poultry producers and refused to have it part of the negotiating agenda (Hobbs et al. 2002). The EU has not yet put import restrictions in place because animal welfare standards are a PPM and such restrictions would surely be successfully challenged. The issue also has still not been put onto the multilateral agenda for negotiation (Kerr 2012). As a result, animal welfare advocates in the EU and elsewhere feel they are being treated as if they are cranks.

[12]They also attempted to justify the ban on the basis of Article XX (b) relating to the *protection of human, animal or plant life or health*. As this is now largely supplanted by the SPS it was not considered in the exemptions outlined above.

Fourth, some members of civil society in developed countries have strongly lobbied politicians to have goods produced using child labor denied access to their markets through the use of trade barriers. Developed country governments cannot, however, go the WTO and say '*our consumers will not accept goods produced using child labor'*. They cannot discriminate between products produced using child labor and those not produced using child labor because the type of labor used is a PPM. Developing countries have reacted to developed country concerns by suggesting that it is simply a protectionist ploy to provide economic protection for domestic industries (Bakhshi and Kerr 2008). The international organization responsible for labor standards is, however, the International Labor Organization (ILO) not the WTO. Unlike the WTO, the ILO has no enforcement mechanism. Those opposed to child labor have had some success in getting firms to impose private labor standards for the products that move through their supply chains. The WTO appears to child labor advocates as an uncaring institution that treats them as if they are cranks.

Finally, the issue of how trade in GM products is to be regulated has grown to dwarf these other concerns (Kerr 2014). In a similar vein to the examples above, some environmentalists and consumers object to the import of GM products or, at a minimum, want imports to be appropriately labeled. Given strong resistance among environmentalist and consumers to GM crops, the EU implemented a moratorium on domestic licensing of GM crops and banned GM crop imports in the late 1990s. The ban was to be temporary until the EU could devise a new domestic regulatory regime for GM crops. Arriving at a set of new domestic regulations proved extremely difficult and the ban on imports remained in place for a considerable period. The long delay prompted a challenge by the US, Argentina and Canada at the WTO in May 2003. After much delay, the WTO Panel reached its decision on September 29, 2006. Although the EU ban had already been lifted (replaced by the long awaited new domestic policy), the Panel nevertheless ruled strongly in favour of the complainants (Isaac and Kerr 2007). The EU had not argued at WTO that '*Some of our environmentalists and consumers have a problem with GM crops and we want to impose trade barriers*' because that would have rested on arguing that against GM technology as a PPM; instead the EU attempted to use SPS concerns as the rationale for the import restrictions—but to no avail. In absence of any compelling evidence in the context of a formal risk analysis, the SPS-based argument failed. Again, it appeared as if that those opposed to GM crops were being classified as cranks whose concerns could simply be dismissed by the WTO. It is worth noting that the EU import ban was perceived by many in North America as simply a way to protect the EU's lagging GM research sector.

These are just a few examples of a multitude of environmental, consumer and other non-producer requests for protection that have come to the fore over the last two decades—others include ecological labeling, organic standards, testing every animal for Bovine Spongiform Encephalopathy (BSE) , labor standards, pelts from Canada's seal hunt, furs produced from animals caught in leg-hold traps and imports of products from China. Some of these have become formal disputes but in most cases governments have declined to acquiesce to the requests for protection due to their existing WTO commitments. From the perspective of all those requesting protection but failing to get action, they are viewed as cranks by international trade institutions.

## Consequences of the Inability to Accommodate New Protectionists

The major consequences of the failure to deal directly with the issue of governments faced with non-producer requests for protection is that the prestige and credibility of the WTO has been damaged among a previously largely unengaged and indifferent public. While support for multilateral trade remains overall positive in most OECD countries, support for further liberalization has been tempered in many countries (Kerr 2010). There are many reasons for the apparent failure of the Doha Round of WTO negotiations, but the ambivalence of civil society to further liberalization is certainly a contributing factor.

In absence of new agreements, governments have sought ways to circumvent WTO disciplines in response to sustained protectionist pressures from non-producer groups. Environmental and labor standards side-agreements are increasingly incorporated into preferential trade agreements. While these side-agreements have not yet had a major impact on altering the international trade architecture, their existence is a recognition of legitimate public concerns that governments feel they need to respond to. Certainly in the US, legislators have been attempting to increase the efficacy of these side agreements in newer preferential agreements. Given that regional trade agreements must conform to WTO norms, and often allow for settlement of disputes at the WTO, these side agreements fail to grapple with the central issue.

In some cases, attempts are being made to circumvent the existing international trade architecture entirely. The Cartagena Protocol on Biosafety (CPB) , a multilateral environmental agreement that has been accepted by 170 countries, which deals exclusively with rules for trade in genetically modified organisms, is a case in point. According to Isaac (2002), eNGOs, frustrated by the WTO had a major influence on the design of the CPB.[13] The mega-adopting countries that produce the

[13]"A request for draft protocols resulted in an Ethiopian submission in October 1996 submitted on behalf of the African delegation and written by the Third World Network (TWN). This draft protocol, considered representative of the views of many developing countries, used as a framework the Basel Convention on the Control of Transboundary Movements of Hazardous Wastes and Their Disposal. As a result, the draft protocol treated shipments of LMOs (Living Modified Organisms; the CPBs term for genetically modified products) with the same degree of prescriptive regulation as shipments of toxic or nuclear waste. Further, this draft protocol placed an enormous burden upon the Party of export (the exporting country) and the exporter to ensure biosafety and to gain approval before any shipment of LMOs" (Isaac et al. 2002, p. 38).

bulk of GM crops have not ratified the CPB.[14] Thus, there is the potential for there to be two sets of rules of trade for GM crops, leading to questions such as: Which agreement takes precedence?; What can be done when membership is not totally overlapping? A host of other issues are discussed later in this book. The CPB is an extremely protectionist instrument which gives importing countries a virtual carte blanche to restrict imports without any effective mechanism for exporters to object (Hobbs et al. 2005; Holtby et al. 2007; Kerr et al. 2014). If it were to become a model for international trade architecture in the future, many of the gains in trade liberalization made in the more than half century since the coming into force of the GATT in 1947 would be in jeopardy. Thus, the consequences of a failure to deal with the issue of government response to non-producer requests for protection could be far greater than a few protesters dressed up as turtles or GM bananas outside WTO gatherings.

## The Need for a New Economic Model for the Governance of International Trade

Given the underlying economic model upon which the current international trade architecture is built, one should probably not be surprised that no progress has been made in addressing the new issues confronting the governance of international trade. Trade officials understand these are difficult issues. It is easy to characterize them as being ostrich-like, preferring to put their heads in the sand and wishing the problems would simply disappear. However, this is too harsh of a depiction. Economists have not given them anything to work with—the economic paradigm that underpinned trade policy in 1947 and remains dominant today is inadequate to address modern day issues. International trade theorists in the economics profession need to turn their efforts to devising a more appropriate model upon which to base the architecture for governing trade going forward.

Devising an improved model presents some significant challenges. The model would have to, for example, formally incorporate consumers with heterogeneous tastes, some of which have very strong preferences. Decision rules relating to what would constitute a sufficient expression of preferences to justify the imposition of trade barriers would have to be developed. It would also need to explicitly incorporate the preferences of individuals who perceive negative consequences associated with imports, but who are not significant market participants—e.g. vegetarians who have strong preferences for high animal welfare standards for imported products. Economic models normally assume that consumers are *sovereign* and

[14]The US cannot be a party to the CPB, which is legally under the broad umbrella of the 1992 Convention on Biodiversity (CBD), because the US Congress has not ratified the CBD. Even though it was clear that the US would not be a party to the CPB, it was allowed to take an active part in the negotiations given the important place it holds in biotechnology development, production and trade.

have perfect information. Given that producers gain when trade barriers are imposed to protect consumers (or others), they may have an incentive to attempt to influence (or capture) consumers—clearly violating the assumption of consumer sovereignty. Campaigns by the organic industry to encourage consumers to support embargos on imports of GM products are one example of an attempt to capture consumers. The new model will need some mechanism to discern if consumers are sovereign or captured, and, hence, whether consumer requests for protection can be considered legitimate. For all of these issues, better means of calculating expected changes in welfare will be required. These are not easy questions for economists to deal with. Trade economists may be able to draw on and or adapt methods from other branches of economics (e.g. valuing non-market amenities from environmental economics) and behavioural theorists (e.g. assessing the stability and resilience of stated preferences through experimentation) but must also always keep in mind that their analysis must be careful and rigorous, for fear that what they devise may be open to exploitation by traditional producer interests.

Devising a new trade model to replace or augment the paradigm upon which the GATT/WTO is based will be a major undertaking for members of the economics profession. Without such an effort, however, little can be expected from trade policy makers. In absence of any significant movement by economists, one can anticipate further erosion of support for international trading institutions and considerable foregone gains from trade. Agricultural biotechnology will be at the centre of the development of a new architecture for the governance of international trade. Until progress is made there are likely to be considerable trade benefits forgone and the rate of improvements in agricultural productivity diminished. Both of these outcomes have major implications for future global food security.

## References

Alfnes, F. 2004. Stated preference for imported and hormone-treated beef: Application of a mixed logit model. *European Review of Agricultural Economics* 31 (1): 19–37.

Bakhshi, S., and W.A. Kerr. 2008. Incorporating labour standards in trade agreements: Protectionist ploy or legitimate trade policy issue? *International Journal of Trade and Global Markets* 1 (4): 373–391.

Belcher, K. 2007. Trade agreements and multilateral environmental agreements. In *Handbook on International Trade Policy*, ed. W.A. Kerr, and J.D. Gaisford, 427–438. Cheltenham: Edward Elgar.

Cordon, W.M. 1957. The calculation of the cost of protection. *The Economic Record* 33: 29–51.

Fox, G. 2002. Evaluating environmental trade disputes in the post-Seattle world. In *Public Concerns, Environmental Standards And Agricultural Trade*, ed. F. Brouwer, and D.E. Ervin, 325–347. Wallingford: CABI Publishing.

Furtan, W.H., R.S. Gray, and J.J. Holzman. 2005. Regulatory approval decisions in the presence of market externalities: The case of genetically modified wheat. *Journal of Agricultural and Resource Economics* 30 (1): 12–27.

Gaisford, J.D., and W.A. Kerr. 2001. *Economic Analysis for International Trade Negotiations*. Cheltenham: Edward Elgar.

Gaisford, J.D., and W.A. Kerr. 2003. Deadlock in Geneva: The battle over export subsidies in agriculture. *International Economic Journal* 17 (2): 1–17.

Gaisford, J.D., and A. Hester. 2007. Why are there trade agreement? In *Handbook on International Trade Policy*, ed. W.A. Kerr, and J.D. Gaisford, 57–70. Cheltenham: Edward Elgar.

Gaisford, J.D., J.E. Hobbs, W.A. Kerr, N. Perdikis, and M.D. Plunkett. 2001. *The Economics of Biotechnology*. Cheltenham: Edward Elgar.

GATT. (1947). The General Agreement on Tariffs and Trade (GATT 1947)—Legal Texts, On-line. http://www.wto.org/english/docs_e/legal_e/gatt47_01_e.htm.

Haberler, G. 1936. *The Theory of International Trade*. London: Dodge and Co.

Hobbs, J.E. 2001. Labeling and consumer issues in international trade. In *Globalization and Agricultural Trade Policy*, ed. H.J. Michelmann, J. Rude, J. Stabler, and G. Storey, 269–285. Boulder: Lynne Rienner Publishers Inc.

Hobbs, J.E., and W.A. Kerr. 2006. Consumer information, labelling and international trade in agri-food products. *Food Policy* 31: 78–89.

Hobbs, A.L., J.E. Hobbs, G.E. Isaac, and W.A. Kerr. 2002. Ethics, domestic food policy and trade law: Assessing the EU animal welfare proposal to the WTO. *Food Policy* 27: 437–454.

Hobbs, A.L., J.E. Hobbs, and W.A. Kerr. 2005. The biosafety protocol: Multilateral agreement on protecting the environment or protectionist club?". *Journal of World Trade* 39 (2): 281–300.

Hobbs, J.E., W.A. Kerr, and S.J. Smyth. 2014. How low can you go? The consequences of zero tolerance. *AgBioForum* 16 (3): 207–221.

Holtby, K.L., W.A. Kerr, and J.E. Hobbs. 2007. *International Environmental Liability and Barriers to Trade*. Cheltenham: Edward Elgar.

Isaac, G. 2002. *Agricultural Biotechnology and Transatlantic Trade: Regulatory Barriers to GM Crop*. Wallingford: CABI Publishing.

Isaac, G.E. 2007. Ethical issues in trade. In *Handbook on International Trade Policy*, ed. W.A. Kerr, and J.D. Gaisford, 404–423. Cheltenham: Edward Elgar.

Isaac, G.E., and W.A. Kerr. 2007. Whose vision of the future? The entrenched international conflict over genetic modification. *The Geneva Post Quarterly* 2 (1): 87–107.

Isaac, G., M. Phillipson., and W.A. Kerr. (2002). International regulation of trade in the products of biotechnology, Estey Centre Research Papers No. 2. Saskatoon: Estey Centre for Law and Economics in International Trade, http://law.usask.ca/documents/estey-journal/Isaac-Phillipson-Kerr%20-%20Biotechnology%20Regulation%20-%20Estey%20Study%202%20-%20Full%20Report.pdf.

Kerr, W.A. 2000. A new world chaos?—International institutions in the information age. *Journal of International Law and Trade Policy* 1 (1): 1–10.

Kerr, W.A. 2003. Science-based rules of trade—A mantra for some, an anathema for others. *Journal of International Law and Trade Policy* 4 (2): 86–97.

Kerr, W.A. 2007a. International trade education: Do we need a new model for the global market? *Journal of International Law and Trade Policy* 8 (1): 1–11.

Kerr, W.A. 2007b. Introduction to trade policy. In *Handbook on International Trade Policy*, ed. W.A. Kerr, and J.D. Gaisford, 1–8. Cheltenham: Edward Elgar.

Kerr, W.A. 2010. What is new in protectionism? Consumers, cranks and captives. *Canadian Journal of Agricultural Economics* 58 (1): 5–22.

Kerr, W.A. 2012. Taming the bear: The WTO after the accession of Russia. *Journal of International Law and Trade Policy* 13 (2): 150–159.

Kerr, W.A. 2014. Food security and trade: Some supply conundrums for 2050. *Journal of International Law and Trade Policy* 15 (2): 115–132.

Kerr, W.A. 2016. Low Level Presence Under the WTO. In *The Coexistence of Genetically Modified, Organic and Conventional Foods*, ed. N. Kalaitzandonakes, P.W.B. Phillips, J. Wesseler, and S.J. Smyth, 311–319. New York: Springer.

Kerr, W.A., and J.E. Hobbs. 2005. Consumers, cows and carousels: Why the dispute over beef hormones is far more important than its commercial value. In *The WTO and the Regulation of International Trade*, ed. N. Perdikis, and R. Read, 191–214. Cheltenham: Edward Elgar.

Kerr, W.A., and S.L. Hall. (2003). Mandatory country-of-origin labeling (MCOOL), Its economic and trade policy implications. In *Proceedings of a conference entitled WTO: Competing policy issues and agendas for agricultural trade sponsored by the Farm Foundation and USDA Economic Research Service*, Washington, D.C., Sept. 17, www.farmfoundation.org/projects/documents/KerrPaper.pdf.

Kerr, W.A., S.J. Smyth, P.W.B. Phillips, and M. Phillipson. 2014. Conflicting rules for the international trade in GM products: Does international law provide a solution? *AgBioForum* 17 (2): 105–122.

Liu, H., W.A. Kerr, and J.E. Hobbs. 2009. Product safety, collateral damage and trade policy responses: Restoring confidence in China's exports. *Journal of World Trade* 43 (1): 97–124.

Lusk, J.I., J. Roosen, and J.A. Fox. 2003. Demand for beef and cattle administered growth hormones or fed genetically modified corn: A comparison of consumers in France, Germany, the United Kingdom, and the United States. *American Journal of Agricultural Economics* 85 (1): 16–29.

Magee, S.P. 1979. Twenty paradoxes in international trade theory. In *International Trade and Agriculture: Theory and Policy*, ed. J.S. Hillman, and A. Schmitz, 91–115. Boulder: Westview Press.

Miner, W.M. 2007. Modern history of trade policy. In *Handbook on International Trade Policy*, ed. W.A. Kerr, and J.D. Gaisford, 29–43. Cheltenham: Edward Elgar.

Roberts, D. 1998. Preliminary assessment of the effects of the WTO agreement on sanitary and phytosanitary trade regulations. *Journal of International Economic Law* 1 (3): 377–405.

Samuelson, P.A. 1969. The way of an economist. In *International Economic Relations: Proceedings of the Third Congress of the International Economic Association*, ed. P.A. Samuelson, 1–11. London: Macmillan.

Samuelson, P.A. 2004. Where Ricardo and Mill rebut and confirm arguments of mainstream economists supporting globalization. *Journal of Economic Perspectives* 18 (3): 135–146.

Samuelson, P.A., and A. Scott. 1966. *Economics—An Introductory Analysis*. Toronto: McGraw-Hill.

Schmitz, A. 1979. Research in international trade: Methods and techniques—with emphasis on agricultural trade. In *International Trade and Agriculture: Theory and Policy*, ed. J.S. Hillman, and A. Schmitz, 273–294. Boulder: Westview Press.

Skully, D. 2007. Tariff rate quotas. In *Handbook on International Trade Policy*, ed. W.A. Kerr, and J.D. Gaisford, 258–266. Cheltenham: Edward Elgar.

Viju, C., M.T. Yeung, and W.A. Kerr. 2014. Zero tolerance for GM flax and the rules of trade. *World Economy* 37 (1): 137–150.

Webster's Ninth New Collegiate Dictionary. (1985). Springfield: Merriam-Webster Inc.

# Chapter 6
# Governance Challenges from Transformative Technologies

**Abstract** This chapter provides a thorough review of various international regulatory initiatives implemented or under negotiation to develop the architecture for regulating the production, trade and marketing of biologically-derived crops, bioproducts and foods. The results of these global governance efforts are compared and contrasted to provide for a detailed assessment of how transformative technology barriers have been identified and addressed within the existing and emerging institutions.

## Introduction

While the current international trade architecture struggles to cope with new forces that are attempting to influence trade policy, our discussion in chapter five presented the case that serious theoretical work will be required to provide an appropriate underpinning for its fundamental redesign. There is little evidence of such an effort among trade economists. The international trade system cannot, however, be put on hold until the theoretical work has been completed. It has been and will be necessary to attempt to move forward within the existing architecture. This chapter examines some of those efforts.

In the absence of significant theoretical progress, the evolution of multilateral regulatory regimes is a slow process of consensus-building among a large number of participants. It suffers from the structural challenges associated with the need to allow input from each of the participants in a timely fashion and to make space for the interplay of argument, persuasion and the blunt political 'horse trading' that comprises the negotiation process. One complication is that any compromise reached during international state-to-state negotiations must also be accepted by the domestic political institutions of the countries involved and the companies and other members of civil society in those markets. The World Trade Organization (WTO), for example, has in excess of 160 member states, representing the overwhelming majority of the 160 million companies and 6.5 billion consumers in the global economy. In such a densely packed world, consensus-based decision-making



S.J. Smyth et al., *Biotechnology Regulation and Trade*, Natural Resource Management and Policy 51, DOI 10.1007/978-3-319-53295-0_6

generates significant potential for *hold-up* and hold-outs. Moreover, it is impossible to remove the influence of the broader international political environment on any deliberations. Given all this, it is somewhat surprising that any progress can be made.

There has also been a rising concern over the last few years about whether developing countries, the largest single bloc in any international dialogue, can fully engage in multilateral negotiations, both because their negotiating teams lack the analytical support to deal effectively with complex issues and their governments are unwilling to commit sufficient resources to assure full participation in the multitude of multilateral decision-making venues. The slow and uneven process of building capacity has, hence, limited the ability to build international consensuses.

The inherently slow pace of reaching any consensus means that international organizations function best in relatively stable environments. They are least effective when rapid change is taking place. Transformative technological advances, such as the advent of agricultural biotechnology, however, lead to a state of disequilibrium and pressures for institutional adaptation and/or the development of new institutions (Kerr 2016a).

While technological advancement is generally thought to increase society's well-being, the process of arriving at a new equilibrium creates both winners and losers.[1] Those who perceive they are likely to be losers can be expected to employ any available means to protect their interests, including the enlistment of existing institutional arrangements in support of their position. If those institutions are faced with a loss of influence (and resources) in the wake of changing circumstances, there may be a natural alliance between vested interests facing loss of economic opportunity and the bureaucracies of increasingly irrelevant institutions. These natural alliances can be extremely effective in slowing the rate of technological change even if, historically, they have not been particularly effective at stopping it. It has never been possible to *stuff the genie of new technology back into its bottle*. Attempts to slow changes brought by technological progress, however, can mean increased tensions in the economy and society as potential winners are frustrated by their inability to act on obvious commercial opportunities or consumption alternatives.

Of course, the potential winners from the arrival of a new technology can also use existing institutional arrangements to enhance their position. If the regulatory regime is biased toward the acceptance of a new technology or has failed to anticipate all of the effects the new technology brings, then existing regulatory processes may not be robust enough to provide an appropriate balance between facilitating the introduction of the technology and the exercise of a judicious degree of caution (i.e., raising the spectre of an error that causes harm).[2] In such cases,

---

[1]As was discussed in Chap. 2.

[2]Commonly known as a type 1 error whereby a decision to go ahead should not, in hindsight, have been made. Type 2 errors are decisions not to go ahead that should, in retrospect, not have been made.

those whose vested interests are served by existing international institutions can be expected to resist reform. As a result, those who are served less well by those same institutions may be tempted to create new institutions that are more amenable to their interests. The result is competing international institutions and regulatory confusion that inhibits those wishing to engage in international transactions involving goods arising from the use of the transformative technology. More importantly, it also increases the level of risk for investors, thus inhibiting investment in further advances in, or applications of, such a technology (effectively a type-2 error where inherently safe and efficacious applications are not adopted) (Kerr 2002).

When politicians and policy makers perceive that their country's future relative prosperity depends to a considerable degree on the successful use of transformative technologies, constraints that inhibit investments in those technologies become highly politicized issues that can set the stage for major international confrontations (Isaac 2002). Many developed countries have made the expansion of the knowledge economy a central component of their industrial policy. Biotechnology, through its commercial use of genetic information, is a central component of the knowledge economy, along with the transaction cost reducing opportunities for sharing information created by the internet and the widespread use of computers for information processing. This strategic focus of government is likely to continue for a considerable period (Boyd et al. 2003). As a result, disputes over international regulatory jurisdictions or reforms will likely take place at the highest political level. They will not be consigned to the realm of professional diplomats. In short, international institutions are likely to be particularly ineffective during the disequilibrium arising from a transformative technological change. Consensus-building will be difficult, if not intractable. Compliance with previously agreed international arrangements is likely to decline, meaning considerable stress for relations between the sovereign states that are members of these varied international organizations.

## Regulatory Innovation

A transformative technology represents a fundamental change. Transformative innovations create shocks with dynamic dimensions and are much more expansive than the more common iterative technological advances. Transformative change involves disjointed, structural adjustments in productive and institutional capability. It destabilizes or overturns existing systems (Kerr 2016a). They are accompanied by new epistemologies, they offer important complementarities (within and across sectors) and they can often involve convergent, recombined or hybrid technologies. While they are often enthusiastically announced by their creators or commercial developers, they seldom remain in the domain of any single sector or market as they tend to gravitate toward and/or influence other activities over long periods and through variable rates of progression. The pervasive and wide ranging scope of transformative change creates uncertainty. It is seldom clear until well after a launch where, when, how, by whom and why the technology spreads to other users and uses. The result is often significant angst, controversy and debate (Phillips 2007).

Transformative changes inevitably result in major gaps in both the domestic and international regulatory architecture that govern technologies and their products. This missing or inappropriate architecture may span a plethora of issues, ranging from the absence of a common vocabulary that can be used to describe the technology's many aspects, to customs classifications for the new goods created, to procedures for assessing the risks and, ultimately, to methods of measuring the benefits. As a result, in the period that follows the initial commercial application of a new transformative technology there will be considerable international activity that attempts to fill the gaps in the regulatory architecture. As the architectural redesign progresses, major areas of contention often become apparent. Initially there may be considerable duplication of effort, jurisdictional confusion and conflicting outcomes as national and international institutions take independent initiatives to make the architecture complete. There may also be private sector initiatives in international standard setting and trade facilitation (Phillips and Kerr 2002; Phillips 2007).

At least seven international institutions have been vying to co-ordinate and govern the food and environmental safety aspects of biotechnologically-developed bioproducts, foods and crops (see Table 6.1). They represent a progression from institutions that are predominately science-based—i.e., the International Plant Protection Convention (IPPC) and the *Codex Alimentarius* Commission (Codex) and the World Organization for Animal Health (OIE)[3]—to those that have broader objectives such as trade facilitation, environmental protection and other social and political goals—the Organisation for Economic Cooperation and Development (OECD), regional initiatives, the WTO and the Cartagena Protocol on BioSafety (CPB)—and finally to institutions that simply attempt to define the boundaries of ownership and thereby set the stage for private negotiation and initiative—the World Intellectual Property Organization (WIPO) and the treaties it houses. As suggested in Chap. 5, almost all of this architecture[4] was developed and implemented prior to the commercialization of biotechnology. The OIE began in 1924, and some of the precursor international treaties related to intellectual property, now under the aegis of the WIPO, go back into the 1800s.

The IPPC and the OIE are multilateral treaties whose purpose is to protect plants and animals from the spread of pathogens through international trade. The IPPC protects natural flora, cultivated plants and plant products while the OIE protects animals and fish. In collaboration with both regional and national plant and animal protection organizations, they provide a forum for international co-operation,

[3]The World Organization for Animal Health is the new name for the long standing institution—the Office International des Epizooties or OIE. When the institution was renamed in 2003 it kept its historic and well known acronym.

[4]The exception is the Cartagena Protocol on Biosafety.

**Table 6.1** International institutions regulating international trade in GM crops

| Institution | Date | Coverage | Parties/ratified | DSB | Orientation |
|---|---|---|---|---|---|
| IPPC | 1952 | Pests and pathogens of plants and plant products | 182 | Non-binding; sets WTO standards via SPS S.3.4 | International standards for plant measures involving quarantines |
| Codex | 1962 | Food labeling and safety standards | 188 | Non-binding; sets WTO standards via SPS S.3.4 | International standards to provide guidance for the food industry and protection for consumer health |
| OECD | 1961 | Harmonization of regulatory requirements, standards and policies | 35 | None | Consensus documents; special commissions and events to seek common ground |
| Regional initiatives | 1990+ | Harmonization of the science of regulation | Various | None | Regional side agreements, MOU, MRA, formal dialogues and joint research projects |
| WTO/ GATT | 1947 | Trade in all goods and most services | 164 | Binding | Establish rules for transparency and dispute settlement through TBT and SPS agreements |
| Biosafety protocol | 2003 | Transboundary movements of genetically modified organisms | 170/103 | None | Treaty creates rules for TBM of GMOs and for liability redress (under development) |
| WIPO | 1967 | Intellectual property policies and treaties | 188 | Non-binding | Treaty creates policies and mechanisms to protect intellectual property |

*Note* Dispute settlement body (DSB); memorandum of understanding (MOU); members of regional associations (MRA)
*Source* Buckingham and Phillips (2001); revised and updated by the authors as of October 2016

harmonization and technical exchange of plant and animal protection information. The IPPC has addressed the international regulation of GM crops through several International Standards for Phytosanitary Measures (ISPMs). The IPPC has its own non-binding dispute avoidance and settlement systems, but its most important role in international trade policy is through the WTO's Agreement on the Application of Sanitary and Phytosanitary Measures (SPS). The SPS is designed to use the IPPC standards as the base for evaluating WTO disputes on plant and environment related issues. National measures based on international standards from this organization are generally not open to challenge under the WTO dispute resolution process. In addition, the IPPC nominates experts for SPS dispute panels at the WTO and provides technical and background information to be used by the panels. Hence, it has far-reaching economic influence on the international commerce pertaining to the crops and bioproducts stemming from biotechnology.

The *Codex Alimentarius* Commission provides a similar service related to food products. Codex develops international food standards that define the product and its essential composition and quality factors, identifies additives and potential contaminants, establishes hygiene requirements, specifies labeling requirements and sets out the scientific procedures to be used to sample and analyze the product. Normally a standard takes six or more years to develop. Once a Codex standard is agreed, member countries are encouraged to incorporate it into any relevant domestic rules and legislation, but they may unilaterally impose more stringent food safety regulations for consumer protection, provided the different standards can be justified scientifically. Codex plays a central role in agri-food trade because its standards, guidelines and recommendations are acknowledged in the SPS and the Agreement on Technical Barriers to Trade (TBT) of the WTO. As with those of the IPPC, domestic standards based on Codex standards are expected to be sustained if trade disputes arise. While there are currently no Codex standards in place for products of biotechnology, there has been significant effort in Codex to develop a standard for labeling food products derived from biotechnology. The Codex Committee on Food Labelling was tasked in 1993, with Canada in the chair, to work to develop a standard on labeling of GM derived foods. The process normally involves eight well-articulated steps. At one point in the middle of the past decade it looked like the committee had achieved step 7 with part of the standard, but disagreements emerged about the definitions of terms being used and the process reverted to step 2, where it has remained gridlocked for years in spite of ongoing efforts. However, in 2011, the US relented on its opposition to the labeling of GM food products and in 2012, Codex adopted the principles for a risk analysis of foods derived from biotechnology, which establishes that if a risk is identified, labeling is an appropriate management strategy. Codex stresses that any risk analysis of biotechnology derived foods has to be science-based and that these principles do not address "environmental, ethical, moral and socio-economic aspects …" (Codex 2012, p. 1). It is important to note that this is a Codex principle on risk analysis of foods derived from biotechnology and not the standard on the labeling of GM foods that the Committee was tasked with 20 years ago.

Since 1995, at the dawn of GM crops, the Organisation for Economic Co-operation and Development (OECD) has been actively engaged in the process of harmonizing international regulatory requirements, standards and policies related to biotechnology. The OECD has engaged in a number of projects to ensure regulatory processes are more transparent and efficient, to facilitate trade in the products derived from biotechnology, to provide a venue for information exchange and as a point of dialogue with non-OECD countries. The OECD leads efforts to provide Consensus Documents, a set of scientific background documents mutually recognized by member states that set out the biology of the crop plant, introduced trait, or gene product and to provide a common base to be used in the regulatory assessment of an agricultural or food product derived through modern biotechnology. As of October 2016 the OECD had concluded 63 consensus documents that are becoming formalized and incorporated in the risk analysis framework of OECD members.

A number of bilateral or multilateral regional initiatives have played, and could increasingly play, an important role in the regulation of trade in goods and services. The proposed Trans-Atlantic Economic Partnership between the United States (US) and the European Union (EU), for example, committed to discuss improving regulatory processes and scientific co-operation in their negotiations. Other commitments to negotiate regulatory reform have been included in other preferential agreements (e.g. the EU-Canada Agreement, the Comprehensive Economic and Trade Agreement and the TransPacific Partnership that has been negotiated) but, thus far progress has remained elusive (Kerr and Hobbs 2015; Kerr 2015). Memoranda of Understanding, mutual recognition agreements, formal dialogues and joint research projects are mechanisms that can be and are being used to create the new vocabulary, processes, practices and norms that should act to decrease bilateral regulatory barriers to GM food trade. Using these mechanisms often assists countries in achieving the greatest degree of market access that is possible at a particular point in time.

The WTO does not have any specific provisions relating to trade in products of biotechnology. As a result, trade in these products is governed, and trade disputes adjudicated, under general WTO principles and commitments, including most favoured nation and national treatment of like-products, policy transparency and science-based standards for addressing *legitimate objectives*. The agenda for the last successful round of WTO negotiations—the Uruguay Round—was established in 1986, long before any GM products had been commercialized. The Uruguay Round, however, did endow the WTO with a mandatory and binding dispute settlement mechanism—something that is unique among the institutions that are vying to regulate GM products at the international level. As a result of this unique institutional endowment, the WTO has become a major, and contentious, venue for resolving disputes involving trade in the products of biotechnology. While the WTO's GATT permits national standards or regulations for the classification, grading or marketing of commodities in international trade (Article XI), and the adoption or enforcement of measures necessary to protect human, animal or plant life or health (Article XX(b)), it sets some limits. The SPS specifies that: (1) measures should not discriminate among countries; (2) standards which conform to international standards developed by international organizations (i.e. Codex, OIE, and IPPC) are presumed to be consistent with the obligations contained in the SPS Agreement; (3) standards that are in excess of established international standards or where no international agreement exists must be based on scientific principles and the completion of a risk assessment; and (4) measures shall not constitute a disguised restriction on international trade.

If a country believes another party has unjustly restricted market access through a SPS measure, it can use the WTO dispute settlement system. To date, WTO panels have decided three important cases concerning the validity of national SPS measures: the EU hormones case; the Australian salmon case; and the Japanese agricultural products case. In all three cases, the contested domestic SPS measure was struck down on the basis that there was no risk assessment completed to support the SPS measure or the risk assessment was improperly done. The salmon case decision sets out that a proper risk assessment: (1) must identify the disease(s) a Member wants to prevent within its territory and the potential biological and economic consequences associated with the entry or spread of the disease(s); (2) must evaluate the likelihood of disease entering and causing damage without the measure; and (3) must evaluate the likelihood of the entry or spread of the disease with the SPS measure (WTO 1998, par. 121). If the risk assessment does not even refer to the SPS measure, it would appear to be doomed to failure before a WTO Panel.

The Cartagena Protocol on Biosafety represents a more recent effort to provide a comprehensive international structure to ensure the protection of biodiversity. It represents an attempt to provide an alternative to the WTO's *exporter-friendly* approach to the regulation of GM products and, in particular, the science-based orientation of the SPS rules. Hence, it has incorporated consideration of non-scientific concerns in the process of justifying the imposition of trade barriers (Ludlow et al. 2014). The CPB was negotiated specifically to deal with trade in the products of biotechnology. The Protocol, the formal legal designation of the CPB, concluded in Montreal in January 2000, provides rules for transboundary movements of genetically modified organisms (GMOs)[5] intended for environmental release and for those destined for the human food chain. The Parties to the CPB are tasked with negotiating the terms of a Supplementary Protocol on liability and redress. These negotiations provide a forum for the world's nations to express concerns about the use of biotechnology in the environment and to write their own national laws to manage any harms to the environment, human health or other forms of loss. The consultation, discussion and negotiation process was to have been finalized within four year of the CPB coming into effect in September 2003.

Article 27 of the CPB involves an effort to devise a set of international laws establishing specific liability for harm to biodiversity arising from GMO imports. It focusses on one particular plant breeding process without taking on the other myriad of threats to biodiversity. This represents a specific reaction to the disequilibrium wrought by the transformative nature of the technology. Article 14 of the Convention on Biological Diversity (CBD), as part of the parent agreement to the CPB, provides for Parties to examine the issues of liability for harm to biodiversity from all causes. Although efforts to take a comprehensive approach to all causes of harm have proved elusive, negotiations on biotech specific liability have made

[5]The CPB refers to products as LMOs, living modified organisms. For the purposes of consistency, we will refer to the products as GMOs or GM products.

some progress through a long negotiation process under the Biosafety Protocol. Nevertheless, the Parties to the CBD are far from reaching any consensus about a liability and redress regime under Article 14, paragraph 2. Even if the Parties can reach a settlement, it is far from clear that any resulting system will be effective. The NAFTA Chap. 13 case referred to the North American Commission for Environmental Cooperation relating to GM maize in Mexico illustrates the challenge of finding cause and defining remedies in complex ecosystems found in most agro-ecologies (Phillips 2016).

Most importantly, the US and a number of important producers of GM products are not signatories to the CPB (e.g. Argentina, Canada and Australia). Hence, a debate has begun about which international organization takes precedent when international law conflicts arise. These issues have not been satisfactorily addressed, leaving countries unclear as to which set of internationally agreed provisions apply to GM products. This topic is explored more fully later in this book.

The World Intellectual Property Organization (WIPO) is a specialized agency of the United Nations. It is the primary international organization dedicated to developing a balanced and accessible international intellectual property (IP) system that can reward creativity, stimulate innovation and contribute to economic development while safeguarding the public interest. WIPO members have ratified and adhere to one or more of the Paris Convention for the Protection of Industrial Property (1884), which provides international conventions related to patents, trademarks and industrial designs, and the Berne Convention for the Protection of Literary and Artistic Works (1886), which relates to copyright. Since 1986 the mandate of WIPO has included: work to harmonize national intellectual property legislation and procedures; legal and technical assistance to developing countries; facilitating resolution of private intellectual property disputes; and seeking to develop a normative international framework for IP.

Despite significant initiative and resources invested in these varied efforts, the distance between what has been achieved and the potential fully-realized international regulatory architecture remains large. Research into all but the largest area crops has slowed in the face of regulatory gaps and inconsistencies and production of GM crops and bioproducts lags almost all of the enthusiastic forecasts of two decades ago, at least partly due to the regulatory and trade barriers between nation states. Similarly, the international marketplace for GM crops and food stuffs has split into GM-indifferent and GM-avoiding markets, with attendant higher costs and lost opportunities for producers and forgone gains in social welfare (Kalaitzandonakes et al. 2016).

## The Risk Analysis Framework

Central to the governance and regulatory systems for biotechnology is the minimization of risks. As they are new and their consequences cannot be fully known, transformative technologies have a degree of uncertainty associated with them. It is an issue for all transformative technologies. Over time, the framework for dealing with risk in relation to new technologies has become standardized. This is especially the case following the technological innovations since the mid-point of the 20th century (Phillips et al. 2006). It was during this period that regulators developed a structured format for risk analysis as a regulatory response to public policy problems.

This process officially became part of government regulation with the American National Research Council's (1983) report to the US federal government. This report has become known as the 'Redbook' on the risk analysis framework (RAF). The report identifies the mission as (National Research Council 1983, p. 1): "… a search for the institutional mechanisms that best foster a constructive partnership between science and government, mechanisms to ensure that government regulation rests on the best available scientific data and judgments in the unavoidable collision of the contending interests that accompany most important regulatory decisions." While this is a fairly broadly interpreted objective, the mandate was more focused, in that it sought to (NRC 1983, p. 17): "… examine whether alterations in *institutional arrangements or procedures*, particularly the organizational separation of risk assessment from regulatory decision-making and the use of uniform guidelines for inferring risk from available scientific information, can improve federal risk assessment activities." This process was done within the scope of increased risk of cancer from exposure to chemicals in the environment.

The Redbook definitions have become the cornerstone for all RAFs. The Redbook defines risk assessment as (NRC 1983, p. 3): "… the use of the factual base to define the health effects of exposure of individuals or populations to hazardous materials and situations."

The definition of risk management is (NRC 1983, p. 3): "the process of weighing policy alternatives and selecting the most appropriate regulatory action, integrating the results of risk assessment with engineering data and with social, economic, and political concerns to reach a decision."

Risk assessment, as defined in the Red Book and standardized through Codex, is a scientifically-based process consisting of the following steps: (i) hazard identification; (ii) hazard characterization; (iii) exposure assessment; and (iv) risk characterization. In the first instance, a hazard is defined as a biological, chemical and physical agent capable of causing adverse effects. Hazard identification involves examining whether the agent could cause an adverse effect, and is informed by laboratory and field observations of adverse effects from similar agents. Hazard characterization involves asking about the relationship between the dose of an agent and an adverse response in the environment, animals or humans. Data collection and analysis, descriptive characterization and dose-response modeling and experimentation are required. The goal is to determine the relationship between the magnitude of exposure (i.e. dose) to a chemical, biological or physical agent and the severity and/or frequency of any causally associated adverse health effects or environmental response. Subsequently, exposure assessment investigates what exposures are currently experienced or anticipated under different conditions. Field measurements, estimated exposures and characterization of populations all contribute to the exposure assessment. Risk, a (NRC 1983, p. 3) "function of the probability of an adverse health effect and the severity of that effect, consequential to a hazard" is then characterized, articulating to the best of the ability of science (applying qualitative or quantitative estimation techniques). This involves identifying and characterizing both, the probability and severity of known or potential adverse health effects in a given population and any uncertainties. While no formal evaluation and decision flows from the assessment stage, the choices made and evidence gathered provide the context for the decision undertaken in the risk management phase.

The development of the RAF based on these factors was derived from a variety of US federal risk assessment guidelines developed by numerous federal regulatory agencies or institutions, dating back to the early 1960s. Some of these guidelines met with greater success than others. In the late 1970s, efforts were undertaken to reduce inconsistencies and duplication among the various federal agencies, while working to improve co-ordination (NRC 1983). This work was done by the Interagency Regulatory Liaison Group (IRLG). In 1979, the IRLG developed guidelines for carcinogens that were then adopted by the President's Regulatory Council. These guidelines were used as a starting point for the development of the RAF under the premise that properly structured uniform guidelines should be used for risk assessment by all federal agencies.

The report of the IRLG served to develop the uniform risk analysis framework, which has gone on to be widely adapted and adopted around the world. The RAF that was developed from the Redbook has served as a template for risk assessment in a most countries. The framework of the RAF used in other jurisdictions is adapted and modified to fit with their scientific rational and protocols to create a comprehensive model for risk assessment.

## The Regulatory Void

While individuals have accepted or rejected GM products, or are trying to sit on the fence, societies have also collectively arrived at different positions regarding the technology that makes biotechnology products possible. These different collective conclusions have led to domestic regulatory systems that diverge considerably. This divergence leads to international trade conflicts. Given the slow international efforts to address the impacts of transformative technologies, national governments have been forced to move at a more rapid pace to devise domestic regulatory regimes. All other things being equal, governments prefer order over chaos. Having arrived at divergent domestic regulatory outcomes in response to emerging transformative technologies, however, national governments tend to seek out international organizations that most closely mirror their domestic regimes and then try and shape them so that they closely reflect their regimes.

A *first mover advantage* may allow a country, or group of countries, to capture an international organization. Once an international regulatory architecture is configured in a particular way, it is difficult for other countries to have it reversed because of the consensus nature of decision-making. If the transformative technology is a sufficiently divisive issue, faced with an international regime that does not reflect their domestic policy conclusion, governments may seek to establish an alternative international institution whose architecture can be shaped to more closely reflect their own.

This is clearly the case with biotechnology, where the US finds the WTO closer to its domestic model for regulating biotechnology. The US has been able to successfully block all EU attempts to have the WTO reformed; refusing to agree to have EU and other countries' issues put onto the agenda for negotiations. The EU, in particular, has tried to have the SPS opened for renegotiation to embed the precautionary principle and to have biotechnology dealt with in the negotiations on agriculture, both to no avail. The EU has been clearly frustrated with their inability to have the WTO modified and dismayed (but probably not surprised) with the results of the Panel decision in the beef hormone case, which is often viewed as a test case for the products of biotechnology because the same political difficulty with members of civil society faces the European Commission (Kerr and Hobbs 2002). As a result, the EU has set its sights on the CPB as an alternative international regulatory architecture. International law is sufficiently imprecise on the issue of conflicting international treaty obligations, meaning it is still unclear whether precedence among organizations can be established (Kerr et al. 2014).

Governments have an interest in having international organizations closely reflect their national regulatory architecture on important issues such as transformative technological change. Having to conform to an international norm when the existing domestic regulatory regime does not align can pose considerable political risks. If the transformative technology is controversial and the national legislation put in place (at least, in part, to allay the fears of civil society groups worried about its chaotic effects) forces a government to *lower* domestic standards to conform to externally imposed international norms, then a government is open to claims that it has surrendered too much sovereignty to *faceless bureaucrats* in far off places and is putting its citizens at risk. This can be a particularly emotive issue when food safety or the environment is involved—as in the case of GMOs. On the other hand, if the international regime is strict relative to the domestic commercial environment, conforming to the international norms may reduce profitable opportunities and create friction in commercial channels.

While the WTO is an attractive site for dispute resolution for many countries, it is far from perfect. As discussed in Chap. 5, the SPS Agreement does not permit non-science concerns such as consumer preference, animal welfare, or non-quantifiable environmental risks to be considered in the determination of whether a SPS measure is acceptable. WTO dispute cases show that the organization is likely to value unrestricted trade and scientific evidence above other factors such as environmental protection or socio-economic considerations; there are few provisions to allow protection based on consumer preference. As a result, there is a very real risk that Panel decisions that are contrary to domestic concerns may be ignored, which would discredit the WTO disputes process and might threaten the WTO as a whole, jeopardizing the other operations of the WTO fostering trade liberalization. In the beef hormone case between the US and Canada on one side and the EU on the other, the EU (which lost the case) has chosen to ignore the WTO panel and accept retaliation. Accepting retaliation signals a breakdown in the political compromise and is a compelling reason for further negotiation (Kerr and Hobbs 2002). The EU has signaled that it wishes to renegotiate the SPS to take account of consumer preferences (Kerr 1999). If an international regulatory regime is going to evolve to accommodate contested life-science products, the conflict between a science-based regime and civil society concerns like the *consumer's right to know* will have to be resolved (Perdikis and Kerr 1999).

It should be noted that the WTO's scientific rationality approach predates the point in time when biotechnology became an important international issue. In the Uruguay Round of General Agreement on Tariffs and Trade (GATT) negotiations, there was a major initiative to bring trade in agricultural products into conformity with general GATT principles. Prior to this, most agricultural trade was conducted under *waivers* that allowed the use of non-tariff measures to restrict imports and the unrestricted use of subsidies. With stronger disciplines on the tools of agricultural protectionism likely to arise from the Uruguay Round negotiations, there was a worry that demands for protection for agricultural products would be satisfied through the nefarious use of border regulations justified on human, animal and plant health grounds. To forestall this circumvention of the increased disciplines on trade in agricultural products, the new SPS Agreement was negotiated. It was given a scientific-rationality basis to remove it from the political process, thus attempting to isolate it from the protectionist influences of producer-based vested interests. Both the US and the EU accepted the principle of scientific rationality (Kerr 2003) but the EU has chafed under its SPS commitments almost since the SPS came into effect.

The CPB has been able to develop in the form it has because, due to a quirk of history, the US is not able to become a signatory and, thus, cannot formally participate in its framing. As outlined above, the CPB comes under the auspices of the Convention on Biological Diversity as part of the United Nations Environment Programme. The CBD arose out of the UN Earth Summit in Rio de Janeiro in 1992. The US did not sign or ratify the CBD and, hence, cannot formally be part of initiatives under its auspices such as the CPB. As a result, the European Union was able to influence the Protocol design without official US opposition.

While international governance mechanisms are operating, there is a considerable trans-Atlantic dichotomy. The EU has strongly indicated a preference for greater inclusion of socio-economic factors in the international regulation of biotechnology, while North America remains firmly committed to science-based regulations. A truly functional international governance mechanism for biotechnology will not be successful until the two halves of this dichotomy are brought together.

## Intellectual Architecture

Development of an effective international governance framework, or revision of the existing framework, must encompass three crucial components: risk analysis and the use of the precautionary principle; international trade; and intellectual property.

### *Risk Analysis and the Precautionary Principle*

The Risk Analysis Framework was developed to deal with the regulation of advanced technology products (which were characterized by a large information gap between the producers of the innovation and the intended consumers) where the goal was to credibly inject science into public policy development (National Academy of Sciences 1983). The language of risk analysis is found in regulatory guidelines for the research, development and commercialization of advanced technology products, including biotechnology, in many countries (including both the EU and the US) and in various multilateral agreements, treaties and organizations. These include the World Health Organization (WHO) and the Food and Agriculture Organization (FAO) of the United Nations.

Significantly different perspectives on how to actually operationalize the RAF in North America compared to the EU arose for biotechnology. As suggested above, the result has been that two distinct and somewhat conflicting regulatory trajectories exist for biotechnology—the scientific rationality and the social rationality trajectories (Isaac 2002; Isaac and Kerr 2003). The differences between the scientific rationality perspective and the social rationality perspective begin with a fundamental difference in belief about the appropriate role of science and technology in society. According to the former, technology yields innovations and enhances efficiency, ultimately producing economic growth. Rising incomes lead to demands for stringent social regulations in areas such as food safety and environmental protection. The result is a regulatory race to the top fostered by technological progress (Isaac and Kerr 2003). The goal of the regulatory regime for science and technology is the maximization of technological gain, subject to achieving acceptable standards of safety.

The social rationality perspective views science and technology as normative activities that by their very nature bring change to what is a delicate social balance of the preferences and concerns of all constituents (Isaac 2007). As change disrupts the balance, the social rationality perspective supports regulatory policies that ensure technological precaution, so that all of the impacts arising from a change are dealt with in a proactive, socially-responsive manner (Isaac and Kerr 2003). Progress in science and technology cannot be left to market forces given their disruptive nature—which is, of course the nature of a transformative technology.

Differing perspectives regarding precaution are central to the divide between the regulatory approaches based on scientific and social rationality. While both interpret the precautionary principle to mean that in the face of scientific uncertainty, regulators must employ precaution, the interpretation and operationalization of precaution differs in fundamental ways.

The scientific rationality perspective views the precautionary principle as a risk assessment tool. It is the scientists entrusted with risk assessment that decide when sufficient information justifies approval or when inadequate information warrants a go-slow approach with a new technology. When evaluating a new technology there is usually a dearth of data on the risks and, hence, risks are constructed using causal-consequence models that draw inferences from the accumulated peer-reviewed scientific literature (Isaac and Kerr 2003). Hence, as a scientifically-rational risk assessment tool, the exercise of precaution is grounded in *sound* science, where risk assessors who hold a required amount of scientific expertise are charged with making decisions. A rules-based approach is the result.

The social rationality perspective goes further and uses precaution as both a risk assessment and risk management tool (Isaac et al. 2002). According to this approach, risk assessors can use the precautionary principle in the same way as in the scientific rationality approach but it can also be legitimately employed by risk managers to ensure precaution in response to non-scientific perceptions and concerns. Using the precautionary principle as a risk management tool potentially increases the social responsiveness of regulations but also increases the discretionary nature of regulations. Given its broader social mandate, it includes socio-economic and ethical considerations. This leads to the possibility that regulators may use the process to prevent a technology from imposing negative economic consequences on those who cannot or will not use the new technology. Implicitly, then, it opens the door to restrictive technology policies to limit access for new technology or for compensation for those who will lose as a result of technological change. When international trade is involved, this has the potential to lead to trade barriers that are not fundamentally different from traditional economic protectionism; in practice many measures implemented as precautions are similar to import quotas or bans, which are viewed by economists and policy makers as the most egregious form of protection (Kerr and Perdikis 2014).

Central to the social rationality approach to precaution is that the ultimate responsibility for its invocation lies with the political leadership rather than being vested in a scientific body, as is the case with science-rationality-based regulatory systems (Kerr 2003). Science can inform the decision in the case where social rationality is applied, but the ultimate decision is a political one (Isaac and Kerr 2003). From a trade policy perspective, there is no guarantee that political decision-makers will deal only with questions of scientific uncertainty—they can also be swayed by traditional protectionist interests or new civil society-based protectionists (Kerr 2004).

The EU has tried to establish norms for how politicians should engage with and use the precautionary principle. While some eNGOs and other civil society groups opposed to biotechnology hope that what van den Belt (2003) calls the strong version of the precautionary principle will be the operational norm, Van den Belt asserts that the strong version of the precautionary principle is logically inconsistent. To date, the EU has not accepted the strong version of the precautionary principle. In 2000 the Commission articulated a set of operational principles under which it can be applied. The EU asserted its use should be governed under three specific principles: the fullest possible scientific evaluation, the determination, as far as possible, of the degree of scientific uncertainty; a risk evaluation and an evaluation of the potential consequences of inaction; and the participation of all interested parties in the study of precautionary measures, once the results of the scientific evaluation and/or the risk evaluation are available. The precautionary approach was also expected to follow three general principles of risk management: proportionality between the measures taken and the chosen level of protection; non-discrimination in application of the measures; consistency of the measures with similar measures already taken in similar situations or using similar approaches; examination of the benefits and costs of action or lack of action; and review of the measures in the light of scientific developments (European Commission 2000). While the principles are articulated, the procedures and practices remain untested. This creates a challenge as that earlier version of the precautionary principle had been enshrined in the CPB before the EU, its major proponent, has been able to determine how to use it effectively. This poses major difficulties, as the CPB and its 170 member states have ascribed to a policy tool that has not yet been normalized and used effectively.

## *International Trade*

The WTO is fundamentally a trade agreement. Its primary purpose is to restrict the ability of governments to impose trade barriers so that investments in international commercial activities are not threatened by the capricious use of trade barriers. It also has a long-term goal of trade liberalization. As suggested in Chap. 5, the economic model that underlies the WTO is not well suited to the modern issues pertaining to biotechnology. In particular, it assumes that consumers have perfect and costless information and, hence, they always benefit from trade liberalization because prices fall as a result of the removal of trade barriers. The model also assumes they benefit from open markets because technological change is amplified through trade, translating internationally into lower prices. In this paradigm, as outlined in Chap. 5, consumers have no reason to ask for protection and, thus, their asking for protection was not anticipated by the framers of the WTO. In contrast, declining international prices unambiguously have a negative impact on import competing producers, so they can almost always be expected to seek protection. Thus, the entire structure of the WTO is aimed at controlling producer-based attempts to obtain protection from their governments against foreign competition.

The result is that the WTO does not recognize the market failure that can arise from incomplete information regarding a new technology (Gaisford et al. 2001). It has no mechanism to accommodate consumer or environmentalist requests for protection from new technologies such as biotechnology. Beyond the failure to recognize that such a market failure exists, there are considerable practical problems that must be solved before the WTO can incorporate consumer and other societal demands for protectionism, including how such claims would be legitimized and how to prevent them from being captured by producer protectionist interests (Perdikis and Kerr 1999). From the point of view of an international trade agreement, the potential harm of extending its policy reach ineffectively is probably greater than the potential gains from doing it successfully. Thus, the WTO exhibits considerable caution (some say inertia) on this issue. This stance is now championed by the US in an effort to prevent consumer issues from being used to inhibit its exports of products of biotechnology. It has refused to allow the issue to be put on the Doha Round negotiating agenda despite EU requests to open up the SPS for re-negotiation of these issues. While the SPS may not be the best place to deal with these types of market failure (Perdikis and Kerr 1999), they have not even been allowed to become an agenda item in the SPS or in any other WTO negotiating forum.

## *Intellectual Property*

The regulatory and commercial system for GM bioproducts, crops and foods is fundamentally premised on well-defined ownership and control of the technologies and products of those technologies. The regulatory system is proponent—or owner-driven—while the private commercial system is most effective when the ownership and control of a technology or product is clearly delimited. The difficulty is that the number of life science patents and other intellectual property mechanisms (e.g. trade secrets, plant breeders' rights and trademarks) have increased substantially over the past 20 years and, as the capacity for patent rights increases, the transaction costs associated with diverse ownership (aka freedom-to-operate concerns) have increased considerably, sometimes to the point of significantly inhibiting innovation. Intellectual property rights (IPRs) underlie the development of new technologies, particularly in the biotechnology sector where innovation is highly dependent on work by others. Much of the earlier work used is subject to IPRs, and much of the output has the potential to generate new IPRs. The use of these IPRs relies on licensing.

The 1990s witnessed a consolidation within the agricultural biotechnology industry through a series of mergers and acquisitions as agrochemical/pharmaceutical firms pursued life science strategies (Hobbs 2014). At that time, there was concern that the increased concentration within the industry would have negative impacts on both research and development and the resulting commercialization of new technology. Despite these concerns, the industry continued to make significant research investments and commercialization of new technology has continued.

Nevertheless, there is concern that proliferating IPRs will create a new class of problems. Recently, there has been a new movement in the agricultural biotechnology industry, one that has seen an increase in gene trait cross-licensing agreements (Smyth and Gray 2011). While these agreements facilitate the much needed sharing of IP across research platforms, they raise additional concerns with respect to market concentration and the impact on transfer of technology to less developed nations (Smyth et al. 2014). A report by the International Expert Group on Biotechnology, Innovation and Intellectual Property (2008) concluded that the increasing numbers of patents is having a counterproductive effect in some aspects of biotechnology, where there has been a decline in innovation. Innovative advances in the health and agricultural sides of biotechnology appear to be frustrated by ever increasing proliferation of IP holders. The report suggests that the degree of concern is important because the high transaction costs associated with IP are repeatedly cited as barriers to delivering badly needed medicines to developing nations. Finally, a report by the OECD (2007, p. 34) asserts that the "role that intellectual property rights play in driving innovation, impeding innovation, and disseminating innovation is poor and unlikely, in the short term at least, to get better."

Clearly, any international governance mechanism for biotechnology, or any other subsequent transformative technology, will need to address the difficulties that the current IP system has created for the innovative process. Intellectual property, while crucial in terms of motivating investment in innovative research and investment in commercialization, is a double-edged sword, as it places barriers and restrictions on the ability of innovative products and processes to be widely disseminated.

Further, the circumstances whereby intellectual property in genetic traits that have value in biotechnology is far from uncontentious (Kerr and Isaac 2005). As the multilateral agreement to protect intellectual property—the WTO's Agreement on Trade Related Aspects of Intellectual Property (TRIPS) has been poorly accepted generally by developing countries, protection of foreign intellectual property, including biotechnology, tends to be poor in those countries (Gaisford et al. 2002; Cardwell and Kerr 2008). The likely result is smaller markets where the benefits from R&D expenditures can be recouped and, hence lower rates of investment in improvements to GM technology (Boyd et al. 2003). Further, it likely biases investments in R&D to crops where the likelihood of garnering returns to intellectual is high, meaning little development of crops agronomically uniquely suitable to developing countries. Intellectual property rights over plant properties considered traditional knowledge is also an unresolved issue with bio-prospecting activities of biotechnology countries being characterized as bio-piracy by some NGOs and developing country governments (Isaac and Kerr 2004).

## Is There a Way Forward?

There was a flurry of regulatory development in a period that stretched from the early 1980 s through to the mid-1990s, when the initial governance frameworks for biotechnology were being developed and implemented. There has been, however, little in the way of substantive progress following this initial period.

The initial flurry of institutional activity regarding an international regulatory architecture for biotechnology can probably be attributed to attempts to fill the regulatory void. This largely took place within the scientific and diplomatic communities. The subsequent slow pace of regulatory activity has arisen in response to blow-back from two sources: (1) increased information regarding biotechnology and the appearance of commercial products has motivated the general population; and (2) potential winners and losers self-identifying or being identified by others have engaged a wider set of interested actors. In some cases very strong preferences relating to biotechnology have developed. Those with a lot to win or lose and those with strong preferences have been able to capitalize on the vulnerability of the consensus-based decision-making processes of international organizations to slow the process of establishing a consensus. This slowing of the process is a tactical rather than a strategic victory. Some of the major actors have had tactical victories —the US by keeping the status quo at the WTO, the EU with the inclusion of the precautionary principle in the CPB and those with strong preferences disrupting international meetings and influencing national negotiating strategies. There has, however, been no strategic victory or winning of the war. Moreover, there has not, as yet, been a concerted attempt at reaching a consensus.

Some had hoped that the WTO ruling on the North American challenge to the EU moratorium on GM crops would resolve the roadblock. It did not (Viju et al. 2012). The ruling came at the same time as the EU initiated coexistence legislation that required all member states to develop domestic requirements for the mutual production of both GM and non-GM crops. As a result, while the verdict in the dispute was seen as a precedent setting victory for GM producers, its effect was limited due to the EU abandoning its moratorium in favour of its new domestic regime. The coexistence plan has not, however, resolved the issue. The EU regulatory regime remains in flux as a number of individual member states have refused to comply with various aspects of the EU Commission's regulatory reforms. Subsequently, there have been initiatives by the EU Commission to devolve aspects of GM-regulation to the member states.

While domestic regulatory systems evolve (albeit slowly), international negotiations seem to have lost momentum. The current Doha Round of WTO negotiations have basically ground to a halt and it is possible that a controversial issue such as a formal challenge to the new EU mechanism at the WTO disputes system could lead to a further discrediting of the multilateral system. No significant progress has been made in major preferential trade initiatives such as the CETA (Kerr and Hobbs

2015) and the TPP (Kerr 2016b). Meanwhile, the CPB negotiations on the liability and redress regime have been mired in controversy, leaving them in negotiating limbo.

With the prospects for of an international governance capacity for transformative technologies seemingly stalled and greater divergences between national systems developing, it may be time to step back from the current processes and think about underlying questions and responses that may help identify options for moving forward. One major problem, the absence of an appropriate theoretical model to underpin the international trade architecture, was identified in Chap. 5. Like much in life, however, there is a simple rubric for sorting through complicated issues—asking the W5 + H questions, including why, who, what, when, where and how.

As discussed above, the foundational questions one must ask is *why might one wish to invest time and energy in the system?* If the simple goal is to gain ones position at all costs, then there is little room for negotiation. Compromise is part of negotiation, although strong preferences sometimes increase the difficulties in finding a mutually acceptable resolution. There are two ways of casting the issue that offer potentially fruitful avenues of compromise. First, if one views the GM regulation and trade issue as isolated and unique, then there may be fewer trade-offs, making it hard to construct comprehensive packages where losses by some parties on the GM issue can be made up by real and measurable gains in other areas. Second, characterizing the issue as a principled dispute involving two irreconcilable views (scientific rationality vs. social rationality) championed by two intractable and implacable foes (the US and EU) may be problematic. The reality is more nuanced. Those countries, as with all other states, have multiple interests in and perspectives on the appropriate approach to transformative technologies. The proponents at times change sides in the debate, depending on whether GM technologies will be used to produce more and cheaper food, to produce foods exhibiting differentiated quality, to produce bioenergy and other bioproducts or to produce industrial and pharmaceutical proteins and enzymes that might be used in the nutrition and health system. The difficulty is that when the debate stays at the *40,000 foot level* of principles, the subtle interweaving of interests and positions is indistinguishable. Answering the why question in one of these two ways suggests that parsing the issue differently—either more comprehensively or more finely—might offer some new paths for convergence of interests.

The next obvious question is *what needs to be governed?* As noted above, transformative technologies tend to operate outside the bounds of the existing knowledge and rules systems. In a way, a whole new superstructure is needed. At its core, it involves agreeing on the language of the technology, be it binary code or the genome made up of A, C, G and T.[6] This process often happens well upstream of government, but is foundational to the governance process. Moving downstream to applications and use, the terminology, concepts, methods and processes of developing, testing and producing the products of transformative technologies

[6]A, C, G and T represent the four nucleotide bases of adenine, cytosine, guanine and thymine.

needs to be created and then normalized. The very process of scientific progress does much of this, but governments often need more, or sometimes a different kind of codification and normalization. The many bilateral arrangements between regulators at the operational level in national regulatory systems, the consensus documents at the OECD and the wide and varied national and international expert panels have gone a long way to making transparent what is known, what is unknown and what is unknowable. At the end of the process, the challenge is to ensure that the application of the RAF and the hazard identification, hazard characterization, exposure assessment and risk characterization are as accurate as humanly possible, are repeatable, are transparent and are responsive to new information and knowledge. That part of the system has not been normalized, but more work is underway than many people might realize (Phillips 2008).

Asking *who* should be involved similarly opens up new avenues for rapprochement. If the issue remains the problem for governments, pressed either by rent-seeking producers or by consumers seeking to exert control, then the gridlock looks intractable. There is no reason that governments are necessarily the only or even the most important actors. There are many other avenues by which vocabulary, terminology, methods and processes are standardized, normalized or institutionalized. New institutional economists (e.g. Coase 1960) argue that if the IP boundaries were clearer, then rational private actors could reach a consensus. The difficulty comes, however, from the high costs and uncertain benefits of organizing large groups of individual consumers (Olson 1965). Hence, few practitioners see this as a feasible option. Nevertheless, in many ways private supply chains vertically organize to manage risk and capture returns, and in the process are forced to develop internally the vocabulary, methods, terminologies and processes. Many products and services now regulated by the state were initially controlled exclusively within the private sphere. Even for GM foods, this has happened. When governments were slow to develop quality assurance through regulation, firms stepped in and provided quality assurance services, testing and liability insurance to bridge the gap in regulation. More might be possible in the crops-based part of the biotechnology sector, given the few firms operating in the area. By the same token, governments have no unique capacity to normalize the myriad of values, beliefs and morals of individuals and groups. Collective action often is more effective than public direction when it comes to articulating *voice* (Phillips 2007). If one compares the search for food labeling standards in the GM, organic, Halal and Kosher food sectors, one can see the power of collective action. While there are a wide range of diverging and mostly unhelpful GM food regulations around the world (Phillips and McNeill 2002), the organic industry has recently normalized all of the local and national standards into one international standard at Codex while the Halal and Kosher foods sectors remain governed and controlled by various religious institutions and processes.

The tougher questions are *where* and *when* to engage in a new or renewed process. If we only consider governments, then we are largely left with multilateral, regional or bilateral processes. The multilateral processes seem to be blocked. Even if they were to unblock in any foreseeable future, none of them is a particularly good fit for resolving the issue. The WTO has as members all of the major producing and consuming countries, but as a trade institution it is neither the place to develop standards nor is it open to discussing options related to broader socio-economic goals. The IPPC and Codex, as standards bodies that involve private firms, are well placed to do much of the foundational work on normalizing the elements of the RAF system, but they are not currently a locus for higher level negotiation related to structure and function of governance. The CPB appears to have taken on board the concepts of precaution and socio-economic balance, but so far they have not been able to make much progress on implementing them in any effective way. It also has a major structural flaw—21 percent of national governments and in particular Argentina, Australia, Canada, Chile, the US and Uruguay (which cultivate over three quarters of the global GM acreage) are not members (Table 6.2). Finally, we are left with the OECD, which at times has demonstrated great leadership in key policy areas. In the GM area it has done some good work, but once again, its incomplete membership is a barrier—less than 20 percent of all nations belong, and key developing countries (e.g. Brazil, India and China), which cultivate much of the world's GM acreage, are not members.

Ultimately, whatever options are chosen, *how* the process proceeds will influence what can be done, how fast it can happen and how acceptable it might be to nation states, firms and civil society. Simple *boundary* drawing that leads to private resolution may appear to be the path of least resistance, but major market imperfections exist which would make it difficult to succeed, and even if it did, it may be unacceptable either to consumers or to regions that will be at the mercy of a highly concentrated market. Bilateral and multilateral intergovernmental negotiations often choose one of four basic negotiation processes. First, some work to create guiding and judiciable principles backed up by processes and procedures to adjudicate compliance (e.g. the GATT 1947 ). Second, many set up issue-specific tables where those nations with the most at stake will bid and ask for explicit changes; agreements then become multilateralized via the principle of most-favoured nation (e.g. many of the GATT Rounds between 1947 and 1973). Third, sometimes broad formula can encircle a problem and then specific targets can be set to progressively (and sometimes differentially) reduce the problem (e.g. the Swiss formula in the WTO and the red/yellow/green categories for subsidies in the Agriculture Agreement of the WTO). Finally, some institutions seek to find consensus, which can work well when a clear objective is in sight but can be problematic when too many variables are in play (e.g. CPB). Governments, at times, will go outside the

**Table 6.2** Membership in major multilateral policy institutions related to GM crops, 2015

| | WTO | IPPC | Codex | CPB | OECD |
|---|---|---|---|---|---|
| Number of members | 164 | 182 | 188 | 170 | 35 |
| % of all countries | 80% | 90% | 95% | 84% | 19% |
| % of GM production by members | 100% | 100% | 100% | 27% | 56% |

*Source* Authors' calculation using data from WTO, IPPC, Codex and CPB, OECD and James (2015)

formal policy system and ask *wise persons* to provide expert opinions or turn to the courts or other formal arbitration systems to decide a dispute. In that context, courts or arbitrators can be restricted to picking one of the opposing views or alternatively have the liberty to construct a resolution that settles somewhere in the middle of the opposing views. Finally, when all else fails, states always have the possibility of *punting*, by delegating the authority and responsibility to a special, independent agency of government, to other levels of government or to a private contractor to decide. Some governments make a virtue of this necessity—the EU, for instance, has a principle of subsidiarity that by practice moves issues to the government or institution that is closest to the people affected. None of these approaches is guaranteed to succeed; nor is any fundamentally flawed. The challenge is to match the method to the problem.

There is no simple answer to the challenge of *how* to proceed. The only thing we know is that something must be done. One must keep in mind that the technology is *out of the genie's bottle*. Commercialization and adoption continue, albeit not at a pace desired by some. The market offers strong incentives. When access to the technology is denied, as it was in Brazil prior to approval of GM crops for production, for example, smuggling and illegal adoption are widespread. The food crisis of 2007–8 also forced hard choices regarding the use of GM technology. In the face of transformational pressures, the current regulatory fracture has not produced a result that is frozen in time. Pressure for an acceptable international regulatory compromise is likely to build.

## Only a Few Steps Along the Way

Transformative technologies represent significant challenges. The process of technological change is one that creates economic disequilibrium and foists change on the wider society. It takes many steps to arrive at a new equilibrium. International institutions are no exception—but due to their very structure international institutions are generally very slow to adapt. As membership and compliance in international institutions is voluntary, decisions must often be reached by consensus. Deliberations are often slow because they involve a large number of parties with a diversity of interests. Two competing international organizations appear to be emerging as the primary international regulators of biotechnology—the WTO and the CPB. They are, however, on different regulatory tracks that represent the positions of the major protagonists in the debate over biotechnology. The WTO appears unable to adapt to accommodate the international trade questions, including liability, raised by the transformative technology. In essence, it is simply acting as if no transformation is taking place. The CPB, on the other hand, has a built-in bias against the technology both through the inclusion of the precautionary principle and the absence of a dispute settlement mechanism. While it recognizes that the existing international liability regime will not be adequate for liability to fully play its deterrence and compensation role in the case of biotechnology, it is finding it difficult to devise an international liability regime that deals with the characteristics of a biologically-based transformative technology in a way that is acceptable to all of the parties to the Protocol. Even if an agreement is reached, it looks like it will be done without Argentina, Australia, Canada, Chile, the United States and Uruguay, all which are major adopter and exporters of the basic foodstuffs that use GM technologies.

It will be some time before an international regime for the products of biotechnology brings the regulation of this transformative technology to a new equilibrium. In the interim, less than smooth international relations pertaining to the regulation of biotechnology can be expected. The forces unleashed by this transformative technology have not yet run their course, so the current regulatory fracture will face new and increasing pressures. As its deficiencies become ever more evident, reaching a viable compromise is likely to become increasingly appealing to all the contending parties.

## References


Boyd, S.L., W.A. Kerr, and N. Perdikis. 2003. Agricultural biotechnology innovations versus intellectual property rights—are developing countries at the mercy of multinationals? *The Journal of World Intellectual Property* 6 (2): 211–232.

Buckingham, D., and P. Phillips. 2001. Hot potato, hot potato: Regulating products of biotechnology by the international community. *Journal of World Trade* 35 (1): 1–31.

Cardwell, R., and W.A. Kerr. 2008. Protecting biotechnology IPRs in developing countries: Simple analytics of a levy solution. *Journal of Agricultural Economics* 59 (2): 217–236.

Coase, R. 1960. The problem of social cost. *The Journal of Law and Economics* 3 (1): 1–44.

Codex Alimentarius Commission. 2012. Principles for the Risk Analysis of Foods Derived from Modern Biotechnology.

European Union Commission. 2000. *Communication from the Commission on the Precautionary Principle*. Brussels: European Commission Press.

Gaisford, J.D., J.E. Hobbs, W.A. Kerr, N. Perdikis, and M.D. Plunkett. 2001. *The Economics of Biotechnology*. Cheltenham, UK: Edward Elgar Publishing Ltd.

Gaisford, J.D., R. Tarvydas, J.E. Hobbs, and W.A. Kerr. 2002. Biotechnology piracy: rethinking the international protection of intellectual property. *Canadian Journal of Agricultural Economics* 50 (1): 1–14.

GATT. 1947. The General Agreement on Tariffs and Trade (GATT 1947)—Legal Texts. On-line. http://www.wto.org/english/docs_e/legal_e/gatt47_01_e.htm.

Hobbs, J.E. 2014. The private sectors: MNEs and SMEs. In *Handbook on Agriculture, Biotechnology and Development*, ed. S.J. Smyth, P.W.B. Phillips, and D. Castle, 56–70. Cheltenham: Edward Elgar.

International Expert Group on Biotechnology, Innovation and Intellectual Property. 2008. *Toward a New Era of Intellectual Property: From Confrontation to Negotiation*. Montreal: McGill University Press.

Isaac, G.E. 2002. *Agricultural Biotechnology and Transatlantic Trade: Regulatory Barriers to GM Crops*. Wallingford, UK: CAB International.

Isaac, G.E. 2007. Sanitary and phytosanitary issues. In *Handbook on International Trade Policy*, ed. W.A. Kerr, and J.D. Gaisford, 383–393. Cheltenham: Edward Elgar.

Isaac, G.E., and W.A. Kerr. 2003. GMOs at the WTO: A harvest of trouble. *Journal of World Trade* 37 (6): 1083–1095.

Isaac, G.E., and W.A. Kerr. 2004. Bioprospecting or biopiracy?—Intellectual property and traditional knowledge in biotechnology innovation. *The Journal of World Intellectual Property* 7 (1): 35–52.

Isaac, G.E., M. Phillipson, and W.A. Kerr. 2002. International Regulation of Trade in the Products of Biotechnology, Estey Centre Research Papers No. 2. Saskatoon: Estey Centre for Law and Economics in International Trade. http://law.usask.ca/documents/estey-journal/Isaac-Phillipson-Kerr%20-%20Biotechnology%20Regulation%20-%20Estey%20Study%202%20-%20Full%20Report.pdf.

James, C. 2015. 20th Anniversary (1996 to 2015) of the Global Commercialization of Biotech Crops and Biotech Crop Highlights in 2015. ISAAA Brief No. 51. Ithaca, NY: ISAAA.

Kalaitzandonakes, N., P.W.B. Phillips, J. Wesseler, and S. J. Smyth (eds). 2016. The Co-existence of Genetically Modified, Organic and Conventional Foods: Government Policies and Market Practices. New York: Springer.

Kerr, W.A. 1999. International trade in transgenic food products: A new focus for agricultural trade disputes. *The World Economy* 22 (2): 245–259.

Kerr, W.A. 2002. *The international trade regime for biotechnology—a costly muddle*, 26–29. Life Sciences Technology January: Business Briefing.

Kerr, W.A. 2003. Science-based rules of trade—a mantra for some, an anathema for others. *Journal of International Law and Trade Policy* 4 (2): 86–97.

Kerr, W.A. 2004. The changing nature of protectionism: Are "Free Traders" up to the challenges It presents? *Journal of International Law and Trade Policy* 5 (2): 91–101.

Kerr, W.A. 2015. Governance of international trade in genetically modified organisms: Is future global food security at risk? *Journal of International Law and Trade Policy* 16 (2): 60–77.

Kerr W.A. 2016a. Disequilibrium, trade and the consequences of adjustment. *Journal of International Law and Trade Policy* 17(2): 59–75.

Kerr, W.A. 2016b. Worlds apart on GMOs—Can trade agreements bridge the gap. NABC Report 27, Proceedings of the 27th North American Agricultural Biotechnology Council Conference, Pennsylvania State University, June 2–3, 2015, 133–145. Ithaca: North American Agricultural Biotechnology Council.

Kerr, W.A., and J.E. Hobbs. 2002. The North American-European Union dispute over beef produced using growth hormones: A major test for the new international trade regime. *The World Economy* 25 (2): 283–296.

Kerr, W.A., and J.E. Hobbs. 2015. A protectionist bargain?: Agriculture in the European Union-Canada Trade Agreement. *Journal of World Trade* 49 (3): 437–456.

Kerr, W.A., and G.E. Isaac. 2005. The international treatment of biological material as intellectual property. *Journal of International Biotechnology Law* 2 (3): 105–111.

Kerr, W.A., and N. Perdikis. 2014. *A Guide to the Global Business Environment*. Cheltenham: Edward Elgar.

Kerr, W.A., S.J. Smyth, P.W.B. Phillips and M. Phillipson. 2014. Conflicting rules for the international trade in GM products: Does international law provide a solution? *AgBioForum* 17 (2): 105–122.

Ludlow, K., S.J. Smyth, and J. Falck-Zepeda (eds.). 2014. *Socio-Economic Considerations in Biotechnology Regulations*. New York: Springer Publishers.

National Academy of Sciences. 1983. Risk Assessment in the Federal Government: Managing the Process. Committee on the Institutional Means for Assessment of Risks to Public Health, Commission of Life Sciences. Washington: National Academy Press.

National Research Council. 1983. *Risk Assessment in the Federal Government: Managing the Process*. Washington, D.C.: National Academy Press.

Organisation for Economic Co-operation and Development. 2007. Intellectual Property. Issues in Biotechnology: Health and Industry. Paris: OECD.

Olson, M. 1965. *The Logic of Collective Action: Public Goods and the Theory of Groups*. Cambridge: Harvard University Press.

Perdikis, N., and W.A. Kerr. 1999. Can consumer-based demands for protection be incorporated in the WTO? The case of genetically modified foods. *Canadian Journal of Agricultural Economics* 47 (4): 457–465.

Phillips, P.W.B. 2007. *Governing Transformative Technological Innovation: Who's in Charge?.* Oxford: Edward Elgar.

Phillips, P.W.B. 2008. The kaleidoscope of risk analysis: The state of the art 25 years after the Red Book. Commentary on the proceedings of the 1st International Conference on Risk Assessment, Brussels, November 18–19, at www.agbio-management.org.

Phillips, P.W.B. 2016. GM Maize in Mexico: The Challenge of Coexistence in a Centre of Origin. In *The Co-existence of Genetically Modified, Organic and Conventional Foods: Government Policies and Market Practices*, ed. N. Kalaitzandonakes, P.W.B. Phillips, J. Wesseler, and S. J. Smyth. New York: Springer.

Phillips, P.W.B., and W.A. Kerr. 2002. Frustrating competition through regulatory uncertainty—International trade in the products of biotechnology. *World Competition Law and Economics Review* 25 (1): 81–99.

Phillips, P.W.B., S.J. Smyth, and W.A. Kerr (eds.). 2006. *Governing Risk in the 21st Century: Lessons from the World of Biotechnology*. Hauppauge, NY: Nova Science Publishers.

Phillips, P.W.B., and H. McNeill. 2002. Labeling for GM foods: Theory and practice. In *Market Development for Genetically Modified Foods*, ed.V. Santaniello, R.E. Evenson, and D. Zilberman, 245–261. Wallingford, UK: CAB International.

Smyth, S.J., P.W.B. Phillips, and D. Castle (eds.). 2014 *Handbook on Agriculture, Biotechnology and Development*. Cheltenham: Edward Elgar.

Smyth, S.J., and R. Gray. 2011. Intellectual property sharing agreements in gene technology: Implications for research and commercialization. *International Journal of Intellectual Property Management* 4 (3): 179–190.

Van den Belt, H. 2003. Debating the precautionary principle: "Guilty Until Proven Innocent" or "Innocent Until Proven Guilty"? *Plant Physiology* 132 (July): 1122–1126.

Viju, C., M.T. Yeung, and W.A. Kerr. 2012. The trade implications of the post-moratorium European Union approval system for genetically modified organisms. *Journal of World Trade* 46 (5): 1207–1238.

World Trade Organisation. 1998. Appellate Decision, Australia—Measures Affecting Importation of Salmon—Complaint by Canada, AB-1998-5 (Appeal from WT/DS18), at http://wto.org.

# Chapter 7
# The Impact of Barriers to Trade on Investment

**Abstract** International trade rules based on science are not functioning efficiently. Considerable effort was put forth in the 20th century to remove politics from the frameworks that governed international trade. Some degree of success was achieved as numerous institutions were founded or their roles expanded. These institutions were established on the premise that science-based frameworks were essential to the efficient functioning of international commerce. The first decade and a half of the 21st century would seem to suggest that these institutions are floundering and that the role of science as the basis of international trade rules is on the decline. This chapter focuses on the fundamental causes for the disruption of international trade and its implications for investment.

## Introduction

As reported above, the official website of the World Organization for Animal Health (OIE) confirms that the problem of countries using trade barriers supposedly justified on scientific criteria for nefarious purposes long predates the establishment of the current international institutional architecture in the post-Second World War period:

> The ratification of the 1924 Agreement creating the OIE reflects a desire clearly expressed by the Secretary General of the League of Nations that year. He invited various governments to designate veterinary experts "to examine the health guarantees that could be provided by cattle-exporting countries, the facilities that importing countries could accord on the basis of these guarantees and, in general, to determine the most effective means of enabling statutory veterinary measures to be applied, taking into account the economic interests of exporting countries and without prejudicing the interests of countries wishing to protect themselves against animal diseases".
>
> … the Economic Committee of the League of Nations thus proposed to facilitate international trade in animals and animal products to try and reverse the *often highly overt tendency of numerous countries to use sanitary arguments purely for the purpose of economic protection* (emphasis added) (OIE 2000).

 
S.J. Smyth et al., *Biotechnology Regulation and Trade*, Natural Resource Management and Policy 51, DOI 10.1007/978-3-319-53295-0_7

While the institutional architecture of the OIE and various other science-based organizations have certainly been added to and evolved over the intervening 70-odd years since the end of the Second World War, the underlying structure has remained largely intact (Kerr 2010). As outlined previously, the major rule-making institution for the governance of international trade, the World Trade Organization (WTO), evolved out of the General Agreement on Tariffs and Trade (GATT) which came into being in 1947. The current WTO is an organization that has been temporarily empowered, by international standards, with a considerable degree of sovereignty by its Member States. In particular, it has a binding disputes settlement mechanism and the power to approve the imposition of economic sanctions on those countries that fail to comply with their WTO commitments. Over time, other international institutions have been given the right to govern aspects of international trade but have not been endowed with a binding disputes settlement mechanism or the right to approve economic sanctions.

The GATT, and latterly the WTO, have had considerable success in reducing tariffs and eliminating the use of other commonly-used trade barriers. This has had two effects. First, governments have been seeking alternative ways of extending protection from foreign competition in the wake of having agreed to limit the imposition of new tariffs and the reduction of existing tariffs. Many of the domestic regulations that inhibit trade are justified on health, sanitary and phytosanitary grounds, which have a scientific basis—but were not crafted with the minimization of trade effects in mind. As they were largely developed independently, they are not harmonized among countries. The resulting incompatibilities can act as inhibitors of international commerce. Purposeful nefarious use of health, sanitary and phytosanitary measures to restrict imports is difficult to prove and, hence, has an attraction for policy makers wishing to extend economic protection to vested interests when they have agreed internationally to reduce or eliminate more traditional avenues of providing protection. This has led to a second effect, namely that exporting firms have found the access to foreign markets promised by lower tariffs in these new trade deals blocked by importing countries' domestic regulations that were put in place behind what had been very restrictive trade barriers.

Prior to the establishment of the WTO, the GATT rules pertaining to health, sanitary and phytosanitary measures were relatively weak. They were, however, in keeping with the GATT's original institutional arrangements which made it relatively easy to obtain *waivers* from GATT commitments—for example, waivers were widely granted for trade in agricultural products—and where dispute settlement judgments required a consensus that included the accused. This made it easy for countries to grant protection for biologically-based products such as those arising from agriculture and to maintain trade barriers in the face of complaints from exporters. As a consequence, there was little need for GATT members to make use of health, sanitary and phytosanitary barriers for protectionist purposes—it was simply easier to use tariffs. The waivers granted for agriculture, however, led to high barriers to market access for most agricultural products and high subsidization in the sector—significant distortions that by the 1980s began to spill over into other sectors and sour broader international relations between major trading partners (Gaisford and Kerr 2001).

In an attempt to diffuse the tensions surrounding the large and increasing distortions of trade in agricultural products is was agreed that their trade should be moved under general GATT disciplines when the agenda for the Uruguay Round was set in 1986—notably prior to the commercialization of any genetically modified (GM) crop. Further, the consensus-based disputes system was viewed as increasingly unworkable for the greatly increased number of GATT Member States. Hence, a binding dispute settlement system was negotiated as part of the Uruguay Round agreement. Of course, obtaining protection for farmers and other vested interests in agri-food systems has been one of the most effective lobbies. The efficacy of these efforts is widely evident in both developed and developing countries. The negotiators during the Uruguay Round were very cognizant of the effectiveness of agricultural lobbies in obtaining protection and fretted about the impact of bringing agriculture under general GATT disciplines and, hence, closing off the ability to grant the kind of protection provided by the long-standing waivers. Furthermore, negotiators knew full well that eliminating the ability under the consensus-based disputes system to ignore complaints brought by trading partners would considerably reduce governments' flexibility to respond to protectionist pressures. The vested interests in agriculture were not going to *fold up their tents and melt away* and policy makers would still be faced, at times, with protectionist requests that they might wish grant. The nefarious use of health, sanitary and phytosanitary measures was one obvious avenue for vested interests seeking protection to pursue.

To close this potential means of circumvention, the negotiators chose science as the decision-making criteria for justifying barriers to trade on health, sanitary and phytosanitary grounds. The Agreement on the Application of Sanitary and Phytosanitary (SPS) Measures was agreed as part of the Uruguay Round. Science is enshrined as the principle decision-making criterion. The new WTO with a binding disputes settlement system and the SPS as one of its sub-agreements came into force in 1995. There were high hopes for the efficacy of the SPS and its science-based focus (Roberts 1998).

Twenty years later there is little doubt that science-based decision-making has not proven to be an effective way to isolate SPS-based trade measures from protectionist influences. While not confined solely to issues pertaining to GM products, the commercialization of the products of modern agricultural biotechnology has brought to the fore the difficulties associated with attempting to use science to govern trade matters when biologically-based products are involved. As suggested in the previous chapter, biotechnology represents a transformative technology that has led to a threat of disequilibrium in the agricultural sector and angst for many civil society groups—both of which can be expected to lead to an increase in protectionist pressure (Phillips 2007). Given that both the fear of disequilibrium and the angst is not distributed evenly among major trading partners, this has led to differing demands for protection. At one extreme, this has precipitated a major rift between the United States (US) and the European Union (EU) over the basis upon which the rules for trade in GM products are to be premised (Isaac and Kerr 2007a). As suggested above, this rift has led to attempts to have the WTO supplanted by alternative, only partially science-based, trade institutions such as the Cartagena Protocol on Biosafety (CPB) to the Convention on Biological Diversity (CBD) (Hobbs et al. 2005; Holtby et al. 2007). Further, there has been widespread debate as to whether the products of modern biotechnology might be sufficiently unique that they require separate institutional arrangements for their governance, both domestically and internationally. As outlined in Chap. 6, given an apparent institutional void, a number of international entities have attempted to fill the gap in the institutional architecture but none has been able to assert its primacy. A major stumbling block has been how to operationalize science as a decision-making criteria. While the operationalization of any new decision-making system might be expected to have some teething problems, in the case of science-based rules for the imposition of trade barriers there is little, if any, indication of substantial progress in the past 20 years.

If the only effect of the ongoing absence of a well-functioning science-based system was short-run disruptions to trade flows, then the economic losses involved might simply be chalked up to the enduring and fundamental trade compromise between protectionism and trade liberalization. There never has and never will be *free trade* (Gaisford and Hester 2007). The international rules of trade are replete with inhibitors to trade flows (e.g. tariff use is widespread) and decision rules that have been operationalized in sub-optimal ways (e.g. anti-dumping) (Barichello 2007; Kerr 2006). One problem with the current science-based system is that it is opaque. The result is that, in the era of globalization, the international business environment is very risky for those that have invested in biotechnology. In short, there is no assurance that a new product based on biotechnology can be traded. If a country is a major exporter of a product where a biotechnology innovation is reaching the stage of commercialization, it may not wish to jeopardize its exports by licensing the innovation domestically. All of this uncertainty inhibits innovation (Episcopos 1995; Leahy and Whited 1996; Bond and Cummins 2004). Market uncertainty regarding the commercialization of GM wheat in the early 2000s, which led to a freeze on further development, contributed to a decline in wheat breeding research for both conventional and new GM wheat varieties. Thus, the economic cost of the current opaqueness in the system extends far beyond that associated with the disruption of current trade flows. Given the potential of biotechnology to enhance human welfare (Gaisford et al. 2001; Holtby et al. 2007; Phillips et al. 2006; Smyth et al. 2004), developing a more transparent decision-making mechanism would seem imperative—the opportunities forgone are simply too great. Thus, a hard look at the 20-year experience with science-based decision-making in the area of SPS-based measures is timely. The system needs either to be abandoned as a failed experiment or fixed. The time for procrastination or tinkering around the edges has, arguably, passed.

## Science-Based Decision-Making and Trade Barriers

From the perspective of those interested in the institutional architecture of international trade policy, the overarching concern with the decision-making mechanisms put in place is whether they can be captured or otherwise manipulated by narrow vested interests when the intent of the trade rules is to serve the general good. Protectionists should never be underestimated. Experience teaches international trade professionals that protectionists are tenacious, resourceful, inventive and tireless—after all there is usually a great deal at stake (Kerr and Foregrave 2002; Kerr 2003). The idea that there are those who do not have the general good as their objective is often a surprise to those in the scientific community and others in the regulatory system who do not have an appreciation of the social sciences. This is patently obvious in, for example, multilateral environment agreements (MEAs) where the trade rules have been crafted by and for environmental scientists (Hobbs et al. 2005; Kerr and Hall 2004). In contrast, trade policy professionals often appear to non-specialists as paranoid while trade policy professionals feel that trade-policy novices are particularly naïve. Assuming for the moment that the trade policy professionals have a point, two questions arise within the context of science-based decision-making in trade policy: (1) who has an interest in harnessing science-based decision-making to their vested interest?; and (2) how is the current system open to capture or manipulation?

Producers in import competing industries faced with strong competition from imports have been perceived as the primary seekers[1] of economic protection since the days of Adam Smith (1776).[2] In part, this is because the organizational costs associated with rent-seeking activities tend to be small for producers relative to those of widely dispersed groups such as consumers or environmentalists (Olson 1965). Further, as pointed out in Chap. 5, the economic model that underpins international trade policy predicts that consumers will be losers if trade restrictions on imports are imposed—prices rise, making consumers worse off—and, hence, they would never logically request them from their government. Of course, producers win from the imposition of trade barriers—and thus have an incentive to ask for protection. The only other actor in this simple model is the government that is either the recipient of tax revenues or a dispenser of subsidies.[3] The model assumes perfect knowledge, meaning that there are no environmental externalities and thus no need for environmental advocates to be included. Similarly, perfect knowledge

[1]Other producer-based rent seeking activities such as requesting export subsidies and trade distorting domestic subsidies are also manifest at times but relief from competition from imports, one suspects, is the overwhelming source of requests for protection.

[2]An accessible version of Smith's classic treatise on economics is Smith (1994).

[3]In some cases governments have a vested interest in these revenues—for example developing countries may be dependent on tariffs for the majority of their tax revenues and, thus, may have a vested interest in resisting the reduction of tariffs. Given that in all developed countries and many developing countries tariffs are such a small portion of total tax revenues, it is common to ignore this trade policy motivation (see for example Zhang and Kerr 2009).

means that consumers understand any ethical aspects of the products they are faced with and factor any concerns into their purchasing decisions, and thus there is no need for ethical advocates (e.g. NGOs advancing animal rights and child labor laws). It also means there is no need for labeling requirement for imports. The model upon which the trade architecture is based also assumes an industrial structure of perfect competition so that market power is not a concern. It was only long after the current international trade architecture was put in place that economists began to incorporate aspects of imperfect competition and market failures into their trade models (Benarroch 2007).

Thus, the underlying model predicts that producers, and no one else, will request protection. As a result, the entire international trade architecture is based on the fear that governments need to respond to producer requests for protection (Kerr 2010). Trade agreements arise from governments understanding there are gains from turning beggar-thy-neighbour, non-cooperative trade wars into cooperative games through negotiation (Gaisford and Hester 2007; Gaisford and Kerr 2003). Trade agreements also recognize that, at times, governments may need to respond to protectionist pressures from producers. The rules represent a compromise between governments' ability to respond to producer requests for protection and the costs associated with choosing to exercise those options. Governments have not negotiated that ability to respond to any other groups asking for protection (i.e. there are no provisions for responding to consumer or other interested parties requesting protection from imports in major trade agreements) (Kerr 2010). This is fundamental to the question of science-based decision-making.

Once the Uruguay Round was completed, the international trade architecture and the new science-based SPS rules were effectively frozen until the completion of the next round of negotiations.[4] The difficulty is that new demands for trade barriers quickly emerged. As discussed in Chap. 5, some consumers began asking that a range of products be excluded from the market: beef produced using growth hormones; goods produced using child labor; goods produced from animals caught in leg-hold traps (Kerr and Hobbs 1994); goods produced from seal pups; meat produced under certain animal welfare standards; and a host of others. By relaxing some of the assumptions of the trade model that underpins the international trade architecture it can be shown that, under certain circumstances, both a sub-set of consumers and governments concerned with the general good may wish to exclude products from their markets (Gaisford et al. 2001).

In a similar fashion, environmental advocates have become much more prominent in seeking new trade barriers since the end of the Uruguay Round. As discussed in Chap. 5, one example was US government measures against tuna from countries whose tuna fishing practices were considered to be detrimental to dolphins. There have also been initiatives to require ecological or green labeling of

[4]The Doha Round was launched in 2001 but, as yet, it has not been successfully concluded. Thus, the Uruguay Round outcome is the core of the current international trade architecture—although some of the architecture lies outside of the WTO structure.

imports, to be applied to imports of tropical timber from endangered rainforests, to limit imports of petroleum products derived from the oil-sands in Canada, and a broader and ongoing debate over whether international trade contributes to environmental degradation (Belcher et al. 2003; Copeland 2007).

While little is actually known about the proportion of consumers that feel trade barriers are needed to keep products out of their markets or how widespread the anti-trade sentiment is in the environmental movement, what is clear is that these sentiments are sufficiently widespread that politicians have, in some circumstances, felt that they could not be ignored. Certainly, some consumers and environmentalists have strongly held beliefs and some politicians have been willing to acquiesce to their requests for protection. As suggested above, the problem, however, is that there is no mechanism in the WTO that allows governments to directly respond to these requests. Given that many of the products that consumers and environmentalists would like to restrict are biologically-based, there is the potential to specify SPS justifications to respond to such requests for protection. In fact, governments have few alternatives.

In some countries, the pressure to limit imports of products of biotechnology has been overwhelming. This is because the transformative technology has attracted the interests of four distinct groups in civil society that tend to have strong preferences. No other issue has produced such a convergence of interests. Among some consumers there is unease about modern food production and distribution, with concerns ranging from pesticide residues, antibiotics in animal production, preservatives and additives, hormones and transgenes. Demand for organic or natural foods are manifestations of these preferences. Genetic modification represents just one more *unwanted* tampering with the food supply for consumers with this particular set of preferences. While many consumers do not share these concerns, those that do have such concerns (and often have strong preferences) have lobbied hard to have GM products banned from, their markets.

Those with strong preferences related to the protection/preservation of the natural environment are often concerned about industrial and agricultural pollution and the market failures associated with them. They tend to be suspicious about claims that introducing new substances into the natural environment will not be harmful. The introduction of GM seeds into the natural environment was considered by such individuals to be far too risky—environmentalists tended to think that insufficient testing had been done in the case of this new and transformative technology. Again, some environmental non-governmental organizations (eNGOs) lobbied hard to keep GM products out of their environment, in part by denying international suppliers market access.

For some in civil society, GM products raise considerable ethical issues (Gaisford et al. 2001; Isaac 2007a). For example, the ability to use modern biotechnology to transfer genes between species in ways that cannot happen through natural selection (e.g. the metaphoric insertion of a 'fish gene' into a tomato) is considered by some to be using science outside the boundaries of what is ethically acceptable—literally *playing God*. For others it was just the first step on the *slippery slope* to human cloning or unnatural extensions to the human lifespan.

Those who raised these and other ethical objections to biotechnology also made it clear to decision-makers that they did not want such products in their market regardless of the source—either domestic or foreign.

Finally, there are those in civil society that believe that large corporations have too much economic power in the economy and/or have too much influence over consumers. These concerns long pre-date the commercialization of GM products. Having most of the techniques associated with the production of genetic improvements and the genetic material itself held on a proprietary basis as intellectual property by large agribusiness firms has amplified this concern that multinational corporations have far too much control over the food supply (Isaac and Kerr 2004; Isaac et al. 2004). Limiting market access is one way to reduce the influence of the large agribusiness firms that develop and own much of the existing biotechnology that has been commercialized. These voices were added to the others wishing to limit market access for GM products.

The convergence of these four groups comprised of individuals with generally strong preferences has made biotechnology the lightning rod for arguments over science-based decision-making in international trade. Agricultural producers in countries where the imports of GM products have been restricted have, for the most part, been silent over the issue of market access. Nevertheless, if trade barriers are put in place to placate the demands of consumers, environmentalists or others, they still act to benefit producers—restricting imports raises the price domestic producers receive. Thus, there is suspicion that consumers are not sovereign in their decision-making—in effect, there is a real fear that they have been captured by traditional producer-based vested interests.

Thus, there are a wide range of vested interests that would like market access restricted for a range of products—vested interests for which no provisions have been made in the international trade architecture. Given that many of the issues pertain to biologically-based product to which SPS trade provisions may apply, there would seem to be ample incentive to attempt to harness science-based decision-making for protectionist purposes. Until such time as the international trade architecture can directly accommodate government actions that respond to requests for protection from non-traditional sources such as consumers and environmentalists—an accommodation that is likely to be very difficult (Perdikis et al. 2001)—science-based decision rules will be subject to a high degree of attention that is aimed at harnessing them to protectionist causes.

On reflection, the decision to ensconce science as the decision-making criteria in the SPS was probably naive and accompanied by a considerable degree of ignorance regarding the feasibility of operationalizing such decision-making. The naivety may have stemmed from a common perception among social scientists—who likely comprise the bulk of trade negotiators—that the physical sciences (i.e. the *hard* sciences) provide definitive, irrefutable answers.[5] Despite this commonly

[5]Relative to the (*soft*) social sciences, which, by comparison, are presumed to provide only *fuzzy* answers to questions.

held perception among social scientists, the reality is that the *hard* sciences are actually constituted in almost a diametrically opposite configuration, both at any particular point in time and dynamically. The scientific method is based on challenging the accepted wisdom. Science does not provide definitive answers. It is based on theories and empirical testing. Thus, any scientific conclusion is, in fact, defined by explicit or implicit probabilities. For example, it is not possible to reach an outcome that something is *100% safe* from using the scientific method. It may be 99% safe or 99.9% safe or 99.99% safe but that means that there is a one percent, or 0.1% of 0.01% probability that it will be unsafe. Thus, no scientific conclusion is definitive. As a result, science as a mechanism upon which decisions are to be based is weakened.

In the absence of *definitive answers*, what could be and is used as a decision-making criterion is a *scientific consensus*. This would appear to be what the framers of the SPS may have had in mind. That they put considerable faith in the ability to arrive at a *scientific consensus* is evidenced by the provisions in the SPS that recognize international standards devised by the Codex, the OIE and the IPPC and the allowance for temporary *precautionary* trade restrictions when existing scientific information is insufficient (Isaac 2007b). In the latter case, evidence of their confidence in the ability to reach a *scientific consensus* is found in SPS Agreement Article 5:7 which states: "In such circumstances, Members shall seek to obtain additional information necessary for a more objective assessment of risk and review the sanitary or phytosanitary measure accordingly within a reasonable amount of time." It seems clear that the expectation was that objective answers could be obtained relatively quickly and not that scientific uncertainly would continue for long periods or possibly indefinitely. Objective measures were required, and expected, for decisions to be made.

The reality, of course, is that there is no obvious reason a *scientific consensus* will arise. There will always be scientists who do not accept the conventional wisdom and, in fact, challenging the conventional wisdom is central to the dynamic process of scientific progress (Khachatourians 2001). Thus, on any issue there will likely be scientists that can be found to question any claimed *scientific consensus*. While international standards organizations such as the Codex, the OIE and the IPPC may arrive at international standards for food safety, among other matters, this does not mean that there is an actual *scientific consensus*.

As science does not provide definitive answers and a scientific consensus is an illusion, it means that policy makers must draw what often appear to be arbitrary *lines in the sand*. This opens the door for manipulation of those *line in the sand* decisions to arrange protectionist outcomes. For example, the imposition of a trade barrier justified on SPS grounds requires that the barrier (the measure) have a scientific justification and that there be a risk (Buckingham and Phillips 2001). Individual countries, however, are allowed to set their own levels of what are considered acceptable risks—because science cannot provide zero risk nor any definitive objective guidance regarding acceptable levels of risk. If individual countries set their own standards for acceptable risk, those decisions are open to protectionist influence. At one extreme, countries can set zero as their acceptable level of risk—which is operationally impossible for exporting firms or countries to achieve (Hobbs et al. 2014). Even if operational tolerances are set at non-zero, but very low, levels the costs associated with testing and segregating supply chains can be as effective as any tariff in disrupting trade (Maltsberger and Kalaitzandonakes 2000). While it might be possible for governments to negotiate mutually-agreed tolerance levels, there is no requirement to do so. Governments faced with strong protectionist pressures have been strategically reluctant to do so.

Beyond the arbitrary nature of deciding on the acceptable level of risk, there are a number of other aspects of science that fail to provide definitive answers and, hence, are open to capture or manipulation for protectionist purposes. One vexing question is *how much science is enough?* The safety of food additives or the use of pharmaceuticals is a complex field. Groups of individuals are not homogeneous in their genetic makeup and environmentally determined tolerances. Further, the substances that individuals imbibe, inhale and otherwise take into their bodies vary tremendously. Conceptually one might imagine testing for most probably possibilities, but safety testing is not costless. Thus, determining the safety of a product is undertaken using sample populations and sub-populations to test for known or expected adverse effects. The reality is that there are always additional sub-populations that can be tested (e.g. pregnant women taking pharmaceutical X and who smoke; men over sixty with a history of heart disease and using insulin). Thus, even if a product has been determined as being safe in a number of markets and uses, an importing country can always argue that tests on further sub-populations should be conducted. This was, for example, one of the arguments used by the EU to maintain its import ban on beef produced using growth hormones when faced with reports from its own scientific experts that there was no evidence that this product represented an unacceptable risk (Roberts 1998). In the end, they justified maintaining the import ban because not enough science had been done. While it is certainly possible for countries to negotiate common standards in this area, science provides no definitive answers. Thus, countries can chose not to conclude such negotiations or to unilaterally determine when sufficient science has been done—all which can work to extend protection to target groups.

Even if there is international agreement on scientific standards, there is a wide array of institutional arrangements that are open to manipulation. Scientific training and certification varies among countries. Science itself cannot determine if the training of scientists in exporting countries is sufficient or that the certification procedures for, as an example, medical doctors are sufficiently rigorous. Again, these are arbitrary lines in the sand that can be established by importing governments. Similar issues can surround the efficacy of testing laboratories, the certification of testing facilities, the quality and level of staffing in laboratories and the oversight of these facilities. Bent on responding to requests for protection, the institutional hurdles that can be put in place are only confined by the inventiveness of the bureaucrats charged with devising them. The central point, however, is that there is no objective science that can be used to make such choices.

Probably nothing better illustrates the difficulties associated with science-based decision systems than the long wrangle over operationalizing precaution. The central question related to precaution is: How should decisions be made when scientific uncertainty exists? The prudent response is that a new technology should not be licensed under situations of scientific uncertainty. That idea is not controversial and, in fact, it has been incorporated in a range of international treaties and agreements including the WTO. It has also become an important element of EU domestic policy in a number of areas. In this sense, precaution sounds like it is science-based. The problem comes with trying to operationalize when precaution might be appropriate. The crux of the matter is defining when scientific uncertainty has been reduced sufficiently for a new technology to be allowed into the domestic market or the natural environment. Of course, the problem is that scientific uncertainty cannot be reduced to zero. If scientific uncertainty exists and precaution is accepted as a legitimate criteria, then the technology should not go forward into the market. The result would be that no new technology would ever gain acceptance. Given that humankind's progress is based on the adoption of new technologies, precaution can carry a very cost in opportunities forgone over the long run. The cost of these type 2 errors can be very high.

Protectionists (and those who wish to limit new technologies domestically) have seized on precaution as the mechanism to allow the achievement of their ends. It is often called the *precautionary principle*—and in some civil society circles it approaches a mantra. Given its inherent simplicity, and hence appeal, it has been formally incorporated into EU policy and a range of international agreements, such as the CPB—but without really addressing how it is to be operationalized to make decisions (Issac and Kerr 2007b). Those that seek to deny new technologies can always trot out evidence of scientific uncertainty, pulling the precautionary trigger to stifle new technologies and to inhibit further investment in innovation. In its internal deliberations on the precautionary principle the EU has suggested that scientific decision-making be abandoned—that science should only inform the decision and that decision-makers should be allowed to take other criteria, including economic criteria, into consideration when invoking the *precautionary principle*. Allowing economic criteria (among others) takes such decisions away from science-based groundings, effectively re-politicizing the decision process (Smyth et al. 2015). Environmentalists (and others) have pushed hard and continue to push to have the *precautionary principle* incorporated as the fundamental decision-making criteria in international agreements. They understand that removing scientific uncertainty is not feasible—thus embedding the principle would give them the ability to thwart any new technology they do not like. Of course, if there is a technology they find acceptable then they simply need not trot out evidence of scientific uncertainty. Thus, *the precautionary principle* becomes a powerful instrument in the protectionist's tool kit.

As noted in Chap. 6, van den Belt (2003) has asserted that the *strong version* of the precautionary principle is logically inconsistent. In essence, if it is not possible to rule out a catastrophic event arising from adopting the new technology it is equally impossible to rule out a catastrophic event arising from not adopting a new technology. Given this logical inconsistency, using precaution as a criterion will require a weaker version of the precautionary principle where decisions are made under conditions of non-zero risk. As a result, one is back to *arbitrary lines in the sand* drawn by governments. Hence, it is open to protectionist influence.

There are other institutional rigidities that make scientific decision-making difficult. Given that delay can be valuable to protectionist interests—for example, delay in arriving at an agreed scientific standard in bodies like the Codex, OIE or IPPC reduces the incentives for new product developers—institutional constraints may inhibit timely decision-making. In the absence in internationally agreed standards for, as an example, food safety, countries have developed their own domestic standards and procedures. As scientists in different countries are trained somewhat differently, the mandates related to food safety differ and the development of domestic standards will have followed alternative trajectories. Devising international standards, however, requires that standards need to be harmonized. The bureaucrats—often scientists—tasked with negotiating common standards often want their domestic rules to become the standard to which others harmonize. This is not necessarily only hubris—there may be good reasons for such an ego-centric approach to harmonization negotiations.

If a bureaucrat agrees to harmonize to a foreign (or compromise) standard they may be accused of one of two things. First, if they claim they are agreeing to a different standard because it is higher than their domestic standard, then they can be accused of previously not doing their best to protect their citizens—they allowed a lower standard to be used when they knew a better standard existed.[6] On the other hand, they cannot admit that the new standard is lower than their existing standard because they could be accused of trading off the safety of their citizens to gain a trade (or some other) benefit. Admitting to lowering safety standards to satisfy foreigners is akin to political suicide. It may be possible to claim that the new harmonized standards are equivalent to the existing domestic standard even if differences exist. Equivalence, however, can be difficult to defend when differences between the old and new standards are pointed out by those wishing to maintain existing standards for protectionist purposes. Given the perils associated with agreeing to harmonized standards, delay is frequently the preferred strategy for those responsible for negotiating harmonized standards. Certainly, the development of harmonized standards at the Codex, OIE and IPPC seems prolonged, often taking decades (MacKenzie 2001). Of course, given the differences in training and the domestic institutional culture, bureaucrats may well believe that their existing domestic standards are superior and see no reason to accommodate international trade obligations (Kerr 1997). These inherent biases toward delay play into the hands of those that garner rents from delaying the development of harmonized international standards.

[6]If they claim that they did not know about the new, better standard, they could be accused of incompetence.

Thus it seems as the old cliché suggests that both *motive and opportunity* exist—and protectionists appear to have been adept at acting on them. Twenty years of tinkering has not been able to effectively operationalize science-based decision-making pertaining to SPS matters. More fundamental changes are likely required.

## The Costs of Incomplete International Regulations

The international trade for conventional grain and oilseed commodities uses agreed upon thresholds for standard crop production aspects such as low level presence of other commodities and weed seeds. Thresholds also exist for less desirable features of crop production such as insect fragments, stones and manure (Hobbs et al. 2014). While one, or some, could dispute the safety of consuming some of the items that are comingled into crop commodity exports, at some point, trade negotiators realized that the establishment of these thresholds were required to facilitate trade. Grain trade experts suggest that international trade in conventional grain and oilseeds functions at about the 95% purity level. The ability to export at higher purity levels does exist, but importers must pay a premium to secure those products. Organics is an example of a higher purity level, which in that case has a co-requisite higher price.

The commercialization and wide adoption of GM crop varieties around the world has, at times, resulted in the manipulation of thresholds. As was highlighted above, safety is the most common trigger mechanism. For example, the US and Canada commercialize new crop varieties using science-based regulations, meaning that all risks pertaining to safety, be they environmental, food or feed related, are addressed prior to regulatory approval being granted.

One instance where international commodity trade is affected is in the EU's decision-making process relating to both the import and production of GM crop varieties. To ensure that there is no ceding of sovereignty, the European Food Safety Agency (EFSA) is responsible for approving GM varieties for food and feed use and production. While the US and Canada are attempting to move towards a policy of synchronised approval for GM crop varieties, the EU is steadfastly insisting that safety of GM crop varieties is a crucial issue, so much so, that it insists on conducting scientific assessments with ever more finely defined population subsets. Between June 1999 and August 2003, the EU implemented a moratorium on the approval and import of GM crops and food products. This moratorium was ultimately ruled by the WTO to be inconsistent with the EU's commitments under the SPS. But rather than comply, the EU Commission in 2004 called for all member states to begin developing frameworks for coexistence.

There have been a number of instances where seed varieties in Europe were found to contain trace amounts of GM varieties. The first report came from Switzerland, which has similar concerns as the EU with agricultural biotechnology.

Smyth et al. (2002) reported that in May 1999 the Swiss Department of Agriculture announced that two Pioneer Hi-Bred non-GM corn varieties, imported and distributed by Eric Schweizer Samen AG, had been found to contain trace amounts of GM varieties. The level of comingling ranged from 0.1 to 0.5% (Fürst 2002). Pioneer distributed sufficient seed to plant an estimated 400 ha, of which, about half had been seeded at the time of detection. The GM traits that were identified were not approved for import or commercial release in Switzerland. As a result, the fields that had been planted were burnt or controlled with herbicides and the importing firm agreed to pay compensation of 700 Swiss francs per hectare. The second incident, in the spring of 2000, involved a breeders lot of canola seed imported by Advanta that contained 0.4% unapproved GM traits (Agnet 2000a). Advanta quickly determined that the unexpected presence of GM canola was caused by gene flow from GM foundation seeds that had been planted in a neighboring field. Canadian seed growers had followed isolation rules but the genes still moved into the conventional foundation seed. The total acreage in most countries (except Britain) was insignificant, Sweden and Germany had 300 ha and France had 600 ha. France ordered all 600 ha to be ploughed down and Sweden allowed the canola to be harvested but prohibited the canola from entering the domestic or EU market. In Britain, over 15,000 ha were planted and had to be destroyed. It was estimated that the cost of compensation from Advanta would exceed one million pounds (Agnet 2000b). In August 2002, the UK Department for Environment, Food and Rural Affairs (DEFRA) announced that they had been advised by Aventis CropScience Ltd. (now Bayer CropScience) of some impurities in canola seed that was being used for Farm Scale Evaluation field trials in England and Scotland (Scottish Government 2002). The initial discovery was part of a routine audit conducted by the Scottish Agricultural College and the level of comingling was 2.8%. Given the time of the discovery, the crops were harvested and the resulting seed was destroyed.

Then in March 2003 trace amounts of GM canola were detected in Canadian mustard exports to Europe (Western Producer 2003). A mustard shipment in late 2002 was tested and found to contain trace amounts of GM material. Since there were (and are) no GM mustard varieties in Canada at this time, importers in the EU conducted further tests and determined that the trace amounts of GM material were GM canola. According to export standards, mustard exports are allowed to contain 1% canola and since 75% of the canola produced in Canada at the time was GM, it was not surprising that trace amounts would be comingled. There is no report on what the European importers did with the mustard shipment.

The response of EU regulators in perhaps the most widely known case of comingling—StarLink™ corn—was again one of co-operation (Paarlberg 2001). Prior to the implementation of the moratorium, the EU did approve a small number of GM crops for production within the EU and Paarlberg (2001) suggests that the EU regulators worked with the US to allow small levels of comingling in corn exports so that their actions were not seen to be undermining their own approval of GM crops, nor an admission that the EU regulatory system was not science-based.

Clearly, there is ample evidence that while the EU moratorium was in place, and the EU Commission was revamping its regulations, both import shipments of commodities and crop production in Europe were affected by the detection of trace amounts (as high as 2.8%) of unapproved GM varieties. While this caused some concern, it only triggered a proportionate regulatory and industrial response and, in some instances, led to financial compensation. During this period the EU response to the low level presence of GM varieties conformed to the science-based international rules of trade. Risks were identified, isolated and managed in proportion to their potential effect on the health and safety and the import of the marketplace.

Since the end of the moratorium in 2003, coexistence has been the buzz phrase and has led to more extreme and far-reaching responses. In 2006, trace amounts of an unapproved GM event were detected in US rice exports to the EU. In what has become known as LL601 rice, the widespread presence of this variety resulted in an announcement from the EU on August 20th, 2006, that it would not accept any further rice shipments from the US (Li et al. 2010).

Smyth et al. (2010) discuss the trade implications in US-EU corn trade following the commercialization of Herculex corn. Again, the event was approved in the US, but not in the EU, and in spite of US testing prior to export, trace amounts were discovered in shipments to the EU. The detection of this variety of corn, released by Pioneer Hi-bred in 2006, caused corn gluten feed exports from the US to the EU to drop by 30–40%. What is interesting in this case is that Pioneer submitted notification of import into the EU for Herculex I in 2000 (Europabio 2006). By 2006, Heculex I had received approval for feed and food use, as well as planting. The problem arose when Herculex (R) Rootworm varieties were detected in shipments coming from the US. These varieties were ultimately approved by the EU in 2009 but for the three preceding years corn trade was disrupted.

The first variety of GM flax, CDC Triffid, was developed at the University of Saskatchewan in Canada. Field trials began in 1989 and approval for livestock feed use in Canada was received in 1996, with food approval coming in 1998. Feed and food approval was received the same year for the US market. In the winter of 1998–99 European flax importers raised concerns about the commercial production of GM flax. Given that an average 70% of Canadian flax production was exported to the EU, pressure from both the EU importers and Canadian producers resulted in the decision to voluntarily withdraw GM flax from the market. Genetically modified flax was deregistered in Canada in 2001 and all known quantities were crushed and/or destroyed. Then in September 2009, the European Union's Rapid Alert System for Food and Feed (RASFF) announced the detection of GM Triffid flax in food products in Germany. By the end of September, most Canadian flax in the EU was in quarantine and flax trade between Canada and the EU was suspended, pending identification of the source of comingling and the implementation of testing a protocol that could provide assurance to EU importers that Canadian flax exports would be free of GM flax (Viju et al. 2014). Trade came to a virtual standstill as testing protocols were devised and subsequently put in place. Costs mounted on both sides of the Atlantic. Testing revealed the presence of Triffid flax in the Canadian supply chain leading to diversion of cargoes to non-EU markets

(Viju et al. 2014). Firms in the EU that import flax, largely for industrial uses such as the manufacture of linoleum, had difficulty sourcing alternative sources of supply. Canadian export levels had not returned to pre-2009 levels by 2016 and considerable testing costs are ongoing. With the passage of time, Triffid flax has become obsolete, overtaken by other non-GM varieties (Viju et al. 2014). Thus, approval for Triffid flax will never be sought in the EU so there is no mechanism to remove the ongoing trade costs borne by flax producers and users.

When it comes to the importation of soybeans from the Americas, it is a considerably different issue. Soy in animal feed in EU is considered a staple, with 75% of the soy used in the EU animal feed industry imported from Argentina, Brazil and the US (LEI 2009). The adoption rates for GM soy in these three sources of imports for the EU in 2007 was estimated to be 95, 66 and 91% respectively.[7] EU import regulations allow for up to 0.9% of approved GM material, 0.5% of unapproved GM material that has been positively reviewed by the European Food Safety Authorities (EFSA) and 0% of GM material that has not been approved by EFSA.[8] The challenge for EU soy importers is that it takes considerably longer for EFSA to approve new GM events than approvals take in other countries. A study by the Dutch research institute LEI (2009) estimated that if EFSA reviews took longer than two years and import thresholds remained at 0.9%, then the cost of importing a tonne of soy would rise from 290 Euros to 7747 Euros. With so little availability of non-GM soy globally, it was widely speculated in the grain industry that the EU was illegally importing shipments of soybeans that tested positive for GM soybeans at low levels.

The EU's determination to use only non-GM soybeans led them to seek new sources of supply. The EU announced in 2009 that it had begun to source non-GM soybeans from India. Importing non-GM soy to feed to animals in EU while millions of people suffer from malnutrition and undernutrition in India raises some provocative ethical questions. The reality of the situation was that the EU could not knowingly import soy from any of the Americas and expect it to have less than 0.9% GM content. In response to the mounting economic pressure on the EU's livestock industries and heavy lobbying by the sector, the EU relaxed its regulation to allow the import of GM-based animal feeds such as soy. Imports of unapproved GM varieties that would be destined for direct human consumption, however, are not allowed to be imported. The opening of the market to GM feeds represented a major climb-down for the EU and calls into question its hard line of GM food imports.

International grain and oilseed commodity trade does not, and probably will not be able to, deliver pure shipments; zero percent comingling is economically and probably technically infeasible (Hobbs et al. 2014). Over time, thresholds in this industry have developed and been agreed upon as mechanisms to facilitate trade in

[7]Some less substantiated references have soybean adoption rates in Brazil in the low 80% range.

[8]The EU soon found these tolerances unworkable and revised the tolerance threshold to be 0.1% comingling for events not approved in the EU, but approved in the country of export.

commodity agriculture. The EU's insistence on zero tolerance is neither theoretically optimal nor operationally tractable (Hobbs et al. 2014). The EU system for GM approvals and for denying market access for GM-based imports does not appear to be consistent with the EU's SPS commitments as they are not solely science-based (Viju et al. 2012). They are also slow and costly to use—slow enough that they might be considered a barrier to trade—in WTO parlance *not minimally trade distorting*. Of course, to decide either of these issues a case would have to be brought to the WTO disputes settlement mechanism. As yet, no appropriate case has been brought forward. While there are a number of reasons why no case has been brought forward including the US wanting a preferential agreement with the EU—the Transatlantic Trade and Investment Partnership (T-TIP)—and Canada seeking its own Comprehensive Economic and Trade Agreement (CETA) with the EU, there is also a major concern that the EU would lose the case, be unwilling to open its market to GM products and be faced with a significant retaliatory cost. The size of the retaliation could lead to deteriorating international relations and a further weakening of the WTO. There is a clear need for alternative thinking to resolve the issues in a non-confrontational way.

## A Framework for Analysis

Ultimately, regulatory regimes are specialized forms of knowledge systems. Thus, the challenges of developing and operating regulatory systems can be recast as a case of knowledge management. A major challenge when dealing with something as nebulous as *knowledge management* is that there is no single way to define or examine it. Legal scholars, economists, philosophers, business analysts, political scientists and sociologists, to name a few, have all examined and framed *knowledge management* in the context of both the underlying assumptions about motivations and the institutional focus of their work. Unfortunately, the field is filled with competing views that do not necessarily sum to a coherent analysis. One way to frame an analysis is to explicitly map the institutional systems that are engaged in creating and managing knowledge and then to move to the individual level, where people, alone or in association with others, use a set of norms, rules or processes, in the context of the prevailing institutions, to achieve some purpose.

On the institutional front, Phillips (2007) asserts that our society is fundamentally faced with a complex systems problem when considering how to govern knowledge. Boulding (1970) suggested three distinct domains—the compulsory, the contractual and the familial—yielding three different methods of integration: coercive states that distribute rights and obligations; quid pro quo exchanges in the market governed by supply and demand; and voluntary dealings, where cooperation, reciprocity and solidarity engage community and society (Paquet 2001). Boulding (1970) argues that society can be viewed as a triangle (his 'social triangle'), where all organizations—including the state, the market and civil society—are built on one or a balance of the three relationship systems. The triad of institutions—governments or states, the market and social or familial organizations —each have specific institutional attributes that make them more effective at producing particular types of goods (Picciotti 1995). The government sector is best at producing public goods, such as basic scientific knowledge and public health and safety—low excludability makes privatization infeasible while the low voice component makes it difficult for the collective sector to organize. The private sector tends to dominate whenever clear and enforceable property rights make rival goods excludable. In the context of knowledge, this would be knowledge embedded in specific products and services. The property of exclusion allows private firms to sell at the marginal cost of production. The participatory sector is best at governing common pool goods (e.g. various types of knowledge, such as standards)—the collective group will usually have better information that will enable them to more effectively manage the resource and capture the benefits. At the individual level, different disciplines posit different human motivations to individuals, such as utility maximization, profit optimization, belonging, power, fulfillment, meaning and happiness. None of these fully explains the motives around knowledge creation and management.

Ostrom (2005) offers a *universal framework* to assess how individuals might interact in pursuit of knowledge management (see Chap. 8 for a more detailed discussion). She argues that whether a participant is an individual or an institution, one must make assumptions about three components of human behaviour in order to animate institutional analysis. Her *institutional analysis and development* (IAD) framework suggests that decision systems (in our case read regulatory systems) have at their heart three key elements:

1. the way that participants acquire, process, represent, retain and use *information*;
2. the *valuation* that participants assign to actions and outcomes; and
3. the process (maximizing, satisficing, or using diverse heuristics) that participants use for *selecting* particular actions or strategic chains of actions in light of their resources (Ostrom 2005, p. 103).

This framework, which defines the scope and way relevant information is agglomerated, explicitly identifies the normative value placed on knowledge and then identifies the process of selecting a pathway of management in the context of constraints (e.g. a budget constraint or bounded rationality) that can be used to consider the structure and function of 'science-based regulation'.

In the context of our case, the underlying information, valuation and selecting rules involved in *science-based* decision-making need to be unpacked. While a daunting task, this is not an insurmountable challenge. The discussion above about the discretion exercised by regulators in the system highlights many of the critical control points in the system. The International Standards Organization (ISO) standard model for industry to develop hazard analysis critical control points (HACCP) plans offers a template for the articulation of generally accepted practices and for the identification and normalization of contested practices.

## Whither Science-Based Trade Regulations?

Politics has always distorted and is continuing to distort international trade. It was hoped (expected) that the SPS Agreement would provide a solution to political meddling in SPS trade regulations. The SPS, however, has failed to stem the tide of trade-distorting policies in this area of international commerce. In fact, it would appear that the politicization of international trade, especially in agricultural products, has simply continued to increase over time and that the commercialization of GM crops has only emboldened those wishing to restrict agri-food trade.

Farmers are heavily protected in many countries, with very effective lobby groups working hard to ensure that this system of support is not soon ended. The willingness of governments to abandon or selectively use science-based regulations in order to preserve domestic agricultural markets or to favour one market over another is apparently so commonplace that little to no concern is raised over the resulting trade distortions. This obvious manipulation of science-based regulation ultimately reveals that the system as it is currently formulated is no longer capable of underpinning international trade. As suggested above, however, in the case of agricultural biotechnology it is not traditional producer protectionism that is motivating the move to increased protectionism. It is other groups in civil society. This has altered the dynamic of protectionism and has moved the dialogue away from the traditional hard bargaining over the pace of trade liberalization that was the GATT staple for fifty years.

The reasons behind the abandonment of science-based regulations by some members of the WTO needs to be examined and a fresh approach to facilitating unfettering of trade needs to be offered. Part of the challenge is that some disembodied stock of knowledge, often called *science*, can fully delimit the acceptable from the unacceptable. Our analysis reveals that there is no definitive set of *science* facts that can underpin decision-making for transformative technologies like GM foods. This does not mean that the science-based system should be replaced by policy-makers preferences—rather it highlights the need to more clearly and concretely define the appropriate human institutions that can develop and use the *near-scientific consensus* that is needed to make decisions that further the public interest.

The uncertainty that is triggered by the movement away from science-based regulation has impacted industry investment decisions. When international wheat markets publicly rejected GM wheat varieties, not only did research into new GM wheat varieties cease, but in some markets investment in wheat development as a whole suffered. Market uncertainty has also affected research with GM varieties of flax, potatoes, sugar beets and alfalfa. The loss entails not just the new crop varieties and the subsequent trade of these varieties, but also the environmental benefits that are becoming associated with the production of GM crops.

In short, the effort to develop an effective regulatory and trade system that facilitates optimal market adoption is not complete. The integration of agriculture into the WTO, the adoption of a binding disputes system and the incorporation of the international food standards setting organizations (Codex, IPPC, OIE) in the trade regulatory system were all vital steps. They should not, however, be viewed as the end of the process. Rather, they provide the launching pad for a process to develop a more highly integrated and effective regulatory system for new agri-food technologies.

## References


Agnet. 2000a. EU Member States Must Enforce GM Crop Rules. May 18th.

Agnet. 2000b. GM Crop Compensation Set to Cost Company Millions. June 2nd.

Barichello, R.R. 2007. Administrative procedures; the distribution of costs and benefits, and incentives in anti-dumping cases. In *Handbook on International Trade Policy*, ed. W.A. Kerr, and J.D. Gaisford, 360–367. Cheltenham: Edward Elgar.

Belcher, K., A.L. Hobbs, and W.A. Kerr. 2003. The WTO and environmental sustainability: Is there a conflict? *International Journal of Environment and Sustainable Development* 2 (1): 2–18.

Benarroch, M. 2007. Economies of scale, imperfect competition and market size. In *Handbook on International trade policy*, ed. W.A. Kerr, and J.D. Gaisford, 141–149. Cheltenham: Edward Elgar.

Bond, S.R., and J.G. Cummins. 2004. Uncertainty and investment: An empirical investigation using data on analysts' profit forecasts. Available online at: http://www.federalreserve.gov/pubs/feds/2004/200420/200420pap.pdf.

Boulding, K. 1970. Organizers of social evolution. In *A primer on social dynamics: History as dialectics and development*, ed. K. Boulding, 19–36. London: Collier MacMillan Ltd.

Buckingham, D., and P.W.B. Phillips. 2001. Hot potato, hot potato: Regulating products of biotechnology by the international community. *Journal of World Trade* 35 (1): 1–31.

Copeland, B.R. 2007. Trade and the environment: What do we know? In *Handbook on international trade policy*, ed. W.A. Kerr, and J.D. Gaisford, 414–426. Cheltenham: Edward Elgar.

Episcopos, A. 1995. Evidence on the relationship between uncertainty and irreversible investment. *The Quarterly Review of Economics and Finance* 35 (1): 41–52.

EuropaBio. 2006. Herculex I (1507) Maize: Safety and General Characteristics. Available at: http://www.europabio.org/sites/default/files/pages/1507_factsheet.pdf.

Fürst, I. 2002. Swiss soiled seed prompts tolerance question. *Nature Biotechnology* 17: 629.

Gaisford, J.D., and A. Hester. 2007. Why are there trade agreement? In *Handbook on international trade policy*, ed. W.A. Kerr, and J.D. Gaisford, 57–70. Cheltenham: Edward Elgar.

Gaisford, J.D., and W.A. Kerr. 2001. *Economic analysis for international trade negotiations*. Cheltenham, UK: Edward Elgar Publishing Ltd.

Gaisford, J.D., and W.A. Kerr. 2003. Deadlock in Geneva: the battle over export subsidies in agriculture. *International Economic Journal* 17 (2): 1–17.

Gaisford, J.D., J.E. Hobbs, W.A. Kerr, N. Perdikis, and M.D. Plunkett. 2001. *The economics of biotechnology*. Cheltenham: Edward Elgar.

Hobbs, A.L., J.E. Hobbs, and W.A. Kerr. 2005. The biosafety protocol: multilateral agreement on protecting the environment or protectionist club? *Journal of World Trade* 39 (2): 281–300.

Hobbs, J.E., W.A. Kerr, and S.J. Smyth. 2014. How low can you go? The consequences of zero tolerance. *AgBioForum* 16 (3): 207–221.

Holtby, K.L., W.A. Kerr, and J.E. Hobbs. 2007. *International environmental liability and barriers to trade*. Cheltenham: Edward Elgar.

Isaac, G.E. 2007a. Ethical issues in trade. In *Handbook on international trade policy*, ed. W.A. Kerr, and J.D. Gaisford, 404–423. Cheltenham: Edward Elgar.

Isaac, G.E. 2007b. Sanitary and phytosanitary issues. In *Handbook on international trade policy*, ed. W.A. Kerr, and J.D. Gaisford, 383–393. Cheltenham: Edward Elgar.
Isaac, G.E., and W.A. Kerr. 2004. Bioprospecting or biopiracy?—Intellectual property and traditional knowledge in biotechnology innovation. *The Journal of World Intellectual Property* 7 (1): 35–52.
Isaac, G.E., and W.A. Kerr. 2007a. Whose vision of the future? The entrenched international conflict over genetic modification. *The Geneva Post Quarterly* 2 (1): 87–107.
Isaac, G.E., and W.A. Kerr. 2007b. The biosafety protocol and the WTO: Concert or conflict. In *The international politics of genetically modified food*, ed. R. Falkner, 195–212. Houndmills, UK: Palgrave Macmillan.
Isaac, G.E., N. Perdikis, and W.A. Kerr. 2004. Cracking export markets with genetically modified crops—What is the entry mode strategy? *International Marketing Review* 21 (4/5): 536–548.
Kerr, W.A. 1997. Removing health, sanitary and technical non-tariff barriers in NAFTA. *Journal of World Trade* 31 (5): 57–73.
Kerr, W.A. 2003. The Free-traders win the debates but the protectionists win the elections: The curious case of MCOOL in the US 2002 farm bill, presented at a workshop entitled Farm Policy Developments and Policy Tensions Under NAFTA—Ninth Agricultural and Food Policy Tensions Under NAFTA, Policy Disputes Information Consortium, Montreal, Canada, April 23–26. Available online at: http://pidc.tamu.edu/farmpolicy/kerr.pdf.
Kerr, W.A. 2006. Dumping: Trade policy in need of a theoretical make over. *Canadian Journal of Agricultural Economics* 54 (1): 11–31.
Kerr, W.A. 2010. What is new in protectionism?: Consumers, cranks and captives. *Canadian Journal of Agricultural Economics* 58 (1): 5–22.
Kerr, W.A., and J.E. Hobbs. 1994. A new era of 'Corn Laws': Agricultural trade with the European Community—1992 and beyond. In *The European Community, Canada and 1992*, 91–102, ed. G.M. MacMillan. Calgary: The University of Calgary Press.
Kerr, W.A., and R.J. Foregrave. 2002. The prophecies of the naysayers—Assessing the vision of the protectionists in the U.S.-Canada debate on agricultural reciprocity, 1846–1854. *Journal of International Law and Trade Policy* 3 (2): 357–408.
Kerr, W.A., and S.L. Hall. 2004. Multilateral environmental agreements and agriculture: Commitments, cooperation and conflicts. *Current Agriculture, Food and Resource Issues* 5: 39–52.
Khachatourians, G.G. 2001. How well understood is the science of food safety? In *Governing food: Science, safety and trade*, ed. P.W.B. Phillips, and R. Wolfe, 13–23. Kingston: McGill-Queens University Press.
Leahy, J.V., and T.M. Whited. 1996. The effect of uncertainty on investment: Some stylized facts. *Journal of Credit, Money and Banking* 28 (1): 64–83.
LEI. 2009. *EU policy on GM soy: Tolerance threshold and asynchronic approval*. Available online at: http://edepot.wur.nl/7856. Accessed October 28, 2013.
Li, Y., E. Wailes, A. McKenzie, and N. Thomsen. 2010. LL601 contamination and its impact on U.S. rice prices. *Journal of Agricultural and Applied Economics* 42 (1): 31–38.
MacKenzie, A. 2001. International efforts to label food derived through biotechnology. In *Governing food: Science, safety and trade*, ed. P.W.B. Phillips, and R. Wolfe. Kingston: McGill-Queens University Press.
Maltsberger, R., and N. Kalaitzandonakes. 2000. The hidden costs of identity preservation. *Feedstuffs* 72 (36): 1–2.
Office International Des Epizootics (OIE). 2000. A short history of the office international des epizootics. Paris: OIE Press. Online at: www.oie.int/eng/OIE/en_histoire.htm.
Olson, M. 1965. *The logic of collective action: Public goods and the theory of groups* (Revised ed.). Boston: Harvard University Press.
Ostrom, E. 2005. *Understanding institutional diversity*. Princeton: Princeton University Press.
Paarlberg, R.L. 2001. Shrinking international markets for GM crops? Presentation to the agriculture outlook forum, February 22–23, 2001. Online at: http://ageconsearch.umn.edu/bitstream/33093/1/fo01pa01.pdf.

Paquet, G. 2001. The new governance, subsidiarity and the strategic state. In *OECD, governance in the 21st century*. Paris: OECD. Online at: http://www.oecd.org/dataoecd/15/0/17394484.pdf.

Perdikis, N., W.A. Kerr, and J.E. Hobbs. 2001. Reforming the WTO to defuse potential trade conflicts in genetically modified goods. *World Economy* 24 (3): 379–398.

Phillips, P.W.B. 2007. *Governing transformative technological innovation: Who's in charge?*. Oxford: Edward Elgar.

Phillips, P.W.B., S.J. Smyth, and W.A. Kerr (eds.). 2006. *Governing risk in the 21st century: Lessons from the world of biotechnology*. Hauppauge, NY: Nova Science Publishers.

Picciotti, R. 1995. Putting institutional economics to work: From participation to governance. World Bank discussion paper 304. Online at: http://www-wds.worldbank.org/servlet/WDSContentServer/WDSP/IB/1995/09/01/000009265_3961219115337/Rendered/PDF/multi_page.pdf.

Roberts, D. 1998. Preliminary assessment of the effects of the WTO agreement on sanitary and phytosanitary trade regulations. *Journal of International Economic Law* 1 (3): 377–405.

Scottish Government. 2002. Unauthorised material in GM crop trials. Available at: http://www.scotland.gov.uk/News/Releases/2002/08/2066.

Smith, A. 1994. *An inquiry into the nature and causes of the wealth of nations*. New York: The Modern Library.

Smyth, S.J., G.G. Khachatourians, and P.W.B. Phillips. 2002. Liabilities and economics of transgenic crops. *Nature Biotechnology* 20: 537–541.

Smyth, S.J., P.W.B. Phillips, W.A. Kerr, and G.G. Khachatourians. 2004. *Regulating the liabilities of agricultural biotechnology*. Wallingford: CABI Publishing.

Smyth, S.J., A.B. Endres, T. Redick, and D. Kershen. 2010. *Innovation and liability in biotechnology: Transnational and comparative perspectives*. Cheltenham, UK: Edward Elgar Publishing Ltd.

Smyth, S.J., P.W.B. Phillips, and W.A. Kerr. 2015. Food security and the evaluation of risk. *Global Food Security* 4: 16–23.

Van den Belt, H. 2003. Debating the precautionary principle: "Guilty until proven innocent" or "innocent until proven guilty"? *Plant Physiology* 132: 1122–1126.

Viju, C., M.T. Yeung, and W.A. Kerr. 2012. The trade implications of the post-moratorium European Union approval system for genetically modified organisms. *Journal of World Trade* 46 (5): 1207–1238.

Viju, C., M.T. Yeung, and W.A. Kerr. 2014. Zero tolerance for GM flax and the rules of trade. *World Economy* 37 (1): 137–150.

Western Producer. 2003. GM canola found in mustard load. Available at: http://www.producer.com/2003/03/gm-canola-found-in-mustard-load/.

Zhang, S., and W.A. Kerr. 2009. Revisiting barriers to trade: Do foregone health benefits matter? CATPRN working paper no. 2009-03. Canadian Agricultural Trade Policy Research Network. Available online at: http://www.uoguelph.ca/~catprn/PDF/Working_Paper_2009-03_Zhang-Kerr.pdf.

# Chapter 8
# International Treaty Precedence

**Abstract** The 1969 Vienna Convention on the Law of Treaties was designed to provide direction and guidance regarding the complicated legality of primacy between international agreements. Given that three of the top five producers of GM crops (Argentina, Canada and the US) have not ratified the Cartagena Protocol on Biosafety (CPB), and in fact are not even signatories to the Protocol, this could become a contentious issue sooner rather than later. This chapter provides a detailed assessment of the various legal implications pertaining to the applicability of the Vienna Convention to the WTO and the CPB.

## Introduction

The World Trade Organization (WTO) is the major multilateral organization upon which the rules of international trade are discussed, negotiated and agreed. Most large economies belong to the WTO and its rules cover approximately 97% of global trade (Kerr 2012).[1] There are, however, other multilateral agreements that contain provisions on international trade—largely in specialized areas such as trade in endangered species or hazardous materials (Kerr and Hall 2004). Many of these agreements are multilateral environmental agreements (MEAs), including the Cartagena Protocol on Biosafety (CPB) which provides an alternative set of trade rules for trade in genetically modified organisms (GMOs).[2] While they may have rules that conflict with those of the WTO, in the case of most MEAs there is sufficient international consensus on the problem being addressed having alternative trade rules that it has never become an issue of dispute at the WTO (Belcher 2007). This may not, however, be the case with the CPB as there is neither consensus on the regulation of trade in GMOs nor on which agreement should take precedence. The rules of trade embedded in the CPB are in many ways antipathetic to those of

[1]Preferential trade agreements such as the North American Free Trade Agreement, the European Union, etc. are covered by the WTO and are expected not to contradict WTO agreements.
[2]The CPB commonly refers to GMOS as LMOs—living modified organisms.

 
S.J. Smyth et al., *Biotechnology Regulation and Trade*, Natural Resource Management and Policy 51, DOI 10.1007/978-3-319-53295-0_8

the WTO (Hobbs et al. 2005)—and may have been designed with that specific purpose in mind (Holtby et al. 2007). Further, while many countries are members of both the WTO and the CPB, some have agreed to only one set of rules. In particular, a number of the major countries that have adopted, and routinely export, genetically modified (GM) products are members of the WTO but not the CPB. Hence there are issues of non-overlapping memberships and how they are dealt with in international law that need to be addressed.

While the conflict between the rules of the CPB and the WTO has been recognized since the inception of the CPB (Phillips and Kerr 2000), the issue is coming to the forefront of public policy for two reasons. First, the ongoing process of asynchronous approvals where more and more GM crops are being approved in some countries and not in others is leading to a rising potential for trade disagreements (Viju et al. 2011). Second, the scientific justifications for placing trade barriers on GM products are fading given widespread adoption and no discernible human health or environmental problems subsequently arising. Now countries that wish to impose barriers to imports are increasingly using socio-economic considerations (SECs) as justifications for the imposition of trade barriers. The CPB allows these as justifications but the WTO does not. In fact, the Agreement on the Application of Sanitary and Phytosanitary Measures (SPS) of the WTO was put in place specifically to de-politicize decision-making in this area by making science the sole justification for the imposition of trade barriers (Kerr 2003). The WTO's rules of trade are primarily motived to constrain a government's capacity to extend economic protection to vested interests threatened by foreign competition—in other words SECs (Kerr 2010). Thus, allowing SECs as a reason to impose trade barriers puts the CPB directly at odds with the WTO's central concern—economic protectionism. As a result, there is the likelihood of a direct conflict between agreements. Hence, a discussion of how conflicts between agreements are handled in international law is warranted. Clarification is required because the current opaqueness of the issue impacts on investments to further develop GM technology (Smyth et al. 2014).

## When International Agreements Conflict—What Is the Law?

The WTO and the CPB are highly divergent regulatory regimes, as discussed in Chap. 4. Given this divergence, and the potential for CPB-related measures to contravene WTO rules, what is likely to occur in the event of a clash? What rules, if any, will take precedence?

As a general principle of international law, when two treaties in the same subject area conflict, the latter treaty prevails in the event of a dispute between two states that are parties to both instruments. While this rule appears to be relatively straightforward, it hides a much more complex set of issues. Indeed as Sinclair observed (1973, p. 62):

> With the post-war growth in international co-operation, accompanied by a massive increase in the numbers and range of international agreements of a law-making character, the problem of incidental conflict between successive treaties has become more acute.

The International Law Commission wrestled with this problem in the course of its work on treaties. The result of its deliberations was Article 30 of the Vienna Convention on the Law of Treaties (1969), which states:

1. Subject to Article 103 of the Charter of the United Nations, the rights and obligations of States parties to successive treaties relating to the same subject-matter shall be determined in accordance with the following paragraphs.
2. When a treaty specifies that it is subject to, or that it is not to be considered as incompatible with, an earlier or later treaty, the provisions of that other treaty prevail.
3. When all the parties to the earlier treaty are parties also to the later treaty but the earlier treaty is not terminated or suspended in operation under article 59, the earlier treaty applies only to the extent that its provisions are compatible with those of the later treaty.
4. When the parties to the later treaty do not include all the parties to the earlier one:

   (a) as between States parties to both treaties the same rule applies as in paragraph 3;
   (b) as between a State party to both treaties and a State party to only one of the treaties, the treaty to which both States are parties governs their mutual rights and obligations.

It is important to note, however, that the rules in Article 30 were clearly designed to be of a residuary nature as 30(2) clearly envisions provisions in a treaty relating to its compatibility with other regimes (Sinclair 1973). In other words, they will only operate in situations where the competing treaties in question are silent in relation to the issue of priority. It should be noted that not all situations where the text of two treaties in the same area diverge are regarded as incompatibilities. This divergence is particularly prevalent in situations where a regional regime and a universal regime operate in the same general subject area, as Sinclair (1973) suggests that the European Convention on Human Rights and the United Nations Human Rights Covenants do. With respect to the potential incompatibility of WTO rules and those contained in the CPB, however, these general rules of international law are unhelpful. Caldwell (1998, pp. 28–29) notes:

> The major difficulty associated with relying on the hierarchical treaty argument is that the later-in-time provision of the Vienna Convention will not adequately serve the GATT/WTO regime or future MEAs. For example, the GATT/WTO regime has a tradition of negotiating additional agreements in new areas of trade liberalization over the course of several years. The latest adoption of the Uruguay Round Final Act clearly places the GATT/WTO regime as the later-in-time treaty in relation to many of the current MEAs. Similarly the negotiations of future MEAs may result in their achieving priority over the GATT/WTO regime. The final result would be a patchwork of differing treaty priorities and minimal clarification of the relationship between the GATT/WTP regime and the MEAs.

Given the radically different regulatory regimes of the WTO and the CPB, the two treaties cannot be said to be in the same subject area (although their spheres of application clearly overlap). Even if such a determination were to be made, the residuary character of the rules contained in Article 30 of the Vienna Convention on the Law of Treaties cannot be overlooked. The CPB is far from silent on its relationship with other regimes, although an analysis of these provisions provides little assistance in answering the question of whether it takes priority over WTO rules.

The Preamble to the CPB contains two seemingly contradictory statements that attempt to outline the relationship between the CPB and other international agreements. Initially, the preamble states, "this Protocol shall not be interpreted as implying a change in the rights and obligations of a party under any existing international agreement" (CPB 2000, p. 2). However, the next paragraph continues, "Understanding that the above recital is not intended to subordinate this Protocol to other international agreements" (CPB 2000, p. 2).

These two seemingly contradictory pre-ambular statements represent classic *savings clauses*. Savings clauses are a common feature of many multilateral treaties and they are effectively an enactment of the principle behind Art 30(2) of the Vienna Convention. The effect of a savings clause is to "save the provisions of an earlier agreement that would otherwise be overcome by incompatible provisions of a later agreement" (Safrin 2002, p. 613).

This rather 'confusing' language (Hagen and Weiner 2000) contained in the Preamble of the CPB is the result of a difficult political compromise reached towards the end of the negotiations. As Safrin (2002, p. 614) notes, the parties had split into three camps in relation to the inclusion of a savings clause in the text of the CPB:

1. A group led by the United States and other major agricultural exporters, who were adamant that the CPB not undermine obligations under existing trade rules and agreement.
2. A group led by the EU who, in the wake of the Beef Hormone Decision of the WTO DSB [dispute settlement body], were fearful that the inclusion of such a savings clause would lend further credence to the belief that a hierarchy of international legal agreements and obligations was developing, where trade concerns and disciplines trumped apparently "subordinate" environmental agreements.
3. A group (the "like-minded group") who supported a compromise position which allowed the operation of a savings clause subject to the exception that its operation could not be allowed to "cause serious damage or threat to biological diversity."

The CPB text adopted this third compromise position thus leading to the rather contradictory pre-ambular statements. In essence, this allowed all parties to claim some form of 'victory' but left it virtually impossible to predict the exact outcome of a potential clash between the terms of the WTO agreements and the CPB.

While the Preamble to a convention does not have the same legal force as the main body of the instrument, it is, nevertheless, a significant aid to interpretation. The objects and purposes of a treaty according to the International Law Commission are primarily to be gathered from the text of the Treaty and particularly from the Preamble (Sinclair 1984). The main text of the CPB contains similarly contradictory statements. Article 2 allows parties to take action on biosafety that is more protective than that envisaged by the protocol, *subject* to the proviso that such measures will be "consistent with the Parties' other obligations under international law" (CPB 2000, p. 3). However, Article 26 subsequently authorizes parties to "take into account socio-economic considerations arising from the impact of LMOs [living modified organisms] on … biological diversity" (CPB 2000, p. 19). As suggested above, SECs are regarded as extraneous by the WTO. Thus, the relationship between the proviso in Article 2 and the authorization contained in Article 26 is at best ambiguous, and at worst, utterly contradictory. In their analysis of these contradictory aspects of the text of the CPB and their relationship with other international obligations (including WTO obligations) Hagen and Weiner (2000, p. 713) conclude:

> In some significant instances the Protocol establishes rights and obligations that can be reasonably interpreted as contradictory rather than merely counterbalancing. For example, it is difficult to see how international trade rights and obligations can remain unchanged as the preamble states, if as the Preamble also states, the Protocol is not subordinate to WTO Agreements.

Clearly, the question of the legal priority of the CPB over WTO agreements (or vice versa) is a complex issue that eludes simple explanation and the issue has spawned a literature that uses rather dramatic descriptors such as 'collision' (Safrin 2002), 'conflict' (Neff 2004; Candeira et al. 2009) and 'tension' (Vallely 2004). However, as with most aspects of international law, the legal subtleties of a given situation are less significant than the political realities. Politically, WTO agreements (and the obligations therein) are generally taken more seriously by states than obligations incurred under other instruments perhaps due to the compulsory jurisdiction of its dispute settlement body. Furthermore, the WTO (and in particular) its Committee on Trade and the Environment clearly advocates regimes that negotiate a rapprochement between trade and environmental goals (which could be said to embody 'sustainable development') rather than creating incoherent, contested and consequently unenforceable hierarchies of international legal agreements.

A key factor in determining the future of the CPB, therefore, is its likely treatment by the WTO should any compatibility issues be raised by aggrieved states. In order to provide a plausible answer to the 'conflict' question, an analysis of the likely rules (and forum of review) governing such a dispute must be undertaken. Prior to that, a discussion of the particular issue where there is a conflict in agreement (SECs) is instructive.

## Socio-economic Considerations

There are a number of areas where the WTO and the CPB conflict in the establishment of trade barriers and in how complaints regarding the justification of those trade barriers can be dealt with. These are set out in Table 8.1. Both the WTO and the CPB allow trade barriers to be put in place to protect human health and the

**Table 8.1** Comparison of WTO and CPB

| Justification for trade barrier | WTO | | CBP | |
|---|---|---|---|---|
| | Complaint | Venue for complaint to be adjudicated | Complaint | Venue for complaint to be adjudicated |
| **Sanitary or phytosanitary concern** | | | | |
| Science-based threat | Scientific justification insufficiently substantiated | WTO dispute settlement mechanism | Scientific justification insufficiently substantiated | No avenue for appeal |
| Risk level unacceptable | Risk assessment not properly conducted | WTO dispute settlement mechanism | Risk assessment not properly conducted | No avenue for appeal |
| Insufficient scientific information (precaution) | Scientific evidence is sufficient | WTO dispute settlement mechanism | Scientific evidence is sufficient | No avenue of appeal |
| Socio-economic considerations | Not allowed in SPS | WTO dispute settlement mechanism | Analysis was flawed | No avenue of appeal |
| **Technical requirement** | | | | |
| Novel product | Like product | WTO dispute settlement mechanism | Not applicable (by definition all GMOs can have trade barriers applied) | Not applicable |
| Production and processing method concern | Not allowed as a justification of trade barrier | WTO dispute settlement mechanism | Not applicable (biotechnology is a PPM and the reason for the CBP | Not applicable |
| Socio-economic considerations | Benefits outweigh the costs | WTO dispute settlement mechanism | Analysis was flawed | No avenue of appeal |
| Agreement allows trade barrier | WTO does not have jurisdiction | WTO dispute settlement mechanism | CBP does not have jurisdiction | International Court of Justice |

environment. In the case of the CPB the country that wishes to export applies to a prospective importing country to be allowed market access for its GMO. The importing country is supposed to examine the available scientific evidence and undertake a risk assessment. It then reports its decision to the applicant. It does not have to justify its decision. There is no mechanism for the applicant country to appeal the decision. The CPB does not have a binding disputes settlement. In essence, the importing country is unconstrained in its ability to establish trade barriers under the CPB (Hobbs et al. 2005). In contrast, the WTO has specific requirements when a country wishes to put trade barriers in place. Further, the WTO has a binding disputes settlements mechanism where countries that do not feel that the justification for the trade barrier is valid can complain and have their complaint adjudicated. The disputes settlement mechanism is binding and if a country does not comply with the dispute panel's ruling, then the WTO can authorize the use of trade sanctions by the country suffering the economic loss arising from the trade barrier. The WTO is the only multilateral organization other than the United Nations Security Council that can authorize the use of trade sanctions (Kerr 2000).

There is a provision in the CPB for disputes to be taken to the International Court of Justice (ICJ). An appeal to the ICJ might arise if an exporting country faced with an importing country imposing a trade barrier justified under the CPB but felt that the product in question did not fall under the ambit of the CPB. This might happen if the exporter used a production method that the member country of the CPB defined as a GMO, but the exporting country did not believe the technology used constituted a biotechnology; in this case, the challenge would be that the importing country exceeded the mandate of the CPB.

The WTO does not allow SECs to be considered in the case of SPS justified barriers and in an only limited way in the TBT. In either case, members of the WTO have an avenue for complaints to be adjudicated. In contrast, SECs are explicitly allowed in the CPB but there is no avenue for any complaint to be adjudicated.

Risk and its mitigation, management and eradication, is an increasingly nebulous topic. As wealth continues to grow, it allows many economies to become focused on risks to their societies that were overlooked during periods when the economy was developing. To a large extent, in the post-war period, societies and economies in Europe were fixated on a 20 year period of rebuilding, reducing the risks of unemployment and overcoming dependence on food imports. It might be argued that not until Rachel Carson's *Silent Spring* (1962), did the importance of other risks begin to creep into the consciousness of industrial societies. As economic recovery became more firmly established, however, other important risks gained prominence in the social conscience and with public policy makers.

While there are identifiable benefits from society's increased focus on risk management and mitigation, one needs to be cognizant of the potential for risk mitigation strategies to be used as a front for trade distorting protectionist policies. Over time, protectionists have been very adept at co-opting policies that justify the protection of their individual vested interests in the name of society's welfare (Kerr and Perdikis 2003). One area where the practice of using disguised trade barriers as a means of science-based risk reduction is in the trade of food and agricultural products. A dominant concern in these regards is the growing practice of using SECs as part of the regulatory risk and biosafety assessment for GM products. The incorporation of SECs into domestic regulatory frameworks has been made available to those countries that adopt and ratify the CPB, especially Article 26. While the focus is specifically on biodiversity and impacts on indigenous communities, the envelope has quite literally been pushed to the margins, as countries have incorporated a wide range of SECs under the guise of Article 26 that have no apparent connection with reducing the risks to either biodiversity or indigenous communities (Falck-Zepeda 2009; Falck-Zepeda and Zambrano 2011; Falck-Zepeda et al. 2013; Smyth and Falck-Zepeda 2013; Ludlow et al. 2014). As part of their review of the socio-economic impacts from GM crops, Lusser et al. (2012, p. 34) note that including SECs as part of the decision-making process provides "more and better information about the technological impact for the decision making process and may help to avoid the adoption of inefficient technologies." However, many would argue that such decisions can and should be left to the marketplace. All new technologies create losers as well as winners and denying a technology on the basis of the existence of losers is simply protecting vested interests at the expense of overall welfare.

The principles of international trade around biotechnology—primarily SPS issues—are based on science, which is something that is agreed to by all Parties to international agreements. The WTO includes the SPS Agreement, whose technical standards framework is provided by the *Codex Alimentarius Commission* (Codex), the World Organization for Animal Health (OIE) and the International Plant Protection Convention (IPPC). The IPPC is a multilateral treaty that seeks to protect natural flora, cultivated plants and plant products from the spread of pathogens through international trade. Through collaboration between regional and national plant protection organizations, it provides a forum for international cooperation, dialogue, harmonization and technical exchange of plant information. The IPPC has addressed the regulation of biotechnology and GM crops though several of the International Standard for Phytosanitary Measures (ISPMs) .

The ISPMs established by the IPPC allows for the implementation of trade barriers under very specific science-based conditions. Following a science-based risk assessment, verifiable evidence is created, documenting the increase risk from allowing a specific traded product to be imported, thus trade in that product can be prevented due to the scientific evidence. Every member country of the WTO is allowed to implement these standards into their domestic regulatory framework without fear of challenge. If a WTO member country implements a regulatory standard that contravenes the IPPC standards, then that country may be accused of erecting a trade barrier in any case brought to the WTO DSB by any other member country. Countries may have higher standards than the IPPC, but only if there is a scientific justification and a risk assessment that satisfies SPS commitments. Annex A.4 of the SPS Agreement (WTO 1995) defines risk assessment as:

> The evaluation of the likelihood of entry, establishment or spread of a pest or disease within the territory of an importing Member according to the sanitary or phytosanitary measures which might be applied, and of the associated potential biological and economic consequences; or the evaluation of the potential for adverse effects on human or animal health arising from the presence of additives, contaminants, toxins or disease causing organisms in food, beverages or feedstuffs.

By way of comparison, Annex III of the CPB (2000, p. 28) defines risk assessment as being, "to identify and evaluate the potential adverse effects of living modified organisms on the conservation and sustainable use of biological diversity in the likely potential receiving environment, taking also into account risks to human health". Even though formal, science-based assessments are to be done to determine risks to the environment and human health, some countries are adopting SECs that would appear to clearly violate Article 26's consistency with international obligations. Numerous SECs are being incorporated into regulatory frameworks that are based on the Precautionary Principle. Those doing so cite Annex III of the CPB which states that "Lack of scientific knowledge or scientific consensus should not necessarily be interpreted as indicating a particular level of risk, an absence of risk, or an acceptable risk" (CPB 2000, p. 28). In essence what the CPB is implying is that if scientific certainty of 100% proof of no harm, ever, is not provided, then it is still acceptable to reject the technology, in this case the registration of a GM crop.

Our argument is that if a country's domestic regulatory system is changed to incorporate SECs, such that the new outcome does not improve environmental safety or plant, animal or human health, then it violates the principles of the SPS Agreement. Table 8.2 lists the SECs identified in a CBD Secretariat expert meeting report, providing examples of SECs that parties to the CPB may take into consideration during their decision-making processes.

Under the WTO, the implementation of new regulations based on the above SECs would constitute a trade barrier in all but a few instances. A few of the SECs in Table 8.2 would have the science-based risk assessment justification to be implemented, including herbicide/pesticide applications and impacts on biodiversity. Risk assessments undertaken on SECs such as ethics, gender impacts, labour, migration and consumer choice would have no effect on the environment or animal, plant or human health and, thus, would be considered under the WTO as unjustified trade barriers.

Disputes by Parties to the CPB will be problematic to resolve. Given that the CPB has no dispute settlement mechanism of its own, Article 34 indicates that disputes between CPB parties will be addressed through Article 27 of the Convention on Biological Diversity (CBD). Article 27 suggests that, first, parties will seek to resolve the dispute through negotiation, and, failing that, parties will engage in third-party mediation. If this still does not resolve the issue, then the dispute is to be submitted to the International Court of Justice.

The referral of trade disputes within the CPB to the ICJ is confusing. Since the ICJ's establishment in 1945, it has never dealt with a case involving agriculture or trade in agricultural products. The most apparent reason for this is the 1947

**Table 8.2** Potential socio-economic considerations

| |
|---|
| Coexistence of LMOs |
| Compliance with biosafety measures, including institutional costs |
| Cultural, spiritual and ethical aspects |
| Economic impacts of changes in application rates and effectiveness of pesticides and herbicides |
| Economic impacts of changes in pest prevalence due to changes in farm management practices |
| Farmers' rights |
| Food security |
| Gender impacts |
| Health-related impacts, including those resulting from changes in the use of pesticides and herbicides |
| Indigenous and local communities impacts on livelihoods, knowledge and biodiversity |
| Impact on the conservation and sustainable use of biodiversity |
| Impacts on consumer choice or consumption patterns |
| Impacts on market access and trade at national and international levels |
| Labor and employment |
| Land tenure |
| Macroeconomic impacts, including those on sustainable development |
| Microeconomic impacts at the individual, household or community level |
| Rural-urban migration |

*Source* adapted from UNEP (2010)

establishment of the General Agreement on Tariffs and Trade (GATT) that was purposely created to deal with the problem of how tariffs were being used to affect international trade. Given that GATT's *raison d'etre* was to resolve trade disputes, it would be illogical from the ICJ perspective to duplicate what GATT was doing and hence, the ICJ and GATT (now the WTO) have managed to ensure that duplication of efforts does not occur. Nevertheless, the IJC might rule on whether a policy exceeded the ambit of the CBP.

The challenge of trade in GM products and growing implementation of SECs is that ultimately trade will be negatively impacted. While trade could be disrupted between two countries that have adopted, produce and export GM products, as some adopting countries have included SECs as part of their regulatory framework (Falck-Zepeda et al. 2013), it is more likely there will be trade disruptions between countries which produce and export GM crops and other countries that do not and use SECs as a means of blocking biosafety approval for planting or food application in that country. The fact that the US, Argentina and Canada, that collectively have 59% of the GM production area, are not a party to the CPB, and the US is not even a party to the CBD, suggests that it is not feasible to have a trade dispute resolved via the dispute settlement process outlined in the CPB and the CBD, solely based on membership. Table 8.3 provides an overview of the countries producing GM crops and their institutional affiliations.

While the complexities of a trade dispute over GM products involving Sudan would appear to be substantial, given their lack of WTO membership, the larger concern is about whether a GM producing nation could file a case with the WTO regarding a non-GM crop adopting country implementing a SEC that is challenged

**Table 8.3** GM crop adopting national and affiliations

| GM adopting countries | Adopted and ratified CPB | Adopted and ratified CBD | Member of WTO |
|---|---|---|---|
| USA | No | No | Yes |
| Brazil | Yes | Yes | Yes |
| Argentina | No | Yes | Yes |
| Canada | No | Yes | Yes |
| India | Yes | Yes | Yes |
| China | Yes | Yes | Yes |
| Paraguay | Yes | Yes | Yes |
| South Africa | Yes | Yes | Yes |
| Pakistan | Yes | Yes | Yes |
| Uruguay | Yes | Yes | Yes |
| Bolivia | Yes | Yes | Yes |
| Philippines | Yes | Yes | Yes |
| Australia | No | Yes | Yes |
| Burkina Faso | Yes | Yes | Yes |
| Myanmar | Yes | Yes | Yes |
| Mexico | Yes | Yes | Yes |
| Spain | Yes | Yes | Yes |
| Chile | No | Yes | Yes |
| Columbia | Yes | Yes | Yes |
| Honduras | Yes | Yes | Yes |
| Sudan | Yes | Yes | No |
| Portugal | Yes | Yes | Yes |
| Czech Republic | Yes | Approved | Yes |
| Cuba | Yes | Yes | Yes |
| Egypt | Yes | Yes | Yes |
| Costa Rica | Yes | Yes | Yes |
| Romania | Yes | Yes | Yes |
| Slovakia | Yes | Approved | Yes |
| Viet Nam | Yes | Yes | Yes |
| Bangladesh | Yes | Yes | Yes |
| % total | 83 | 97 | 97 |

*Source* ISAAA (2015), CPB (2016), CBD (2016), WTO (2016)

as a trade barrier and whether the non-GM adopting nation could argue (or successfully argue) that the CPB should, according to the Vienna Convention on the Law of Treaties, be recognized as the binding agreement by which the dispute should be administered.

## Critical Assessment of SEC Regulations

Both the WTO and CPB definitions of risk assessment allow for the evaluation for the potential of the risk being assessed to be harmful to humans, animals or the environment. For CPB countries to be compliant to their international obligations, they will have to ensure that they adhere to the WTO's SPS Agreement (WTO 1995) definition of risk assessment:

> The evaluation of the likelihood of entry, establishment or spread of a pest or disease within the territory of an importing Member according to the sanitary or phytosanitary measures which might be applied, and of the associated potential biological and economic consequences; or the evaluation of the potential for adverse effects on human or animal health arising from the presence of additives, contaminants, toxins or disease-causing organisms in food, beverages or feedstuffs.

Given that Article 26 of the CPB stipulates that all SECs have to be consistent with international obligations, and the WTO requires science-based risk assessments, the following SECs are assessed according to their ability to reduce the risk to the environment or plant, animal or human health due to the commercial production of GM crops. Where SECs are unable to verifiably reduce risks of this nature, then their inclusion as part of a biosafety assessment for GM crops or products would be deemed to be a violation of the country's WTO commitments and obligations.

### *Labor Impacts*

Labor is one of the SECs identified by the UNEP report as shown in Table 8.2. Labor has a long history of being impacted by innovative processes and products. For example, those employed in the production of clothing and its various related products have been impacted by innovation dating back to new inventions in spinning wool and cotton from the latter part of the 1700s. Agricultural production in many developing countries is a labor intensive process. While the amount of land operated by small landholders is very modest compared to those typical in modern market economies, the lack of mechanized equipment for seeding, weed management and harvest results in many of these aspects of farming being done by hand. Figure 8.1 illustrates that several developing countries have over two workers per hectare of land, while at the other end of the scale, countries such as Canada and the US have 0.05 workers per hectare.

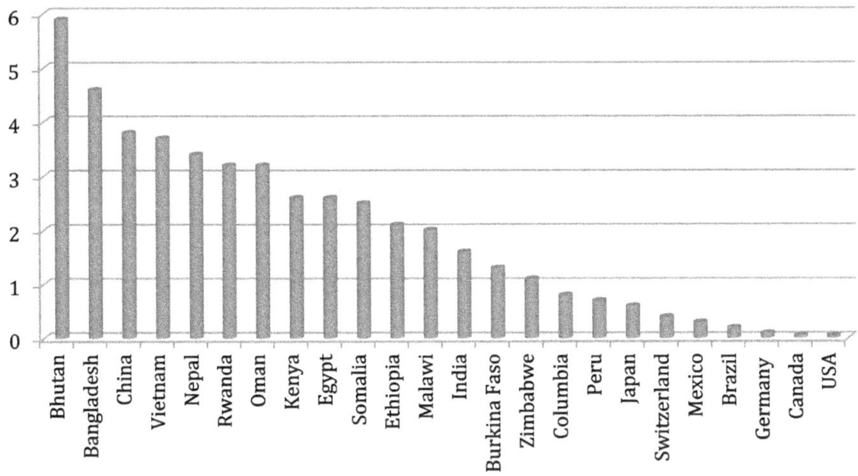


**Fig. 8.1** Farm workers per hectare of arable and permanent crop land. *Source* World Resource Institute (2013)

In South Africa, the average small landholder farm is only several hectares in size. Given the small size of the fields, weeding is done by hand. Gianessi (2009) reports that to prevent weed damage, conventional maize requires 276 h of hand weeding per hectare, sorghum 150 h/ha, cotton 200–400 h/ha and rice 200–418 h/ha. Gouse (2012) found that smallholder maize farmers in South Africa that had adopted GM maize, spent on average 267, 177 and 152 h/ha on manual weeding over three seasons. Weeding is typically done by women and Gouse further found that the female household members indicated that the labor saving in the maize field enabled them to spend more time at home, caring for children or in their vegetable plots.

Gouse (2014) identifies two methods that can be used to assess labor impacts: computable general equilibrium (CGE) models; and a social accounting matrix (SAM). In an ideal scenario, disaggregated time series rural household data would be utilized to assess the labor impacts of GM crops. Gouse acknowledges that the biggest challenge to undertaking these assessments is the lack of data in most developing countries. In instances where data does exist, it frequently lacks a level of detail sufficient enough to perform simulations. To address these insufficiencies, data would have to be gathered to undertake the assessment, requiring 2–4 years of field research.

While an analysis of this kind provides useful information about labor impacts, it does not provide quantifiable information on how risks to the environment or plant, animal or human health can be reduced by undertaking such an assessment. Given that none of these risks can be quantified through a labor impact assessment, any inclusion of this SEC would violate a country's WTO obligations. It represents de facto economic protectionism.

## *Intellectual Property Rights*

Intellectual property rights (IPRs) are also identified in the UNEP (2010) report as a socio-economic issue that should be taken into consideration as part of the decision-making process for GM products. Currently, there are several sources of IP protection that can be applied to GM crops. Plant breeders rights' are available through the Union for the Protection of New Varieties of Plants (UPOV) agreements, while patent rights are protected through the WTO Agreement on Trade Related Aspects of Intellectual Property Rights (TRIPS) , which also provides minimum standards for copyright and related rights, trademarks, geographical indications, industrial designs, layout-designs of integrated circuits and undisclosed information.

Lawson (2014, p. 180) posits that the risk assessment of IPRs under the guise of the CPB "is best framed as whether there are exemptions, exceptions and flexibilities in the international IP agreements that might accommodate decisions involving SECs made under the CPB." Lawson suggests that when the issue is examined from this perspective, there are a range of exceptions and flexibilities where SEC might address IP. Lawson offers a four-step methodology to undertake an IP risk assessment: first, identify what IP is embodied in the LMO; second, identify who owns the IP; third, test to determine if the IP is enforceable in the jurisdiction; and fourth, assess whether the CPB's SECs affect the IP.

Including IP as part of a risk assessment for a GM crop or product would require the ability to access IP data. This data can be found in various online IP databases in developed countries, such as the European Patent Office, the Canadian Intellectual Patent Office and the US Patent and Trademark Office. Searches of these data bases can be conducted to identify what IP is held, by whom and as to how it applies to which products.

Gaining greater knowledge about the IP context for a GM crop or product undergoing a risk assessment may be useful for post-commercialization reasons, but in no way can it obviously reduce the risk to the environment or plant, animal or human health. As such, the inclusion of a SEC based on intellectual property would be in violation of the WTO SPS Agreement and might effectively operate as a TBT that infringes the principle of national treatment.

## *Religious/Cultural*

While ethics is often included with religious and cultural aspects pertaining to biotechnology, Coe (2014) acknowledges that it can be problematic in trying to separate ethics due to it being quite interwoven with religious and cultural aspects. However, Coe (2014, p. 249) suggests that two key themes set religious and cultural aspects apart, "the concepts of sacredness and the desire for happiness and well-being" are intrinsic to religious and cultural meaning. Given that discussions

**Table 8.4** Democratic engagement on biotechnology

| Country | Year | Topic |
|---|---|---|
| Argentina | 2000 | Genetically modified foods |
| Australia | 1999 | Gene technology in the food chain |
| Belgium | 2003 | Genetically modified foods |
| Brazil | Not available | Genetically modified foods |
| Canada | 1999 | Food biotechnology |
| Denmark | 1987 | Gene technology in industry and agriculture |
| Denmark | 1992 | Transgenic animals |
| Denmark | 1999 | Genetically modified foods |
| France | 1998 | Genetically modified foods |
| India | 2000 | Genetically modified foods |
| Japan | 2000 | Genetically modified foods |
| Netherlands | 1993 | Genetically modified animals |
| New Zealand | 1996 | Plant biotechnology |
| New Zealand | 1999 | Plant biotechnology |
| Norway | 1996 | Genetically modified foods |
| South Korea | 1998 | Safety and ethics of genetically modified foods |
| Switzerland | 1999 | Genetic engineering and food |
| United Kingdom | 1994 | Genetically modified foods |
| Unites States | 2002 | Genetically modified foods |

*Source* The Loka Institute (2013)

involving religious and cultural aspects of an innovative technology can be emotionally-laden and subject to misinformation, numerous countries have used citizen consultations to disentangle them. Table 8.4 provides a summary of countries that have undertaken democratic engagement methods as a part of addressing social concerns regarding biotechnology.

Democratic engagement is most frequently conducted through consensus conferences (Medlock and Einsiedel 2014). This method involves advertising for citizens willing to participate in such an event. Once the group of citizens is formed, they are given background information and provided with presentations on various relevant topics, allowing for question and answer periods. Discussions are then held and the citizens write a consensus document that serves as the report of the process. While consensus is not a rigid requirement for the process, it is stressed as an important part of the process and that considerable time should be devoted to trying to resolve differences, before moving on and recognizing that consensus could not be reached on specific issues. Irreconcilable or contentious issues signal to governments those aspects of the greatest challenge for society to adjust to or manage.

Based on the WTO definition of risk assessment, there is no ability within democratic engagement models to quantify any of the ethical aspects of risk, as required by the RAF protocols. Therefore, the inclusion of religious and cultural perspectives as SECs for a biosafety assessment cannot be science-based and would be a violation of WTO obligations.

## *Market Access and Trade*

Market access and assurance that international trade will not be disrupted have become dominant socio-economic concerns for many developing nations and are a major reason for the lack of GM crop adoption in Africa (Paarlberg 2008; Smyth et al. 2013). Inadvertent mixture of bulk agricultural commodities is an inevitable part of international seed and commodity trade, which is why thresholds have been established that allow for the low level presence of undesirable items such as weed seeds, stones, insect fragments and other crop varieties. Meeting norms for levels of comingling is particularly problematic for African countries that have strong export market connections with countries inside of the European Union (EU), as the EU presently has a 0.1% tolerance for GM events not approved for import into the EU. Ensuring that comingling is adequately managed to prevent trade problems for those exports destined for non-GM markets raises a range of SECs, frequently bundled into policies related to coexistence between conventional, GM and organic crops.

Gruere (2014) suggests different methods would be required to examine the domestic coexistence aspect and the international trade component. Viju (2014) extends this discourse at the international level and posits that there are three methods that can be utilized to determine if the commercialization of a GM crop would be welfare enhancing in specific trade markets: partial-equilibrium models; general equilibrium models; and gravity models.

Gruere and Sengupta (2009) offer a model that can be utilized to undertake a market risk assessment. Figure 8.2 illustrates a five-step process that determines whether, and to what extent, market risk would exist when faced with the commercialization of a GM crop. The challenge from a science-based risk assessment

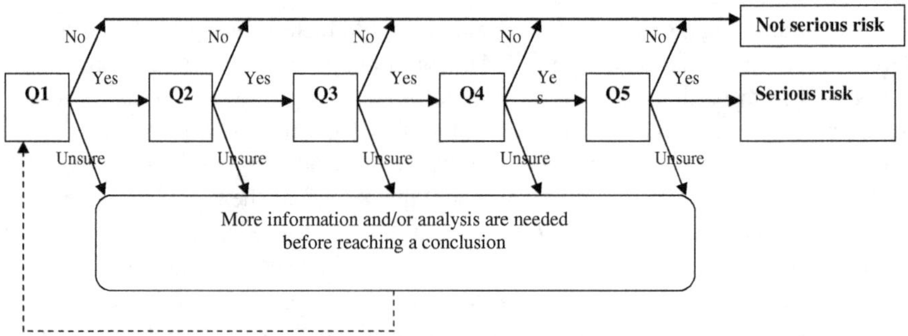


**Fig. 8.2** Market risk assessment decision tree. *Notes Q1* Is the alleged risk substantiated? *Q2* Are export losses likely with the decision? *Q3* Are presumed export losses non-negligible for the country? *Q4* Is the risk unavoidable? *Q5* Is the risk greater than the benefits? *Source* Gruere and Sengupta (2009)

perspective is that the questions posed at each of the five steps are very subjective questions that are susceptible to manipulation such that the results could routinely establish that the market risk of commercializing a GM crop would be dramatically larger than the benefits that would be identified. As with any decision to commercialize an innovation, there will ultimately be those that adopt the innovation and gain a share of the benefits and there will be individuals that chose to not adopt the innovation and, in the classical economic sense, be said to be 'losers' and share the negative externalities of the innovation. No innovation is without externalities that to some extent generate negative impacts on some component of society.

While many would conclude that this style of risk assessment is important prior to commercializing products, the effects can be less about managing environmental and safety risk and more about protecting market positions. The market response to the pending introduction of GM wheat in Canada and the US in 2003, which stalled the commercialization of the technology, illustrates the risk of opening up review beyond strict science-based aspects of risk assessment. Should market access and trade be included in a biosafety regulatory decision-making process, it probably would not comply with the WTO.

## *Consumer Choice*

The heart of the debate pertaining to consumer choice depends on whether a country labels GM foods and products. Canada does not mandate labeling of GM products, whereas the EU mandates that all products containing greater than 0.9% GM ingredients must be labeled as GM. The obvious place to start is with whole foods. The two longest-standing examples are GM sweet corn and GM papaya. GM sweet corn was commercialized in 2012 and has reached an estimated market share of 40% (Non-GMO Project 2013), while GM papaya was introduced in 1998 and has reached 82% market share (Gonsalves and Gonsalves 2014). In 2015–16, Canada and the US approved new GM potatoes and apples. Processed foods are more of a challenge. The reality is that there are virtually no products labeled as GM products available for sale in either Canada or Europe—producers have engineered their processes to minimize the required disclosers—while in the US a few product categories, such as breakfast cereals and soups, have begun to be labeled to comply with new legislation that came into effect in 2016.

The issue of labeling, whether it is mandatory or voluntary, spans the full spectrum of opinions. Environmental groups and critics of biotechnology claim that greater than 95% of consumers responding to surveys asking about GM labeling indicate that they want GM content to be labeled, but other surveys show that only 2% of unprompted consumers ask for GM labeling. The real demand for labeling lies somewhere between the two; determining whether it is greater or less than 50%

should help to determine what type of labeling is optimal. Surveys are needed to determine whether or not consumers would want foods labeled as GM and second whether what consumers would be willing to pay, or conversely, require as a price discount, to purchase GM products.

There are several survey methodologies one might use to gather consumer data regarding acceptance to GM foods and products. Methods can range from in-store surveys that consumers voluntarily complete to more detailed phone surveys regarding grocery purchasing habits. Moses and Fischer (2014) provide a review of the surveys that have been utilized in gather consumer attitudes to GM foods and products. Lusser et al. (2012) discuss the two main methods used to determine consumer willingness to pay: contingent valuation, which poses questions about hypothetical purchase intentions; and revealed preference, which increasingly uses experimental auctions to explore preferences.

Deferring back to the WTO definition of risk assessment, none of these risk assessment methods quantify risk reductions for the environment or plant, animal or human health. In accordance with the WTO SPS Agreement, if a risk assessment cannot document science-based evidence of reducing this nature of risks, then there is no scientific basis for risk assessments of this nature and would not be in accordance with WTO obligations.

These examples serve to illustrate that the inclusion of SECs lacking science-based methodologies into biosafety regulatory frameworks will most likely violate a country's WTO obligations, and, moreover, would be contradictory to Article 26 of the CPB, that asserts that the proposed SECs should be "consistent with international obligations" (CPB 2000, p. 19).

## Policy Implications

Two dominant policy issues stand out. First, what course of action will follow if a leading GM crop producing, non-CPB nation such as Argentina, Canada or the USA decides that a SEC regulation of a non-GM cultivating CPB party nation is a trade barrier? Second, what course of action will follow if a GM crop producing CPB party challenges a non-GM crop producing CPB nation's SEC regulations as a trade barrier?

In the first scenario, the Vienna Convention on the Law of Treaties would establish that since the three GM crop producing nations are not parties to the CPB, then the WTO would be the international agreement that would take precedence. The Dispute Settlement Body (DSB) of the WTO would be the mechanism engaged to resolve this international trade dispute, such as has been the case in multiple previous of trade disagreements. The advantage of the DSB is that any trade dispute is addressed expeditiously and the resolution is definitive, including offering the authority to levy sanctions against the nation in violation.

The scenario where both nations are members of the WTO as well as the CPB raises considerably more interesting circumstances. With both nations being party to the CPB, a newer international agreement on the trade and regulation of GMOs, both nations could choose to use the CPB in an effort to resolve the dispute. The lack of a DSB makes this option considerably more nebulous as negotiation is to first be employed as a means of resolution, followed by third-party mediation. Within this process, there is no binding decision-making process, hence it is essentially a mechanism to extend and draw out a contentious trade issue, without ultimately having a venue to resolution.

This scenario could present itself, even though the use of SEC-based regulation as defined above would violate the SPS Agreement, particularly as nations utilizing SECs have not defined the impact necessary to trigger a SEC assessment. Article 26 allows Parties to employ SECs only when GMOs will impact the conservation and sustainable use of biodiversity. Typically, SECs are being applied as a parallel regulation with science-based regulation, without the identification, let alone the quantification, of an impact on biodiversity.

The potential to engage the CPB as the international agreement to resolve a trade dispute involving GM crops raises several concerns. First, depending on the value of trade between the two nations, pressure could be brought to bear on the nation to which trade is of greater importance to have to engage the CPB as the resolution mechanism. This may limit the ability of nations to act with complete sovereignty if they are required to capitulate to the pressure brought by an important trading partner and engage the CPB rather than the WTO to resolve a GM crop trade dispute. Second, the lack of a defined process, such as that resulting from using the WTO's DSB, results in an ongoing process that can readily be subject to manipulation to delay, prolong and frustrate any attempt to reach a resolution. The negotiation could turn into a lengthy process that might take years to conclude, all the while allowing a protectionist country to reap the benefits of an unfair trade barrier. A final concern is that establishing such a course of action would set a precedence and could work to undermine the credibility of the WTO in trade disputes involving GM crops or products where both nations are party to the CPB.

Ultimately, it would appear that the stage is being set for a showdown of sorts, between those nations adopting GM crops using science-based regulatory frameworks that rely on the WTO for the resolution of disputes and those nations using the socio-economic considerations approved in the CPB to block trade. Historically, the engagement of GATT and now the WTO have provided predictable and generally accepted international governance of trade in an era of high support for science. More recently, trust in and support for science has waned, allowing for competing international governance capacity to develop. Whether these two competing bodies for international governance in the trade of GMOs can coexist remains to be seen. As it stands now, nearly all of the signs suggest they are on a collision course.

## References


Belcher, K. 2007. Trade agreements and multilateral environmental agreements. In *Handbook on International Trade Policy*, ed. W.A. Kerr and J.D. Gaisford, 427–438. Cheltenham: Edward Elgar.

Caldwell, D.J. 1998. Multilateral environmental agreements and the GATT/WTO Regime. Washington, DC: US National Wildlife Federation. Online at: http://www.iatp.org/documents/multilateral-environmental-agreements-and-the-gattwto-regime-0.

Candeira, A.L., A.M. Canellas, M.A. Recuerda, E.A. Garcia, F.T. Botija, P.D. Peralta, L.R. Ghisleri, M.R. Marinez-Larranaga, and A. Anadon. 2009. GMO's and the resolution of conflicts under the WTO. *European Food and Feed Law Review* 4: 282–294.

Carson, R. 1962. *Silent spring*. Boston: Houghton Mifflin.

Cartagena Protocol on Biosafety. 2016. Parties to the protocol and signature and ratification of the supplementary protocol. Online at: http://bch.cbd.int/protocol/parties/.

Cartagena Protocol on Biosafety to the Convention on Biological Diversity. 2000. Montreal: Secretariat of the Convention on Biological Diversity.

Coe, A. 2014. Culture and religion. In *Socio-economic considerations in biotechnology regulation*, ed. K. Ludlow, S.J. Smyth, and J.B. Falck-Zepeda, 247–258. New York: Springer.

Convention on Biological Diversity. 2016. List of parties. Online at: http://www.cbd.int/convention/parties/list/default.shtml.

Falck-Zepeda, J. 2009. Socio-economic considerations, Article 26.1 of the Cartagena Protocol on Biosafety: What are the issues and what is at stake? *AgBioForum* 12 (1): 90–107.

Falck-Zepeda, J.B., and P. Zambrano. 2011. Socio-economic considerations in biosafety and biotechnology decision making: The Cartagena protocol and national biosafety frameworks. *Review of Policy Research* 28 (2): 171–195.

Falck-Zepeda, J.B., J. Wesseler, and S.J. Smyth. 2013. The current status of the debate on socio-economic assessments and biosafety highlighting different positions and policies in Canada and the US, the EU and Developing Countries. *World Review of Science and Technology and Sustainable Development* 10 (4): 203–227.

Gianessi, L. 2009. Solving Africa's weed problem: Increasing crop production and improving the lives of women. Crop Protection Research Institute.

Gonsalves, C., and D. Gonsalves. 2014. The Hawaii papaya story. In *Handbook on agriculture, biotechnology and development*, ed. S.J. Smyth, P.W.B. Phillips, and D. Castle, 642–660. Cheltenham, UK: Edward Elgar Publishing.

Gouse. M. 2012. Farm-level and socio-economic impacts of a genetically modified subsistence crop: The case of smallholder farmers in KwaZulu-Natal, South Africa. Submitted in partial fulfillment of the requirements for the degree Ph. D. in Department Agricultural Economics, Extension and Rural Development, University of Pretoria.

Gouse, M. 2014. Labor impacts. In *Socio-economic considerations in biotechnology regulation*, ed. K. Ludlow, S.J. Smyth, and J.B. Falck-Zepeda, 189–200. New York: Springer.

Gruere, G.P. 2014. National trade interests. In *Socio-economic considerations in biotechnology regulation*, ed. K. Ludlow, S.J. Smyth, and J.B. Falck-Zepeda, 215–227. New York: Springer.

Gruere, G.P., and D. Sengupta. 2009. Biosafety and perceived commercial risks: The role of GM free private standards. IFPRI Discussion Paper 00847. Washington, DC: International Food Policy Research Institute. http://www.ifpri.org/sites/default/files/publications/ifpridp00847.pdf.

Hagen, P., and J.B. Weiner. 2000. The Cartagena protocol on biosafety: New rules for international trade in living modified organisms. *Georgetown International Environmental Law Review* 12: 698–717.

Hobbs, A.L., J. Hobbs, and W.A. Kerr. 2005. The biosafety protocol: Multilateral agreement on protecting the environment or protectionist club? *Journal of World Trade* 39 (2): 281–300.

Holtby, K.L., W.A. Kerr, and J.E. Hobbs. 2007. *International environmental liability and barriers to trade*. Cheltenham: Edward Elgar.

International Service for the Acquisition of Agri-biotech Applications. 2015. 20th Anniversary (1996 to 2015) of the Global Commercialization of Biotech Crops and Biotech Crop Highlights in 2015. ISAAA Brief No. 51. ISAAA: Ithaca, NY.

Kerr, W.A. 2000. Is it time to re-think the WTO?—A return to the basics. *Journal of International Law and Trade Policy* 1 (2): 99–107.

Kerr, W.A. 2003. Science-based rules of trade—A mantra for some, an anathema for others. *Journal of International Law and Trade Policy* 4 (2): 86–97.

Kerr, W.A. 2010. *Conflict, chaos and confusion—The crisis in the international trading system.* Cheltenham: Edward Elgar.

Kerr, W.A. 2012. Taming the bear: The WTO after the accession of Russia. *Journal of International Law and Trade Policy* 13 (2): 150–159.

Kerr, W.A., and S.L. Hall. 2004. Multilateral environmental agreements and agriculture: Commitments, cooperation and conflicts. *Current Agriculture, Food and Resource Issues* 5: 39–52.

Kerr, W.A., and N. Perdikis. 2003. *The economics of international business: A guide to the global commercial environment.* Estey Centre Program in International Trade Education, Saskatoon: Estey Centre for Law and Economics in International Trade.

Lawson, C. 2014. Intellectual property. In *Socio-economic considerations in biotechnology regulation*, ed. K. Ludlow, S.J. Smyth, and J.B. Falck-Zepeda, 177–188. New York: Springer.

Ludlow, K., S.J. Smyth, and J.B. Falck-Zepeda (eds.). 2014. *Socio-economic considerations in biotechnology regulation.* New York: Springer.

Lusser, M., T. Raney, P. Tillie, K. Dillen, and E. Rodrigues-Cerezo. 2012. *International workshop on socio-economic impacts of genetically modified crops: Co-organised by JRC-IPTS and FAO.* Luxembourg: Publications Office of the European Union.

Medlock, J., and E. Einsiedel. 2014. Democratic engagement. In *Handbook on agriculture, biotechnology and development*, ed. S.J. Smyth, P.W.B. Phillips, and D. Castle, 522–537. Cheltenham, UK: Edward Elgar Publishing.

Moses, V., and S. Fischer. 2014. Consumer choice. In *Socio-economic considerations in biotechnology regulation*, ed. K. Ludlow, S.J. Smyth, and J.B. Falck-Zepeda, 67–79. New York: Springer.

Neff, R.A. 2004. The Cartagena protocol and the WTO: Will the EU biotech products case leave room for the protocol? *Fordham Environmental Law Review* 16: 261–288.

Non-GMO Project. 2013. Information on GMO sweet corn. Online at: http://www.nongmoproject.org/learn-more/sweetcorn/.

Paarlberg, R. 2008. *Starved for science: How biotechnology is being kept out of Africa.* Harvard, MA: Harvard University Press.

Phillips, P.W.B., and W.A. Kerr. 2000. Alternative paradigms—The WTO versus the biosafety protocol for trade in genetically modified organisms. *Journal of World Trade* 34(4):63–75.

Safrin, S. 2002. Treaties in collision? The biosafety protocol and the world trade organization agreements. *American Journal of International Law* 96: 606–627.

Sinclair, I. 1973. *The vienna convention on the law of treaties*, 1st ed. Manchester: Manchester University Press.

Sinclair, I. 1984. *The Vienna convention on the law of treaties*, 2nd ed. Manchester: Manchester University Press.

Smyth, S.J., J. McDonald, and J. Falck-Zepeda. 2014. Investment, regulation and uncertainty: Managing new plant breeding techniques. *GM Crops and Food: Biotechnology in Agriculture and the Food Chain* 5(1):1–14.

Smyth, S.J., and J.B. Falck-Zepeda. 2013. Socio-economic considerations and international trade agreements. *The Estey Centre Journal of International Law and Trade Policy* 14 (1): 18–38.

Smyth, S.J., W.A. Kerr, and P.W.B. Phillips. 2013. Accelerating adoption of GM crops through a trade liability regime. *Plant Biotechnology Journal* 11 (5): 527–534.

The Loka Institute. 2013. Tracking danish-style, citizen-based deliberative consensus conferences worldwide. Online at: http://www.loka.org/TrackingConsensus.html.

United Nations Environment Programme. 2010. Summary report on the survey on the application of and experience in the use of socio-economic considerations in decision-making on living modified organisms. UNEP/CBD/BS/COP-MOP/5/INF/10. Online at: http://www.cbd.int/doc/meetings/bs/bsws-sec-01/information/bsws-sec-01-mop-05-inf-10-en.pdf.

Vallely, P.J. 2004. Tension between the Cartagena protocol and the WTO: The significance of recent WTO developments in an ongoing debate. *Chicago Journal of International Law* 5: 369–378.

Vienna Convention on the Law of Treaties. 1969. Online at: http://untreaty.un.org/ilc/texts/instruments/english/conventions/1_1_1969.pdf.

Viju, C., M.T. Yeung, and W.A. Kerr. 2011. Post moratorium EU regulation of genetically modified products—Triffid flax. CATPRN Commissioned Paper, No. 2011–03, Canadian Agricultural Trade Policy and Competitiveness Research Network. http://www.uoguelph.ca/catprn/PDF-CP/CP2011-03-viju-yeung-kerr.pdf.

Viju, C. 2014. Market access and trade. In *Socio-economic considerations in biotechnology regulation*, ed. K. Ludlow, S.J. Smyth, and J.B. Falck-Zepeda, 201–213. New York: Springer.

World Resource Institute. 2013. Workers per hectare by country. Online at: http://www.NationMaster.com/graph/agr_wor_per_hec-agriculture-workers-per-hectare.

World Trade Organization. 1995. The WTO agreement on the application of sanitary and phytosanitary measures (SPS Agreement). Online at: https://www.wto.org/english/tratop_e/sps_e/spsagr_e.htm.

World Trade Organization. 2016. Members and observers. Online at: http://www.wto.org/english/thewto_e/whatis_e/tif_e/org6_e.htm.

# Part III
# Breaking the Gridlock

# Chapter 9
# Refining the Risk Analysis Framework

**Abstract** This chapter investigates the changing methods of risk assessment for agricultural products and notes a deterioration in the likelihood that risk assessment exercises will be completed successfully. Genetically modified products are used as an example. The changing nature of risk assessments is found to be inhibiting international market access, reducing trade and, hence, making investments in productivity enhancing technologies in agriculture less financially attractive. This changing dynamic of risk assessment has negative impacts on food security.

## Introduction

There are three major elements to food security: (1) food availability; (2) incomes that are sufficient enough to acquire food; and (3) increased agricultural productivity. An individual is not food secure if safe food is not available, especially on a daily basis. Even if food is available in an individual's locale they are not food secure unless they have sufficient income to purchase it. As global population continues to expand, ensuring that enough food is available and affordable requires that productivity in food production at least keeps pace with the rising number of mouths to feed. As noted earlier, this undoubtedly requires global trade in food to balance those areas unable to meet their needs locally. Otherwise there will be more and more individuals chasing ever scarcer food—which can only lead to higher prices, lower availability and food insecurity for some. Unfortunately, the circumstances at the start of this century do not bode well for the achievement of food security by mid-century. A report by the Food and Agriculture Organization (FAO 2006) estimated that the number of food-insecure people in 2006 was 845 million. The report forecast that the food insecure population would decline by one-third to just under 600 million by 2015. The stark reality is that food insecurity has grown by over 40% and stood at approximately 1.2 billion people (Nature 2010). More recent estimates place this figure as high as 2 billion. Of course, there are many changes that could be made that would improve how the food currently produced is used and distributed, but ultimately if the Malthusian Trap is to be avoided, agricultural productivity must increase.

 
S.J. Smyth et al., *Biotechnology Regulation and Trade*, Natural Resource Management and Policy 51, DOI 10.1007/978-3-319-53295-0_9

Food security has gained prominence recently due to the rapid increases in food prices globally in 2007 and 2008, which led to a rising awareness regarding the challenges of satisfying a global population forecast to rise by more than one-third to approximately 9 billion by 2050 (Beddington 2010; Evans 2009). Rising incomes combined with population increases lead to projections that global demand for food may double in the same period. Meeting this increase in demand presents a major global challenge for the agri-food system. Indeed, the Deputy-Director of the FAO has observed, "agricultural production needs to increase by 70% worldwide, and by almost 100% in developing countries, in order to meet growing food demand" (Tutwiler 2011). Meanwhile, there is considerable evidence that there has been serious underinvestment in agricultural research and development over recent decades (Alston et al. 2009; James et al. 2008). Even if investment in research and development in agriculture could be increased to eventually backfill the current shortfall, there are considerable lags—of up to 25 years—between when investments are made and increases in productivity are fully manifest (Alston 2010). Climate change and the current reality that a considerable proportion of agricultural land is being farmed in ways which are unsustainable means that even greater investments than in the past will likely be needed to enhance productivity (Davies 2013; Hertel 2011; Kulthreshtha 2011; Tilman et al. 2011). Thus, future food security requires we look at all the impediments to higher agricultural productivity —systemic problems that inhibit advances in productivity should be a priority.

There are a wide range of reasons for underinvestment in agricultural research, including: governments' fiscal difficulties (Gaisford et al. 2001); the inability to capture the full benefits arising from investments due to 'out of country' leakages (Alston 2002); misaligned incentives (Malla and Gray 2005); resistance to technological change (Haggui et al. 2006); poor protection for intellectual property (Cardwell and Kerr 2008); high transaction costs in identifying and acquiring existing intellectual property (Smyth and Gray 2011); and long, complex and costly regulatory processes for new technologies (Smyth et al. 2004). These inhibitors of investment in agricultural research have received considerable academic scrutiny. One further factor that can negatively impact productivity enhancing investments for agriculture is the process used for assessing risk. Prior to receiving permission for commercialization, new technologies and their products must be assessed for the risks they may pose. To move to commercial production, a product must be judged to pose a sufficiently low degree of risk to be acceptable to society (Phillips et al. 2006). Over the last 20 years the process of risk assessment has been evolving and diverging geographically. The major spur for the diverging treatment of risk has been the advent of modern agricultural biotechnology (Isaac 2007). The rift over modern agricultural biotechnology is often portrayed as a disagreement between the European Union (EU) and the United States and, while they be the major champions of the divergent approaches to risk, the rift has global implications for investments in agricultural technologies and food security over the long run (Isaac and Kerr 2007a).

An increase in the likelihood that a new product or process will not be considered safe enough to be commercialized will reduce the appetite—both public and private—to make the required investment. This is true whether or not the concerns have a legitimate rationale. In private decisions relating to investments, expected benefits are usually explicitly weighed against the expected costs associated with the investment (Smyth et al. 2013). The higher the probability of failing to reach an acceptable level of risk, the smaller the expected benefits will appear to be and, at the margin, less investment will be made (Kerr and Yampoin 2000; Smyth et al. 2013). Similarly, if over time, part of a market has a reduced likelihood of achieving an acceptable level of risk, the total value of the expected benefits is reduced (Gaisford et al. 2001). All of this inhibits investment in R&D in agriculture, reducing the rate of technological change just when it is most needed to ensure food security (Smyth et al. 2011). While agricultural biotechnology has been the key technology that has led to changes in risk evaluation, once new methods are accepted there is an increased chance that they will be applied to all new agricultural innovations in products or processes, which could jeopardize efforts to address food security as we move towards 2050.

## International Scope of Science-Based Risk Assessment

The risk evaluation systems operating in most modern market economies have generally been scientifically-based processes that combine the identification and characterization of hazards with assessments of exposure to characterize risk. In essence, they purport to objectively assess the probabilistic outcomes of discrete adverse events, for the most part abstracting from issues related to risk management or risk communications. The practice is that governments establish a risk threshold for products, or classes of products, to enable them to reject new products with unacceptable risks but allow those with acceptable impacts to enter the market.

Traditional risk assessment theory suggests that risk is a combined measurement of the degree of exposure multiplied by the hazard; that is the level of adverse effects of the agent on other organisms. This can be expressed as:

$$\text{RISK}^{\text{scientific}} = \text{HAZARD} \times \text{EXPOSURE}$$

The acceptable threshold for risk for any product is usually set externally from the risk assessment, defined through a process of scientific, political and commercial dialogue. If an assessment is conducted and the level of risk was determined to be higher than was judged to be acceptable, then government agencies would not approve the technology or product for release and use. While the hazard would appear to be quite objectively derived through risk assessment by the global scientific community, the acceptable levels and the estimated relative level of risk for a product can and do vary widely between intended uses (Tengs et al. 1995). Hence, it is not unreasonable to expect to see different levels of risk accepted in different circumstances.

There has been significant effort put into understanding the divergence between objectively assessed risks (the old science-based model) and what many call socially constructed risks. Sandman (1994) believes the old formula under estimated the perceived level of risk because it ignored the public response to a risk, which he termed 'outrage'. He argues that regulators should instead use the following formula for understanding consumer perceptions of risk:

$$\text{RISK}^{\text{socially constructed}} = \text{HAZARD} \times \text{OUTRAGE}$$

Sandman (1994) suggests that public concern is focused on whether the risk is acceptable rather than on the scientifically perceived incidence of that risk. While that model accommodates areas where outrage dominates, it does not fully account for the interaction between expert and public opinion on matters related to exposure.

Perhaps a better configuration of the risk analysis framework is one that incorporates three independent elements, that is, hazard identification and characterization, exposure assessment and consumer/citizen response, or outrage. Thus:

$$\text{RISK}^{\text{modern}} = \text{HAZARD} \times \text{EXPOSURE} \times \text{OUTRAGE}$$

Hazard and exposure would be included in the scientifically derived measure of risk, but the outrage factor could vary widely, at times muting concerns and raising tolerance for risky outcomes and at other times dramatically raising the bar for market entry (Phillips et al. 2006).

Ultimately, the risk assessment system ought to be designed to make socially optimal decisions—that is accepting safe products and rejecting unsafe products. As with any human system, there is potential for error, especially when considering a new class of products where there is little or no empirical evidence. While the system is and should be designed to avoid making type 1 errors, that is, accepting something that is not safe, it has to be mindful of the trap of making type 2 errors, unnecessarily and wastefully rejecting safe products and activities (Table 9.1). While we can tally up the cost of type 1 errors in lost lives or damaged ecosystems, we cannot convincingly estimate the cost of foregone opportunities and the loss of all of the attendant benefits that could flow from them. The difficulty is that social amplification of risk significantly raises the potential of making a type 2 error, thereby diminishing the flow of new and innovative products in a science-based economy. Such type 2 errors in the agricultural industry can have dire effects on the efforts to enhance productivity in aid of achieving food security in the future.

Thus, there is a pressing need to consider the appropriate role for science and society in the evaluation of new risks associated with technological improvements.

**Table 9.1** The typology of errors

| Decision | Product is safe | Product is unsafe |
|---|---|---|
| Accept as safe | Correct | Type 1 error |
| Reject as unsafe | Type 2 error | Correct |

## The Politicization of Risk

Science-based regulations provide the underpinnings of international organizations, agencies and agreements and have done so for decades. While not perfect, the development of science-based regulations has fostered the establishment of frameworks that provide consistent and repeatable decisions to those parties involved in, among other things, the international trade of agricultural products. While disagreements have arisen, been addressed and been resolved, developed countries have traditionally respected the rationale of grounding the regulation of international trade in agricultural products in science. Strong evidence of this is the acceptance by consensus of the World Trade Organization's (WTO) Sanitary and Phytosanitary (SPS) Agreement at the conclusion of the Uruguay Round multilateral trade negotiations in 1994 (Isaac 2007).

Starting about 20 years ago and coinciding with the first commercialization of GM crops, there has been an international effort to move away from international science-based risk assessments. The Cartagena Protocol on Biosafety (CPB) represents a recent effort to provide a comprehensive international structure to ensure the protection of biodiversity and to facilitate consideration of non-scientific concerns. The CPB is a new international institution negotiated specifically to deal with trade in the products of biotechnology. Since it came into effect in 2003, it has provided one set of rules for transboundary movements of GM organisms intended for environmental release and for those destined for the food chain. The Protocol provides a forum for the world's nations to express concerns about the use of biotechnology in the environment and to justify and promote their own national laws directed at addressing potential harms to environment, human health or other forms of loss. Article 27 of the CPB involves an effort to write an international law encoding specific liability for harm to biodiversity caused by biotechnology, in effect addressing only one plant breeding process without materially taking on the myriad of threats to biodiversity posed by other breeding methods or other human activities.

In general, socio-economic assessments provide useful information about the potential impacts of new technologies, but do not contribute to reducing or minimizing the scientific risks of the technology. In the case of a GM crop, this information may include socio-economic impacts at the farmer, household, industry and trade levels. Furthermore, socio-economic assessments may include consideration of up to 16 different non-pecuniary and indirect impacts, including the effect of lower health risks arising from reductions in pesticide use or shifts to less toxic active ingredients, market size of the new crop, possibilities for tracking and tracing, implications for biodiversity, the need for specific regulations in areas close to important ecological zones and changes in farm labor organization (Ludlow et al. 2014). Nevertheless, the results of a socio-economic assessment can almost always be expected to be controversial and may vary depending on the choice of methods, baseline data used, spatial and time focus and even the research team conducting the analysis.

It is posited here that the experience with a WTO challenge under the SPS pertaining to an EU ban on imports of beef produced using growth hormones (Kerr and Hobbs 2005) and domestic resistance to the products of biotechnology (Perdikis 2000; Hensen 2001) dramatically altered the EU's adherence to science-based regulations, encouraging movement towards socio-economic-based regulation of agricultural products and, hence, the politicization of risk assessment. While the change is scattered through a range of EU legislation, it is best illustrated by the regulatory regime put in place to govern agricultural biotechnology in 2003—after a moratorium on regulatory approvals instituted in 1999 (Viju et al. 2012). The new regulatory regime established the European Food Safety Authority (EFSA) to work with established political institutions—both of which must grant approval—resulting in a decoupling of risk assessment and the product approval processes. The EFSA conducts risk assessment using science-based methodologies, however, the product approval process is decided in European Commission institutions, resulting in the politicization of assessment.

Between June 1999 and August 2003, the EU had a moratorium on the approval and import of GM crops and food products. During the moratorium, the regulation of GM crops in field trials and experimental research plots was science-based and remained under the jurisdiction of individual EU Member States. This moratorium was ultimately ruled by the WTO in 2006 to violate international commitments (Isaac and Kerr 2007b). In 2003, however, the EU implemented a new regulatory regime for GM crops and food products. In the wake of the WTO judgement on its moratorium, the EU claimed that the new regime would be WTO compliant, but it needed time to make the necessary adjustments (Viju et al. 2012). By 2004, under its new regulatory regime, the EU Commission called for all member states to begin developing frameworks for coexistence whereby agricultural producers growing organic or conventional crops would not suffer economically from their crops comingling with GM products. Presently, most states have developed, or are still developing coexistence legislation. However, seven member states have refused to do so, visibly signalling their opposition to GM crops.

Comingling or the low-level presence (LLP) of GM material has been detected in shipments of canola, corn and mustard in the intervening period (see Chap. 7). The challenge is that international trade of bulk agricultural commodities never has and realistically cannot function with zero-tolerance, as is currently required in EU regulations pertaining to GM crop varieties that have not been approved. Before the establishment of ESFA and the new rules, science-based regulations underpinning the domestic regulatory systems of Member States effectively managed those few instances GM comingling occurred the incidents were addressed and resolved without the closure of borders and the suspension of international trade. Political interference was minimal.

A number of recent examples illustrated how, since the establishment of the new regulatory regime for the products of biotechnology in the EU in 2003, risk has been significantly politicized.

In 2006, trace amounts of an unapproved GM material were detected in US rice exports to the EU. The widespread presence of what has subsequently become known as LL601 rice resulted in an EU announcement on August 20th, 2006, that it would no longer accept rice shipments from the US (Li et al. 2010). As is noted by Kershen (2009), a 14-month USDA investigation as to how this comingling occurred, costing over US$1 million, offered no conclusive explanation. As the courts rejected a class action lawsuit, over 1000 lawsuits were launched against Bayer CropScience, the developer of LL601 (Redick and Endres 2009). The authors note that some reports indicate that in the end over 6000 lawsuits were filed. The lawsuits sought compensation for comingled crops and for depressed international markets for rice exported from the US. The German food producer, Rickmers Reismüehle, sued two Arkansas defendants—the large grower co-operative Riceland Foods and Producers Rice Mill—alleging that shipments to the company contained unapproved GM rice in breach of several contracts. Riceland Foods and the Producers Rice Mill turned to the developer of the rice variety for an explanation and for compensation. In the spring of 2011, Riceland Foods was awarded US$136.8 million based on their lawsuit against Bayer CropScience. In December 2009, the first of the producer cases were settled with two farmers receiving settlements. One farmer received an award of US $1.95 million, while the second received US$53,000. In the summer of 2011, Bayer offered US$750 million to settle all outstanding producer lawsuits related to the LL601 rice case, on the condition that at least 85% of the total rice acres planted between 2006 and 2010 were to be encompassed by the settlement (Endres and Johnson 2011). Producers were to be compensated on a per acre basis. This offer was accepted by the affected rice producers. Seven years after the initial detection the EU market has still not fully reopened to US long grain rice exports with current exports being less than one-third of previous levels. The liability costs and reduced sales associated with the closure of the EU market will undoubtedly feed into future investment decisions pertaining to GM crops.

Corn was the next product to trigger action. Smyth et al. (2010) reviewed the implications for US-EU corn trade following commercialization of Herculex corn. Pioneer Hi-Bred submitted notification of import into the EU for Herculex I in 2000 (EuropaBio 2006) and by 2006, Heculex I had received approval for feed and food use, as well as planting. The problem arose when the next generation of Herculex, a series of rootworm varieties, was detected in shipments coming from the US. In spite of testing prior to export, trace amounts were discovered in shipments to the EU. The detection caused corn gluten feed exports from the US to the EU to drop by 30–40%. These varieties were ultimately approved by the EU in 2009 but for the 2006–2009 period, corn trade was disrupted.

Then in September 2009, the EU's Rapid Alert System for Food and Feed announced it had detected GM flax in food products in Germany. One GM flax variety was approved in Canada in 1997 and the seed was multiplied from 1997 to 1999, but the variety was withdrawn from commercial production in 1999 due to EU concerns. The variety was deregistered in 2001, without GM flax ever receiving approval in Europe. Hence the problem. Following the detection, Canadian flax in Europe was in quarantine and flax trade between Canada and Europe was suspended, pending identification of the source of comingling and the implementation of a testing protocol that could provide assurance to European importers that Canadian exports would be free of GM flax. Testing of 26,000 flax samples in Canada revealed that 0.05% of samples tested positive (Ryan and Smyth 2012). Shipments of Canadian flax to the EU were embargoed for a three-month period, at the height of Canada's flax export season, resulting in an estimated market loss of C$58 million (Dayananda 2011). Ultimately trade resumed, but without recouping the losses.

These cases illustrate the nature of the problem: the slow and indeterminate EU regulatory system generated asynchronous approvals, which triggered trade responses disproportionate to or disconnected from the underlying scientifically justified risk. As noted in a March 2012 Editorial in Nature Biotechnology, "[i]n Europe, since the mid-1980s, regulators have shifted from evidence-based risk assessments to implementation of rules that specifically discriminate against transgenic products and emphasize the precautionary principle" (Nature Biotechnology 2012). The EU has moved from science-based systems as the underpinning of domestic regulation and international trade to the use of socio-economic considerations in decision-making, illustrating one instance where the principles of the CPB have been incorporated into regulatory frameworks.

In spite of knowing that trade in agricultural products is greatly impeded at a tolerance level of zero percent, the European Parliament moved to enact Directive 2001/18 that any GM variety not approved for import or feed production detected in agricultural imports, or found growing in the EU, would be illegal; hence, the tolerance threshold was established at zero (European Parliament 2001) By 2011, this strict regulatory threshold was proving unworkable and, in a move designed to provide assistance to the European animal feed industry, a threshold of 0.1% was agreed upon for the detection of unapproved GM material that had been approved for production in a non-EU country. But the zero threshold still applies to food imports. This differential threshold suggests that their application is arbitrary and open to political influence.

Political manipulation of EU thresholds is particularly obvious in the EU's treatment of GM soybean imports. In 2009, the University of Wageningen (LEI 2009) reported that if the EU was determined to source non-GM soybeans for use as animal feed, the cost of complying with the EU's 0.9% tolerance would raise the price of a tonne of soybeans from €290 to over €7700. The EU's solution to this pressing problem was to allow GM soybeans to enter Europe for animal feed. The size of this challenge for the EU was quantified by a United States Department of Agriculture report that places the value of GM soybean imports at US$6.5 billion in 2013 (USDA 2013). The report noted that the EU imports 70% of its soybean meal requirements and, of this, 80% of the soybeans imported are GM. Based on this, it is evident that when it is fiscally advantageous for the EU to manipulate their regulations, they will do so.

Prior to the establishment of the revised EU regulatory framework in 2003, regulators and those impacted by LLP, made decisions based upon scientific risk assessments. Before 2003, a few countries allowed farmers to harvest and export crops with LLP affected crops. Post-2003, at the earliest detection of LLP, the EU has automatically closed its borders to imports of the affected product and any domestic production or product has been isolated and destroyed. If science-based coexistence was allowed in the 1999–2003 moratorium period in the EU, it is not obvious why that is no longer feasible afterward. The science is more not less certain and the evidence of safe and efficacious use is mounting. The only explanation one can imagine is that risk management in the EU is no longer about science-based assessment, but is now a political accountability issue—no one within the EU system responsible for assessing and approving GM crops and food products wants to take responsibility for allowing an LLP event to be dealt with through a solely science-based decision process. One result is the closing of borders to international trade. With regulatory accountability residing in the political realm, there is a widening disconnect between regulators and those affected by the regulation. The impacts are not only in the exporting and producing nation: the detection of GM flax and embargo in Europe was estimated to have cost European industry €39 million and over 600 jobs (COCERAL 2010).

## Socio-economic Regulatory Frameworks

The Cartagena Protocol on Biosafety to the Convention on Biological Conservation is a focal point for the development and promotion of socio-economic assessment of genetically modified organisms (GMOs) . The objective of the Protocol is to contribute to ensuring an adequate level of protection in the safe transfer, handling and use of "living modified organisms resulting from modern biotechnology" that may have adverse effects on the conservation and sustainable use of biological diversity, also taking into account risks to human health and specifically focusing on transboundary movements (Article 1 of the Protocol) (Secretariat of the Convention on Biological Diversity 2000, p. 3).

Under the Protocol, parties may include socio-economic considerations (SECs) in reaching decisions on imports, including the planting of GMOs. Some authors, such as Jaffe, argue that the Protocol limits the scope of SECs to those factors affecting biodiversity with an emphasis on those affecting local and indigenous communities (Jaffe 2005). Nevertheless, many countries are using or have considered including SECs in their national legislation. As technological change always creates winners and losers, it will always be possible to identify potential negative socio-economic consequences (Leger et al. 1999).

The key part of the CPB is Article 26, which provides the opportunity for including a socio-economic assessment in national biosafety regulations. The text of Article 26 is as follows:

1. The Parties, in reaching a decision on import under this Protocol or under its domestic measures implementing the Protocol, may take into account, consistent with their international obligations, socio-economic considerations arising from the impact of living modified organisms on the conservation and sustainable use

of biological diversity, especially with regard to the value of biological diversity to indigenous and local communities.

2. The Parties are encouraged to cooperate on research and information exchange on any socio-economic impacts of living modified organisms, especially on indigenous and local communities.

International concern has been raised that SECs will become a mandatory part of approval processes and further complicate the approval of new crops. This will become a particular problem if the SECs are not bounded by the realm of conservation and sustainable use of biological diversity. During preparation for the Meeting of Parties to the CPB in Mexico in 2016, some Member States proposed revising article 26 to make it mandatory to undertake SECs; while that amendment was not approved by the Conference of the Parties, it illustrates the division of opinion on this matter.

Over the past 20 years, risk analysis has in some jurisdictions and in relation to some specific items progressively moved away from science-based assessment. Regulations are no longer solely concerned with the scientific aspects of risk, such as hazard or exposure; rather, they now involve issues such as ethics, labor impacts and consumer choice. The politicization of risk creates an objective regulatory function that might be expressed as a two stage process whereby the scientific risk hurdle might be cleared:

$$\text{RISK}^{\text{scientific}} = \text{HAZARD} \times \text{EXPOSURE}$$

But, subsequently, second stage the political risk assessment, which is isolated from the scientific evidence and deliberations, involves a complex calculation:

$$\text{RISK}^{\text{political}} = f\left(\text{RISK}^{\text{scientific}}, \text{Outrage, Unsubstantiated Information, NGO Advocacy, Institutional Architecture}\right)$$

The two systems are only modestly connected, and in many ways do not overlap conceptually or practically, as a different array of government actors engages in the political risk assessment using a variety of considerations that are more qualitative and judgemental. In effect, governments are influenced by the presence of social outrage from society, which in turn, is a factor of hearsay, allegations and innuendoes promoted by interested parties, including environmental and social justice non-governmental organizations (eNGOs). Ultimately, the outcomes are determined by the nature of the decision architecture, which defines standing, authority and responsibility. According to Isaac (2007, p. 288):

> … risks are defined to include speculative risks which lack experience, data, a causal-consequence mechanism and an accepted analytical method for assessment; they are logical possibilities – irrefutable and untestable. Typically, such risks would have no standing within a science-based framework. Under the … approach science only informs the decision and 'other legitimate factors' are also weighted.

**Table 9.2** Typology of food safety risks

| Food safety issue | RISK$^{scientific}$ | RISK$^{socially\ constructed}$ | RISK$^{modern}$ | RISK$^{political}$ |
|---|---|---|---|---|
| High cholesterol foods | High | Moderate | Moderate | Low–moderate |
| Foods high in sugar | High | Moderate | Moderate–high | Moderate |
| High sodium foods | Moderate | Low | Moderate | Low |
| Decomposing snail in beverage bottle | Low–high | High | High | Low |
| Dead frog in package of frozen vegetables | Low–moderate | High | High | Low |
| E-coli in hamburgers | High | Low | Moderate | Low |
| Salmonella | High | Low | Moderate | Low |
| Mycotoxins | High | Moderate | High–moderate | Low |
| Filth & extraneous materials (insect fragments, stones, twigs, rodent manure) | High | High | High | Low |
| GM foods | Low | High | High | High |

*Source* Authors

The various formulas for assessing risk can be compared in terms of their response to a series of food safety concerns (Table 9.2).

The politicization of risk jeopardizes the movement towards increased food security by utilizing political influence to regulate (and delay or reject) safe new products or technologies (i.e. a type 2 error), while drawing attention and resources away from identified, substantiated and remediable food safety and environmental risks. This trend in risk assessment is actually amplifying the potential for type 1 errors, not so much that unsafe foods are being approved, but rather that existing products and technologies that are less safe than they might otherwise be are not subject to increased regulatory scrutiny and are not contested by newer, safer formulations. For example, food products that have been well documented in the health science literature as having been identified for concern given the presence of high cholesterol, sugar or sodium embodied in the product receive some political encouragement to improve these products or to better label the health problems associated with consumption, but foods that truly present safety concerns often rank quite low in terms of political priority.

Food products that occasionally contain undesirable attributes such as beer with the decomposed remains of a snail in the bottle or the presence of fumonisins in corn-based food products (Missmer et al. 2006) receive low priority under a politicized risk assessment. The presence of decomposing remains in food products is certainly revolting to the consumer and the outrage associated with such events usually triggers increased food safety scrutiny within commercial supply chains has ensured that these risks are minimized. At the political level, however, there is simply not the impetus to recognize this as being a leading health/food safety concern. Less obvious, credence factors (i.e. those that are undetectable through search or immediately on consumption) are especially prone to being ignored.

One way to illustrate this is to juxtapose the EU efforts to manage their market after trace amounts of GM flax were detected in exports from Canada in 2009 (at a cost of C$80 million to both parties and no identified human health risk) with the fact that in Europe from May to July 2011, over 50 people died from the consumption of fenugreek contaminated with e. coli. (World Health Organization 2011). The political risk from the death of consumers from unsafe organic food in the EU was perceived as being low, while the political risk from virtually undetectable trace amounts of GM flax which had been judged as having a sufficiently low risk when a science-based assessment was used—albeit in the US and Canada —was subject to massive regulatory intervention.

The politicization of risk in the EU reduces the expected returns from investing in technological improvements in agriculture—and, hence, threatens the achievement of food security goals. Important technological advances such as GM wheat have already been shelved due to expected difficulties with political risk assessments in the EU, even in the face of evidence that GM wheat would probably receive approval under a scientific risk assessment (Furtan et al. 2005). While not a direct issue for food security, Canadian GM flax was deregistered in both Canada and the US in response to politicized risk assessment in the EU (Ryan and Smyth 2012). Meanwhile, a number of major developers of agricultural technology have moved their R&D operations out of the EU—disruptions that at a minimum slow the pace of technological advancement. In January 2012, BASF announced that it was moving its research division from Europe to the United States due to the lack of timeliness in regulatory decisions in the EU (BASF, 2012). In part, this decision is based on the fact that it took 13 years to receive approval for a BASF developed GM potato variety in Europe (BASF 2010). Further, in July 2013 Monsanto announced that it was scaling back research investments within the EU and abandoning GM crop variety submissions that were in the pipeline (Cressey 2013).

Developing countries also fail to adopt GM technologies due to perceived political risk in the EU and the potential that it would threaten their export markets (Smyth et al. 2013). As some developing countries eschew the acceptance of biotechnology due to political risk factors in their potential markets it reduces the incentive for agricultural technology firms in developed countries to undertake R&D into tropical and other crops specifically suited to their environment (Paarlberg 2002). While significant increases in international trade in agricultural products will be required to reach food security goals by mid-century, increases in agricultural productivity must also make a major contribution. In a globalized economy, the impact of differences in how risk is dealt with can have wide-ranging detrimental impacts on technological progress in agriculture.

## Conclusions

Millions of our ancestors likely died from the consumption of unsafe food. Over time, science-based, food-safety testing and detection mechanisms have been developed and implemented to improve the quality of consumed food, minimizing the risk of consuming unsafe food products. Of course, no system functions at 100% safety. Type 1 errors will occur. The present reality is, however, that consumers die both from eating unsafe food and from famine, but no one yet has died from eating GM food. The politicization of risk does not deliver either safer food or technological improvements.

Science-based risk assessments have been successful in denying the commercialization of unsafe GM foods while politicized risk assessments continue to advocate that consuming GM foods is a danger to one's health or the environment. If this regulatory divergence meant only that consumers in some rich countries have fewer food choices, the making of this kind of type 2 error would not be a particular concern. The EU has also made market access for the products of developing countries under its Generalized System of Preference (GSP-plus) program contingent on accession to the CPB—meaning that the non-scientific risk assessment methods are being proactively spread to developing countries that need productivity enhancing innovations (Ansong 2013). Given the long term negative impacts on technological improvement in a period when concerns regarding future food security are high, a re-assessment of politicized risk is needed.

While the focus of this chapter has been on one particular agricultural technology—genetic modification—that has the promise of being a major contributor to future food security, the real danger lies in the potential acceptance of politicized risk more generally. Once it is applied to one technology and accepted as a guiding principle, it can be extended to other technologies. New technologies will always have their doubters and detractors. They inevitably create potential 'losers' who have a vested interest in having a technology denied. While future food security may not be dependent on fully exploiting the potential of agricultural biotechnology, it does depend on general technological advancement. Formally allowing non-scientific factors to enter into risk assessments gives legitimacy to those pandering to political systems for their own vested interests.

## References

Alston, J.M. 2002. Spillovers. *The Australian Journal of Agricultural Economics* 46: 315–346.

Alston, J.M. 2010. The benefits from agricultural research and development, innovation, and productivity growth. OECD food, agriculture and fisheries papers, no. 31. Paris: OECD Publishing. http://dx.doi.org/10.1787/5km91nfsnkwg-en.

Alston, J.M., J.M. Beddow, and P.G. Pardey. 2009. Agricultural research, productivity and food prices in the long run. *Science* 325: 1009–1010.

Ansong, A. 2013. Creating WTO law by stealth: GSP conditionalities and the EC—Tariff preferences case. *The Estey Centre Journal of International Law and Trade Policy* 14 (2): 133–144.
BASF. 2010. News release: European Commission approves Amflora starch potato. Online at: http://www.potatopro.com/news/2010/european-commission-approves-amflora-starch-potato.
BASF. 2012. News release: BASF to concentrate plant biotechnology activities on main market in North and South America. Online at: http://www.basf.com/group/pressrelease/P-12-109.
Beddington, J. 2010. Food security: Contributions from science to a new and greener revolution. *Philosophical Transactions of The Royal Society B Biological Sciences* 365 (1537): 61–71.
Cardwell, R., and W.A. Kerr. 2008. Protecting biotechnology IPRs in developing countries: Simple analytics of a levy solution. *Journal of Agricultural Economics* 59 (2): 217–236.
COCERAL. 2010. Low level presence of GMOs not authorized in Europe: The linseed CDC triffid case, committee of cereals, oilseeds, animal feed, oils and fats, olive oil and agrosupply, Brussels, trade file on copy with authors.
Cressey, D. 2013. Monsanto drops GM in Europe. *Nature* 499 (7459): 387.
Davies, A.R. 2013. Food futures: Co-designing sustainable eating practices for 2050. *EuroChoices* 12 (2): 4–10.
Dayananda, B. 2011. The European Union policy of zero tolerance: Insights from the discovery of CDC triffid. Unpublished M.Sc. thesis, University of Saskatchewan, Saskatoon.
Endres, A.B., and N. Johnson. 2011. $750 million settlement in GM rice contamination. Available at: http://www.farmdocdaily.illinois.edu/2011/07/750_million_settlement_in_gm_r.html.
EuropaBio. 2006. Herculex I (1507) Maize: Safety and general characteristics. Available at: http://www.europabio.org/sites/default/files/pages/1507_factsheet.pdf.
European Parliament. 2001. Directive 2001/18/EC of the European Parliament and of the council of 12 March 2001 on the deliberate release into the environment of genetically modified organisms and repealing Council Directive 90/220/EEC. Available at: http://eur-lex.europa.eu/LexUriServ/LexUriServ.do?uri=OJ:L:2001:106:0001:0038:EN:PDF.
Evans, A. 2009. *The feeding of the nine billion: Global food security*. London: Chatham House.
Food and Agriculture Organization. 2006. The state of food insecurity in the world: 2006. Online at: http://www.fao.org/docrep/009/a0750e/a0750e00.htm.
Furtan, W.H., R.S. Gray, and J.J. Holzman. 2005. Regulatory approval decisions in the presence of market externalities: The case of genetically modified wheat. *Journal of Agricultural and Resource Economics* 30 (1): 12–27.
Gaisford, J.D., J.E. Hobbs, W.A. Kerr, N. Perdikis, and M.D. Plunkett. 2001. *The economics of biotechnology*. Cheltenham, UK: Edward Elgar Publishing Ltd.
Haggui, F., P.W.B. Phillips, and R. Gray. 2006. Opposition to genetically modified wheat and global food security. In *International trade and policies for genetically modified products*, ed. R.E. Evenson, and V. Santaniello, 175–190. London: CABI Publishing.
Hensen, S. 2001. Consumer perceptions of food safety: Survey research on economics and social psychology. In *Governing food: Science, safety and trade*, ed. P.W.B. Phillips, and R. Wolfe, 91–113. Montreal: McGill-Queens University Press.
Hertel, T.W. 2011. The global supply and demand for agricultural land in 2050: A perfect storm in the making. *American Journal of Agricultural Economics* 93 (2): 259–275.
Isaac, G.E. 2007. Sanitary and phytosanitary issues. In *Handbook on international trade policy*, ed. W.A. Kerr, and J.D. Gaisford, 383–393. Cheltenham: Edward Elgar.
Isaac, G.E., and W.A. Kerr. 2007a. Whose vision of the future? The entrenched international conflict over genetic modification. *The Geneva Post Quarterly* 2 (1): 87–107.
Isaac, G.E., and W.A. Kerr. 2007b. The biosafety protocol and the WTO: Concert or conflict. In *The international politics of genetically modified food*, ed. R. Falkner, 195–212. Houndmills, UK: Palgrave Macmillan.
Jaffe, G. 2005. Implementing the biosafety protocol through national biosafety regulatory systems: An analysis of key unresolved issues. *Journal of Public Affairs* 5 (3–4): 299–311.
James, J.S., P.G. Pardey, and J.M. Alston. 2008. Agricultural R&D policy: A tragedy of the international commons, Staff paper P08-08, Department of Applied Economics, University of

Minnesota, http://digitalcommons.calpoly.edu/cgi/viewcontent.cgi?article=1010&context=agb_fac&seiredir=1&referer=http%3A%2F%2Fscholar.google.ca%2Fscholar%3Fstart%3D120%26q%3Dj%2Bm%2Balston%26hl%3Den%26as_sdt%3D0%2C5#search=%22j%20m%20alston%22.
Kerr, W.A., and J.E. Hobbs. 2005. Consumers, cows and carousels: Why the dispute over beef hormones is far more important than its commercial value. In *The WTO and the regulation of international trade*, ed. N. Perdikis, and R. Read, 191–214. Cheltenham: Edward Elgar.
Kerr, W.A., and R. Yampoin. 2000. Adoption of biotechnology in Thailand and the threat of intellectual property piracy. *Canadian Journal of Agricultural Economics* 48 (4): 597–606.
Kershen, D. 2009. Legal liability and agricultural biotechnology: Ten questions. Available at: http://agribiotech.info/details/KershenFinal%2003%20layout.pdf.
Kulthreshtha, S.N. 2011. Climate change, prairie agriculture, and prairie economy: The new normal. *Canadian Journal of Agricultural Economics* 59 (1): 19–44.
Leger, L.A., J.D. Gaisford, and W.A. Kerr. 1999. Labour market adjustments to international trade shocks. In *The current state of economic science*, vol. 4, ed. S. Bhagwan Dahiya, 2011–2034. Spellbound Publications PVT Ltd.: Rohtak.
LEI. 2009. EU policy on GM soy: Tolerance threshold and asynchronic approval. Available online at: http://edepot.wur.nl/7856.
Li, Y., E. Wailes, A. McKenzie, and N. Thomsen. 2010. LL601 Contamination and its impact on U.S. rice prices. *Journal of Agricultural and Applied Economics* 42 (1): 31–38.
Ludlow, K., S.J. Smyth, and J. Falck-Zepeda (eds.). 2014. *Socio-economic considerations in biotechnology regulation*. New York: Springer Science + Business Media.
Malla, S., and R. Gray. 2005. The crowding effects of basic and applied research: A theoretical and empirical analysis of an agricultural biotech industry. *American Journal of Agricultural Economics* 87 (2): 423–438.
Missmer, S., L. Suarez, M. Felkner, E. Wang, A. Merrill Jr., K. Rothman, and K. Hendricks. 2006. Exposure to fumonisims and the occurrence of neural tube defects along the Texas-Mexico border. *Environmental Health Perspectives* 114 (2): 237–241.
Nature. 2010. How to feed a hungry world. *Nature* 466 (7306): 531–532.
Nature Biotechnology. 2012. Agnostic about agriculture. *Nature Biotechnology* 30 (3): 197.
Paarlberg, R.L. 2002. The real threat to GM crops in poor countries: Consumer and policy resistance to GM foods in rich countries. *Food Policy* 27 (3): 247–250.
Perdikis, N. 2000. A conflict of legitimate concerns or pandering to vested interests? Conflicting attitudes towards the regulation of trade in genetically modified goods—The EU and the US. *Journal of International Law and Trade Policy* 1 (1): 51–65.
Phillips, P.W.B., S.J. Smyth, and W.A. Kerr (eds.). 2006. *Governing risk in the 21st century: Lessons from the world of biotechnology*. Hauppauge, NY: Nova Science Publishers.
Redick, T., and A.B. Endres. 2009. Jury verdict against Bayer for liberty link rice breaks new ground in biotech liability. *Agricultural Law Update* 26: 2–6. 11 and 12.
Ryan, C.D., and S.J. Smyth. 2012. Economic implications of low level presence in a zero tolerance European import market: The case of Canadian Triffid flax. *AgBioForum* 15 (1): 21–30.
Sandman, P. 1994. Mass media and environmental risk: Seven principles. *Risk* 5 (5): 251–260.
Secretariat of the Convention on Biological Diversity. 2000. *Cartagena protocol on biosafety to the convention on biological diversity*. Montreal: Secretariat of the Convention on Biological Diversity.
Smyth, S.J., and R. Gray. 2011. Intellectual property sharing agreements in gene technology: Implications for research and commercialisation. *International Journal of Intellectual Property Management* 4 (3): 179–190.
Smyth, S.J., P.W.B. Phillips, W.A. Kerr, and G.G. Khatchatourians. 2004. *Regulating the liabilities of agricultural biotechnology*. Wallingford, UK: CABI Publishing.
Smyth, S.J., A.B. Endres, T. Redick, and D.L. Kershen. 2010. *Innovation and liability in biotechnology: Transnational and comparative perspectives*. Cheltenham, UK: Edward Elgar.

Smyth, S.J., W.A. Kerr, and P.W.B. Phillips. 2011. Recent trends in the scientific basis of sanitary and phytosanitary trade rules and their potential impact on investment. *Journal of World Investment and Trade* 12 (1): 5–26.

Smyth, S.J., W.A. Kerr, and P.W.B. Phillips. 2013. Accelerating adoption of GM crops through a trade liability regime. *Plant Biotechnology Journal* 11 (5): 527–534.

Tengs, T., M. Adams, J. Pliskin, D. Safran, J. Siegel, M. Weinstein, and J. Graham. 1995. Five hundred life-saving interventions and their cost-effectiveness. *Risk Analysis* 15 (3): 369–390. Reprinted in The Economist, August 2001.

Tilman, D., C. Balzar, J. Hill, and B.L. Befort. 2011. Global food demand and the sustainable intensification of agriculture. *Proceedings of the National Academy of Sciences of the United States of America* 108 (50): 20260–20264.

Tutwiler, A. 2011. Abstract for keynote address to the GMCC-11 conference. Vancouver, Canada, 26–28 Oct 2011.

United States Department of Agriculture. 2013. EU-27 agricultural biotechnology annual. GAIN report no. FR9142. Washington: USDA.

Viju, C., M.T. Yeung, and W.A. Kerr. 2012. The trade implications of the post-moratorium European Union approval system for genetically modified organisms. *Journal of World Trade* 46 (5): 1207–1238.

World Health Organization. 2011. Outbreaks of E.coli O104: H4 infection: Update 30. Available at: http://www.euro.who.int/en/what-we-do/health-topics/emergencies/international-health-regulations/news/news/2011/07/outbreaks-of-e.-coli-o104h4-infection-update-30.

# Chapter 10
# Coexistence Policies and Strategies

**Abstract** Coexistence, as defined by the European Union, involves simultaneous GM and GM-free crop and food production, assuming both products have passed regulatory risk evaluation. Much of the recent discussion about coexistence focuses on how governments can create it, or in some cases how to do it more efficiently and effectively. At one level, the issue of coexistence is a problem of market-making. But economic theory and commercial practice suggest that there are challenges with leaving market-making to private initiative. Meanwhile, given that the detection of the presence of an unapproved GM event results in the closing of borders to the import of commodities, the issue inevitably moves to the WTO. This chapter reviews the economic rationale for involvement by governments and other non-market actors and assesses the evidence of the impact of various interventions on practical coexistence.

## Introduction

GM products are now extensively produced and permeate much of the global agri-food market (James 2015). Coexistence is replacing 'precaution' as the current buzz-phrase defining the relationship between genetically modified (GM) foodstuffs and governments, industry, farmers, consumers and citizens. The most comprehensive definition asserts co-existence is about protecting farmers and all stakeholders in the supply chain from the possible economic consequences of accidental mixing of GM and non-GM crops or derived products. In effect, coexistence involves simultaneous GM and GM-free crop and food production, assuming all products have passed regulatory risk evaluation. In practical terms, this policy addresses segregation and identify preservation of GM, non-GM and organic canola, corn and soybeans in both producing and importing countries, but does not include the adventitious presence of GM crops that have not been fully approved by regulators, such as LLRICE601 and LLRICE604 in 2006 or CDC Triffid in 2009.

The rationale for policy intervention is that "failure in achieving practical coexistence will impact diffusion of new technologies in agriculture and potentially



S.J. Smyth et al., *Biotechnology Regulation and Trade*, Natural Resource Management and Policy 51, DOI 10.1007/978-3-319-53295-0_10

global trade in agricultural products" (GMCC 2007). Coexistence strategies are then defined as a mix of science-based technical measures, political directives and economically feasible organizations, involving scientists, policy-makers and stakeholders from across the entire agricultural supply chain.

Much of the discussion about coexistence in the past decade has focused on how to ensure we have the right science and processes to effectively and efficiently deliver choice. Our discussion on other chapters examines the regulatory challenges of creating and maintaining effective regulatory processes. This chapter looks at the problem from the perspective of market-making and market management. Under stringent conditions one might be able to assume that private markets unmediated by the state or other actors might be able to ensure optimal amounts of GM and GM-free crop and food production to satisfy producers and consumers. The reality, however, is that the market is at times not well suited to undertake this task. The state and/or civil authorities may need to provide support and leadership in order to deliver effective co-existence. Ultimately, the incomplete market rules lead to conflicts in international markets, which then are litigated at the WTO. Hence, this chapter examines both aspects of the market-making system.

## Coexistence in Theory and Practice

The concept of coexistence has been borrowed from a number of other fields, including psychology, international relations and security studies. The concept has been used to describe the outcome of individuals, groups or societies existing together in time or place, essentially existing in mutual tolerance of their inherent differences. This can involve either active or passive coexistence (Khaminwa 2003). Ultimately, the literature suggests a few instructive points for a discussion of technological coexistence—there is a spectrum of outcomes that result from a range of unplanned to well scripted actions. An early industrial application of the concept of coexistence involved discussion about the impact of externalities between different co-located industries, such as forests being used simultaneously by loggers, tourists and indigenous peoples, water supplies being used by industrial factories, power plants, irrigators and native or sport fisheries and land having multiple demands for public or private use. Development history and economic theory are full of examples of these natural conflicts over rival, subtractable resources (e.g. Coase 1960).

Against that backdrop, the European Union (EU) in 2003 proposed a new policy of "coexistence of genetically modified crops with conventional and organic agriculture" as a policy response to perceived concerns by EU producers and consumers about the role of GM technologies in the food system (2003/556/EC http://ec.europa.eu/agriculture/coexistence/index_en.htm). The EU policy defines coexistence as "the choice of consumers and farmers between conventional, organic and GM crop production, in compliance with the legal obligations for labelling defined in Community legislation." Suitable measures were prescribed for cultivation, harvest, transport, storage and processing to achieve 'sufficient' segregation between GM and non-GM production. Importantly, the EU has assumed coexistence always refers to GMOs that have passed EU authorization processes related to health or environmental risks, so that coexistence rules do not address aspects of risk assessment or management. In effect, the goal is to ensure that adventitious traces of GMOs stay below the labeling threshold laid down in Regulation (EC) No 1829/2003 and Directive 2001/18/ (Commission Recommendation 2003/556/EC).

On 23 July 2003, the Commission adopted Recommendation (2003/556/EC) to provide guidelines for the development of national strategies and best practices for coexistence. The guidelines state that any Measures should be efficient and cost-effective (not go beyond what is necessary to comply with EU threshold levels for GMO labeling), specific to different types of crop (based on the probability of admixture by crop) and reflect local and regional variables. By February 2009, 15 Member States had adopted specific legislation on coexistence (Austria, Belgium, Czech Republic, Denmark, France, Germany, Hungary, Lithuania, Luxembourg, Latvia, Netherlands, Portugal, Romania, Sweden and Slovakia). In Austria and Belgium, the competence lies at regional level, and not all regions may be covered by legislation. Draft legislation of three further Member States had been notified to the Commission but not ratified (European Commission 2009). No further reports on progress have been disseminated.

The extent of adoption varies significantly among the EU Member States. A review of the measures adopted suggests that while many of the provisions were appropriately targeted to manage coexistence during seed multiplication and trade, some exceeded what might be necessary for the management of coexistence. A number of measures addressed concerns related to environmental protection (effectively going beyond the provisions laid out in community legislation), some included case-by-case, farm-level approval or notification procedures for GM crop cultivation (which duplicates authorization for the use of GM crops already authorized for EU cultivation), some proposed or adopted measures that aim to reduce adventitious presence of GMOs below the EU thresholds and some set larger isolation distances between GM and non-GM or organic fields than required to meet the thresholds. In terms of implementation, the general practice was that Member States place responsibility for implementing coexistence measures on the farmers who grow GM crops—farmers engaged in non-GM crop production do not have to change established farming techniques following the introduction of GM crops in adjacent fields. But most member states appear to allow neighboring farmers, on a voluntary basis, to decide amongst themselves not to segregate GM and non-GM production, which means that non-GM production would have to be labeled as GM (Communication from the Commission to the Council and the European Parliament —Report on the implementation of national measures on the coexistence of genetically modified crops with conventional and organic farming, COM/2006/0104 final).

In the US, Canada, Brazil and Argentina, the four largest mega-adopters and exporters, there are no specific rules that require or support coexistence. As a result, those seeking to differentiate their crops are left to their own devices to ensure they can meet the tolerances of the market. Both theory and evidence suggest this can be a challenge in the production system (Belcher et al. 2005) and beyond. In short, the challenge of coexistence is asymmetrically managed—with the EU and some other key importers setting up regulatory structures that impose conditions on the market while the key exporters have left this to producers and marketers to resolve. The outstanding question is whether the market can manage the emerging conflicts or whether competition national authorities or international agencies, such as the World Trade Organization (WTO), will need to intervene.

## Efficient Markets and Coexistence

Generally, markets are assumed by economists to be driven by self-interested efforts to deliver optimal amounts of goods and services. Individuals, operating as producers or consumers, buy or sell labor, other inputs, goods or services based on their own calculations of what is in their own best interests, mediated by market prices of the various factors and goods. These markets are driven by voluntary transactions (Benacek 2005). The economic theory of markets suggests that, under certain rigorous conditions, if everyone acts in a self-interested way, then we can optimize our economic activities. Economist Paul Seabright (2005, p. 15) notes that the cooperation that delivers existing or new products often appears effortless because people and institutions embody a kind of tunnel vision, by "play[ing] one's part in the great complex enterprise of creating the prosperity of a modern society without knowing or necessarily caring very much about the overall outcome." Aggregating the actions of individual consumers and producers generally leads to an optimum, where no one can improve their circumstances by further trading. Seabright argues that "one of the great intellectual achievements of modern economics has been to work out very precisely the circumstances under which decentralized systems of market exchange can produce results that are efficient, in the sense of improving the condition of every individual as far as possible whenever this can be done without harming someone else … the pattern of modern life has emerged without ever having been consciously willed by anyone" (Seabright 2005, p. 21).

In a perfect world, then, optimal quantities, qualities and prices of goods and services would be produced and consumed. But the optimal outcome (often called 'perfect competition') is more of a theoretical construct than an observable reality. Theory clearly shows that it can only be realized if five rigorous but uncommon conditions hold. First, all actors must be atomistic—there must be a large number of producers and consumers in a given market, each so small that their actions have no significant impact on others (i.e. both buyers and firms must be price takers and not exert any market power). Second, the goods and services must be homogeneous or perfect substitutes. Third, there must be perfect and complete information—all firms and consumers must have complete information to production technologies, attributes of a product, prices and quantities. Fourth, there must be no barriers to entry or exit for either producers or consumers. Fifth, there must be no external effects of any production or consumption that lie outside the market—both positive externalities (e.g. spillovers from research or public health) and negative externalities (e.g. pollution and congestion) must be known and included in the price of the product.

As different necessary conditions break down, different types of market failures arise. The issue of coexistence presents three specific challenges to markets.

First, the actions of suppliers, farmers, food handlers, food processors, wholesalers and retailers can and often do limit choice in the market. All other things being equal, older technologies and products tend to be priced out of the market by innovative products. This process, what Schumpeter (1954) called 'creative destruction', is a natural outcome of technological change (and usually a good outcome in the long run). As a result, some producers may no longer be able or willing to produce as before and some consumers may have absolutely less choice about what to buy and consume. Some have argued that there is a 'moral' obligation to ensure both producers and consumers can choose from all existing options, but economists tend to apply the strictures of Pareto, Hicks and Kaldor to rationalize the collateral damage of technological change. In short, it is not up to the market to meet all demands—some may simply not be economical.

Second, there are real and documented cases where the actions of one producer could have a material negative effect on other producers. In effect, there is a technical negative externality. For example, the costs and prices observed by one group of producers (e.g. the GM producers) may not fully reflect the full costs of their actions if genetic material adventitiously presents itself in other producer's product and lowers their profits. Ackerlof (1970) offered the theory of lemons to suggest that when a potentially unwelcome trait is introduced into a market without effective product differentiation, it causes the entire product line to suffer lower demand and prices (his example showed that poor quality cars in the resale market and consumers' inability to distinguish between the lemons and higher quality cars worked to depress prices and drive out higher-quality cars). There is significant literature that attempts to explain when and where markets will be able to optimally provide the right quantities and qualities of goods and services (Coase 1937; Williamson 1979; Mahoney 1992). This literature examines the respective roles of individuals, firms and governments in constructing the appropriate rules and institutions that protect against new technologies or products prematurely or unnecessarily destroying value of other product lines.

Third, scientists, industry and many regulators assert that GM and non-GM crops are essentially the same (homogenous) but some consumers and citizens beg to differ. Even if there may be no substantive physical difference between their nutritional value, composition or potential allergenicity, the simple reality that a transgene may have been used in the production of the GM food changes its provenance and implicit value for some consumers. Economic theory posits that the product, per se, does not give utility to the consumer—the characteristics of the product do (Lancaster 1966). While consumers cannot directly purchase units of foods of quality, they can choose to avoid the goods that they perceive to have lower quality attributes. Exposure, attention and perception are important because they affect what consumers comprehend, the attitudes they have and what they remember, which in turn affects the decisions they make. Thus consumer information and knowledge play a major role in the attitude formation process. There is extensive literature in the emerging field of behavioural economics that suggests citizen and consumer cognition and opinion formation are not as clear cut as assumed in the neoclassical economic model. Who defines product attributes, how they are defined, how they are verified and what they signify are not easily sorted out. Firms and markets at times are unable to discern and resolve this issue and hence fail to provide the right quantities of the right qualities at market clearing prices. Consumer economics has spent a great deal of time examining the most appropriate role of the state and markets in assuring full and accurate information is available to assist production and consumption decisions.

In essence, these three theoretical justifications provide a prima facie rationale for states to intervene in markets to secure an appropriate level of coexistence. The unanswered question is whether their interventions will make things better.

## Market-Making Actors

The advent of biotechnology in the agri-food system is not the first and will not be the last example of technology advancing faster than the institutions needed to govern it. North (1991, 1) concludes that "neo-classical theory is simply an inappropriate tool to analyze and prescribe policies that induce development. It is concerned with the operation of markets, not with how markets develop." Mokyr (1997, p. 1) explains "there usually is, at some level, an (sic) non-market institution that has to approve, license, or provide some other imprimatur without which firms cannot change their production method. The market test by itself is not always enough. In the past, it almost never was." Mokyr (1997, p. 2) notes that increasingly:

> The adoption of a wholly new technology is often the target of long debates and public discourse, unlike many other technical and economic choices. The role of persuasion and rhetoric in these decisions is something economists have paid scant attention to, and hence they have not had much success in understanding why, for example, some economies have adopted nuclear power or why some have allowed experimental drugs to be sold and others did not. Furthermore, not all resistance is purely social. There are instances in which the technological 'system' resists a novel and improved component because it does not fit the operation of the whole.

Phillips (2007) asserts governing these types of transformative technologies and their products presents a complex systems problem. Boulding (1970) posits three distinct domains—the compulsory, the contractual and the familial—offer three different methods of integration: coercive states that distribute rights and obligations; quid pro quo transactions in the market governed by supply and demand; and voluntary dealings, where cooperation, reciprocity and solidarity engage community and society (Paquet 2001). Boulding (1970) argues that society can be viewed as encompassed by a 'social triangle', where all organizations—including the state, the market and civil authorities—are built on one or a balance of the three relationship systems.

The triad of institutions—governments/states, the market and social/familial organizations—each have specific institutional attributes that make them more effective at producing particular types of goods (Picciotti 1995). The government sector is best at producing public goods—low excludability makes privatization infeasible while the low voice component makes it difficult for the collective sector to organize. The private sector tends to dominate whenever clear and enforceable property rights make rival goods excludable—the property of exclusion allows private firms to sell at or above the marginal cost of production. The participatory sector is best at governing common pool goods (e.g., standards)—the collective will usually have information that enables them to more effectively manage the resource and capture the benefits.

While in theory there are candidate institutions that could and probably should take the lead to address the three market challenges identified above, in practice they all are self-limited. Governments, while motivated to resolve moral issues, are often incapable of meeting the technical norms of democracy and public good. Markets are obviously imperfect: amorality, externalities and asymmetric information are endemic in the system. Finally, the 'third sector' faces real challenges in appropriately and responsibly mobilizing the public (Olson 1965).

## Critical Assessment of the Challenges of Coexistence

With three fundamental market problems and three core actors (and possibilities of hybrids), there are a range of ways we might effectively and efficiently resolve those issues.

### *The Morality of Sustaining Choice*

The root issue facing GM foods is the fear (some assert reality) that the new, advancing technologies will drive out less economic methods of production. We have seen that many times in the food system. Tractors and combine harvesters, for example, have replaced horses, oxen and large threshing crews, displacing many of those who earned their incomes and livelihoods from those activities. Schumpeter has called that the process of 'creative destruction', where the new relegates the old to the margins or the history books. The economic assumption is that those resources released from the production of less competitive goods and services, and the labor saved through technological innovation, will be reallocated to more productive and compensatory activities. Economists rationalize this outcome through the Pareto optimality criteria, where the economic change unambiguously generates a net gain in economic welfare (the core Paretian criteria), which is large enough to fully compensate all those who would be disadvantaged and still leave some ahead (i.e. Hicks' compensated demand criteria) but, because the event is occurring in a democracy, it is not actually organized (i.e. the Kaldor criteria). Hence, economists abstract from this problem. Sociologists, anthropologists, psychologists and ethicists have a mix of views—while many call for government intervention in support of those losing from change, at least some philosophers assert that there is no moral imperative for the state to intervene to ensure that everyone's preferences are respected (e.g. Thompson 2006).

One difficulty is that few technological transformations are costless. There can be significant costs of changing crops, changing jobs and changing consumption. In those cases, the state may step in and compensate those who face the largest costs of change. We provide unemployment support and retirement bonuses, retraining programs, incentives and subsidies to farmers and firms to change their production systems or to retool their operations and tax and price support for many sectors facing restructuring. While the cost for these interventions usually comes from the common purse, in some instances firms bear much of the cost directly (e.g. workers covered by collective agreements often have clauses that mandate their employer to provide assistance to deal with technological change). For the most part, however, we do not directly force those gaining from a specific technological change to compensate those who lose from that change. The biggest challenge of doing so is to identify those specific people to compensate and to apportion that cost appropriately among the winners. The transactional costs of such an effort usually force governments to do this indirectly through general income and employment programs rather than support targeted at those specific actors affected by technological change.

## *The Lemons Problem*

Full containment of GM traits is not be easily attained—under current conditions even small traces of GM produce can act as 'lemons' in non-GM food markets, driving away consumers and lowering the market clearing prices for all producers. Following Coase's advice, one option is simply to assign the ownership rights and then let GM and non-GM producers negotiate a balance among themselves. This is exactly what some countries have done. In Canada and the US, for example, once GM producers have secured regulatory approval, they have the de facto and de jure right to commercialize their technology and are not required by law to enact any proactive containment policies or programs. In those circumstances, non-GM producers either have to invest in isolation or negotiate some form of buffer between their production and that of GM producers. In one sense, this is analogous to the circumstances of most organic producers around the world, who self-identify as using specific production methods and are responsible for ensuring their compliance with the norms. Generally, where there are price premiums, non-GM farmers have an incentive to make the effort to preserve the unique identity of their product. In this way, the costs and benefits are internalized and balanced in decision-making.

The EU, in contrast, assigns the rights to non-GM producers (REG 2003/556/EC), thereby compelling seed merchants, GM producers, wholesalers and food processors to undertake policies and practices that isolate GM produce from the non-GM and organic production and supply chain. In this case, however, the commercial benefits of sustained identity for non-GM crops (if any) are captured by the non-GM producers but the costs of identify preservation are largely borne by GM producers. This disconnect between costs and benefits creates the possibility that too much will be expended to maintain coexistence.

One product that has prompted significant study was the proposed introduction of Roundup Ready wheat in Canada and the US. Kuntz (2001) examined the potential cost of GM wheat acting as a lemon in the Canadian international wheat trade, concluding that uncontrolled comingling could jeopardize up to 85% of the price premium estimated to be extracted from international markets by the operation of the then Canadian Wheat Board. Huso and Wilson (2005) and Wilson et al. (2008) were more sanguine about the prospective effect on US producers, concluding that given a lower estimated premium for bread wheat in the US the economic effect would be relatively smaller. Furtan et al. (2005) undertook a complementary analysis, incorporating the changes in farmer benefits along with industry and consumer impacts, concluding that in the presence of the lemons effect there would be no clear first-mover benefits for Canadian farmers adopting GM wheat. Furtan et al. (2007) went on to examine whether there were adequate private incentives for market actors to find some intermediate outcome. They concluded that even if GM producers had the right to operate without restrictions, organic producers in Western Canada had large enough premiums to provide them with the incentive to negotiate landscape clubs that would isolate and preserve their organic status and related market premiums.

In a way, the choices made in Canada, the US and the EU apply aspects of three key product differentiation strategies that are often variably applied to the lemons problem: segregation; identity preservation; and traceability (Phillips and Smyth 2004).

Lin (2002, p. 263) defines segregation as the requirement "that crops be kept separate to avoid co-mingling during planting, harvesting, loading and unloading, storage and transport". Segregation systems are mostly used when food safety could be jeopardized by comingling of the segregated product and all other like-products (Table 10.1). Participation is not optional—any producer or firm involved with segregated products has to comply with rules that have been set or approved by a regulatory agency. A private firm often will have responsibility for developing the actual system but a regulator will be the final arbiter of the system itself. The focus of product delivery within a segregated supply chain is mostly downstream. Segregated commodities commonly have industrial value, so these products are supplied to meet the criteria of the processor. Product failure would most definitely see a complete recall of any products suspected of being affected and could result in criminal prosecution in the most severe instances. Testing and auditing are vital

**Table 10.1** Product differentiation strategies

| | Segregation | IPPM | Traceability |
|---|---|---|---|
| Objective | Product safety | Revenue capture | Liability management |
| Status | Mandatory | Voluntary | Voluntary or mandatory |
| Leader | Regulator | Private company | Commodity group, standards organization or regulator |
| Information flow | One-way | One or two way | Two-way |
| Supply chain focus | Downstream | Downstream | Upstream |
| Testing/auditing | 1st party/regulator | 2nd party/brand owner | 3rd party/standards organization |
| Production arrangements | Regulation and contracts | Contracts | Membership in standard |
| Production controls | Buffers/land use/containment | Product quality | Process standards and record keeping |
| Enforcement | Public | Private | Collective |
| Quality criteria based on | Regulations or HACCP | Brands or product standards | Process standards (e.g. ISO) |
| Tolerance levels | Set in law | Defined by market | Performance based |
| Information provided to | Regulator | Consumer | Regulator, retailer or processor |
| Penalties for failure in product market | Criminal prosecution; mandated product recalls | Consumer fraud charges; lost brand value | Consumer fraud charges; exclusion from product category |
| Price premium | None | Yes | None |
| Labeling | None required | Private brands | Quality marquee |

*Source* Adapted from Smyth and Phillips (2003)

features of segregation systems and are usually conducted by agents of the regulator. While this system is the most formal and visible of the three systems, it is not commonly used. There are very few open-field crop segregation systems presently operating in the global agri-food system. The most commonly cited segregation systems in Canada relate to canola. Canada now has systems in place for high erucic acid rapeseed (HEAR), which has industrial value due to the specific oil profile. A series of non-GMO HEAR varieties are actively cultivated under segregation rules in Canada (called contract registration by the regulator) and in 1996–99, two varieties of transgenic, novel oil canola (Calgene Laurical varieties) were also contract registered and produced. All plant-made pharmaceuticals would also require full segregation, and while none have been approved in Canada a few have been cultivated in open fields in the US. As one might expect, the costs of these systems have tended to be relatively high, and by implication, the product premiums must be higher. HEAR earns on average a $60/tonne premium in Canada and unsubstantiated reports peg premiums well over $100/tonne for other specialty industrial and pharmaceutical crops in the past few years.

Identity preserved production and marketing (IPPM), in contrast, is initiated by private firms in the food industry to extract premiums from a marketplace that has expressed a willingness to pay for an identifiable and marketable product trait or feature. IPPM systems are closed-loop channels that facilitate the production and delivery of an assured quality by allowing identification of a commodity from the disposition of germplasm to the processed product on a retail shelf (Buckwell et al. 1999; Lin 2002). The objective of any IPPM system is revenue management. Premiums need to be available to attract participants and the efforts of voluntary participants will be directed towards capturing a share of the premium. The lead stakeholders in IPPM systems are always private firms seeking to capture increased value of special traits. The role of any regulator is to ensure that industry standards are handled in such a way as to prevent consumer fraud. As with many products, information may be asymmetric, as only the product seller can know with certainty what level of cheating, if any, has occurred in the delivery of the product. Moral hazard may be present due to the presence of premiums. Effective IPPM systems that span entire supply chains often have extensive two-way information flows. Information about purity and quality of the product flows downstream and information coming from consumers' preferences flows upstream. While the information flow in IPPM systems is two-way, the focus of the system is downstream. Each participant in the system wants to ensure they extract a portion of the value of the special trait, whether from production, processing or retailing the product. This means that each participant will focus on the needs of the next participant in the supply chain. Market failure can result in fraud charges for improper labeling and also create fears among consumers that certain brand names cannot be trusted. Second parties acting on behalf of the brand owner or developer of the special trait will usually do testing and auditing.

Numerous IPPM systems operate around the world. Some involve only the breeders and the wholesale market or processor while others extend to the retailer. Their structure depends on the attribute they are trying to preserve. Some novel oils, such as low linolenic oils that are more stable in fryers, only have value at the processing level while others, such as high oleic oils, have health attributes that can be marketed to consumers. Identity preserved production and marketing systems are especially important for providing information to consumers about the provenance of a product, as those attributes are not visible or detectable in the product itself. A number of IPPM systems operate in North America. While organic products are perhaps the most noticeable IPPM products, Cargill has an IPPM system in place to export canola to Japan (the variety gives off virtually no odour when used to fry food), General Mills operates an IPPM system for a select variety of white wheat that possesses a special trait for 'flake curling' when processed into breakfast cereal and Dow AgroSciences uses an IPPM for Nexera canola sold into the Japanese specialty oil market. In each of these systems, there are a range of costs and benefits. Essentially, the systems must generate adequate premiums in the intermediate or final markets to justify the added costs. The market actually determines the value of the systems, by revealing what tolerance for off-types it will allow (which sets the benchmark for costs) and then setting a premium for the differentiated trait. A number of studies show the complexity and sensitivity of system costs to the tolerances for adventitious admixtures (Gosnell 2001; Matlzberger and Kalaitzandonakes 2000). A number of case studies have calculated the incremental costs (over the commodity system) to be in the range of US $20/tonne to US$35/tonne; by assumption their continued operation implies that the premiums must exceed those levels.

The third product differentiation system is traceability. The International Organization for Standardization (ISO) defines traceability as the "ability to trace the history, application or location of an entity by means of recorded identifications." The *Codex Alimentarius Commission* has adopted this as their working definition for all Codex standards (Codex 2001). Traceability systems are designed to ensure that products that encounter problems can be identified, isolated and removed from the supply chain, both mitigating harm and reducing potential liability (Smyth et al. 2004a, b). Participation in a traceability system can be voluntary, depending on where in the supply chain the participant is located. The closer the participant is to the start of the supply chain, the more likely it will be that participation is voluntary. The lead stakeholder may be a commodity group demanding greater clarity in or selection of food products, a standards council comprised of representatives from all sectors of the supply chain or the regulator seeking to ensure consumer protection. Traceability systems have information flowing two-ways as these systems are designed to react quickly to food safety concerns. If a product is discovered to exceed any defined tolerance level at any point in the supply chain, traceability will be used to identify the source of the problem and to locate any retail products that may be affected. Information on food safety flows upstream while information on specific product lots is available to trace flows downstream. This results in the focus of traceability systems being upstream. Market failures can result in consumer fraud charges in addition to permanent exclusion from selling into that supply chain. Testing and auditing is conducted according to the standards developed by third-party organizations.

There are an increasing number of traceability systems operating. Perhaps the oldest and most developed system is the manufacturers' lot number system that traces processed foods from plants to consumers. Those were implemented largely to handle liability of contaminated foods. More recently, the UK retail chains have adopted extensive traceability systems (from retailer to farmer), initially for beef but now for a growing list of products, in order to manage their liabilities flowing from the UK Food Safety Act, 1990. The livestock industry in 25 countries has either introduced or announced plans to introduce partial traceback systems (from slaughter house to birth), while there has been significant discussion about if and how a traceback system can be developed for commodity grains and oilseeds. While these systems often add between 1 and 5% to the cost of a product, they do not add any directly identifiable premium to the product—instead they act as a form of insurance against food contamination that could cause the recall of an entire product category. The EU requires in its latest set of labeling rules a traceback system both within the EU and in the exporting countries in order to validate claims of GM or GM-free status in products without detectable rDNA (EU 2001). This is a novel use of traceback, as no other systems operating in the global food system are imposed by the state—all of the others are economic choices adopted by liable firms seeking to limit their exposure to losses.

## *Standards*

There are no clear, unambiguous, transparent or acknowledged norms or standards about what constitutes GM or non-GM foods. The common definition asserts genetically modified organisms have genetic characteristics which are altered by the insertion of a modified gene or a gene from another organism using the techniques of genetic engineering. Individuals have mixed and at times confused the concepts of genes, transgenes and genetic engineering and how they are used. Regulators complicate this by using a wide and divergent range of terminology and definitions. The EU in Article 2(2) of Directive 2001/18/EC defines genetically modified organism (GMO) as an "organism, with the exception of human beings, in which the genetic material has been altered in a way that does not occur naturally by mating and/or natural recombination" which includes recombinant nucleic acid techniques, mechanical techniques (including micro-injection, macro-injection and micro-encapsulation) and cell fusion (including protoplast fusion) or hybridization techniques, where live cells with new combinations of heritable genetic material are formed by means of methods that do not occur naturally. The US uses the term genetic engineering, which involves recombinant DNA techniques to transfer genetic material among donor and recipient organisms. Just to complicate matters, Canada uses for the purposes of regulatory action the concept of plants with novel traits (PNTs), which are plants that contain a trait which is both new to the Canadian environment and has the potential to affect the specific use and safety of the plant with respect to the environment and human health. These traits can be introduced using biotechnology, mutagenesis or conventional breeding techniques. Adding to this already complicated picture is the fact that the technologies are continuing to evolve. A suite of new biotechnologies, sometimes called NBTs—including gene editing systems such as CRISPR-CAS9 and a variety of knock-out methods—are yet to be fully incorporated into regulatory processes.

Meanwhile, there is a small but significant non-governmental effort to develop standards to handle the concerns related to GM/GE foods. The organic industry, for example, has acted in most developed countries to modify their standards to exclude genetically modified/engineered plants or animals. As befits a process-based standard, their standard only deals with planned, intentional use of GM seeds and does not explicitly address adventitious presence of genetically modified traits.

Mancur Olson highlights the fundamental challenge. Given the underlying non-rival, non-excludable but voice-intensive nature of standards, there are few inherent incentives for those who might benefit from clear and uniform standards to invest their time and energy in the expensive and time-consuming process of developing standards. Given the anonymity of the process and the large number of actors, boundedly-rational opportunistic individuals and institutions will simply wait for others to do the heavy lifting—in short, free-riding is the rational process.

The Canadian process to create a standard for GM foods, funded by the government of Canada, offers one strategy for overcoming the common pool problem of standards. In 1999 the Canadian Council of Grocery Distributors (CCGD) asked the Canadian General Standards Board (CGSB) to assist in developing a National Standard for the Voluntary Labelling of Foods that Are, or Are Not, Genetically Engineered. The CCGD, representing 80 percent of Canada's grocery and supermarket operators, sought to give food processors consistent guidance to assist them in developing label claims that are informative, understandable, verifiable, and not false or misleading. The CGSB, an organization in Public Works and Government Services Canada, oversaw the process to develop the standard. The committee that developed the food labeling standard consisted of 53 voting members and 74 information (or non-voting) members who represented consumers, the food industry, and general interest groups, such as governments and universities. As required under the CGSB process, the three categories were balanced so that no one group was able to dominate the voting. The committee first met in November 1999 and over the next four years it met formally 11 times and held three votes to reach consensus. The final vote released on November 6, 2003, showed that of the 53 total votes expected, 40 voted yes, 6 vote no, 5 abstained and 2 did not vote. The Standard for Voluntary Labelling and Advertising of Foods That Are and Are Not Products of Genetic Engineering as a National Standard of Canada of the Standards Council of Canada was announced on April 15, 2004. This standard offered a two-tailed measure, with non-GM being defined as less than 5% of the product while a proactive assertion of GM content (e.g. a particular nutritional feature) would require 95% of the product to be of that type (CGSB 2004). As a standard, this measure then is adjudicated by national consumer laws related to truthful claims.

## *Labeling*

Ultimately, labels are the most direct form of communication between the actors in the supply chain and the consumer. Effective and efficient labeling systems forestall market failures, but the policy area is fraught with challenges. Labels can and probably should be the front line of defence in response to the 'lemons market.' To do that, however, labels must address the needs of individuals.

Governments are aware of the important role of labels in market-making. Most countries currently have laws that require product labels to disclose weights and ingredients and, increasingly, governments have extended this to include nutritional information. In addition, most countries have truth-in-labeling laws that require labeling claims to be accurate and not misleading. In addition to these formal provisions, firms frequently brand and market their products as differentiated goods in order to optimize their value in the marketplace. In some cases these are supplemented by standards that specify what production methods have been used (or not used), such as Kosher, Halal or organic products. Enforcing these systems then involves governments, firms and various civil authorities or their agents.

Virtually all countries agree on one aspect of GM food labeling—namely that GM foods that have substantive changes in nutrition, composition or allergenicity must be proactively labeled. Beyond that, it is not clear which set of actors should do what. The European Union (Art. 12.2 of REG 1829/2003) requires that all foods containing 0.9% or more GM material by content (either present or if the inputs used involved GM traits) must be proactively labeled as GM. This model of legally mandated positive labeling of GM content is used by at least 22 other countries, including Australia, New Zealand, Japan, China, Korea, Thailand and Brazil, only the threshold varies from 1 to 5%. Alternatively, Canada and Argentina do not have explicit rules mandating labeling; they rely instead on pre-existing consumer laws related to truthfulness in labeling to control for unsubstantiated claims (Phillips and McNeil 2001). While this standoff has been stable for more than a decade, recently there has been efforts to develop mandatory labelling in more than 20 US states and nationally. As of July 1, 2016, Vermont enacted labeling laws but that action is being actively litigated on a number of fronts by a variety of actors.

Government labeling rules could go part of the way to bridging the information asymmetries but they are unlikely to be enough—in absence of any clear definitions, norms or standards, labels simply mask the fundamental information gaps. In absence of a proactive regulatory or market solution, disputes will continue to arise. Most of the more contentious problems will involve international trade.

## The WTO and Low Level Presence

The issue of the management of a traded product which has become inadvertently mingled at trace amounts with another product that is not allowed to enter the customs territory of an importing country has not historically been a concern of international trade law. As a result, the issue of 'low level presence' is not specifically dealt with in international trade agreements. One interpretation of this gap in international trade law would be that the ability to set trade policy in this area falls exclusively in the jurisdiction of individual nation states. Another interpretation is that Member States of the World Trade Organization (WTO) can trigger dispute settlement panels to provide interpretations that will fill in gaps.

Dispute panels generally only make rulings that deal with issues that have not come up before within areas where trade agreements apply. It is clear that dispute panels cannot extend trade agreements into areas where no negotiated agreement exists. For example, a disputes panel could not rule on a case involving trade barriers put in place due to a particular labor practice—say the use of child labor—because labor issues are not dealt with in the WTO but, rather, are dealt with by the International Labour Organization (Bakhshi and Kerr 2008). But in the context of LLP, there is room for action. The well-known Tuna-Dolphin case, for example, involved a WTO Panel that ruled on US legislation and effectively extended extraterritorially to the international management of marine mammals (Isaac et al. 2002). The areas of international trade law where a dispute is likely to arise over LLP relate to food safety or risks to the environment—which are covered under the WTO's Agreement on the Application of Sanitary and Phytosanitary Barriers (SPS). Hence, while one would have to await the bringing forth of a formal dispute to the WTO to determine whether a panel would hear the case, the potential for a determination under the commitments undertaken in the SPS cannot be ruled out.

In modern international supply chains for bulk agricultural products, it is prohibitively expensive to ensure that no comingling takes place—the optimum economic level for comingling is non-zero (Hobbs and Kerr 1999). This necessitates agreements on tolerance levels. These agreed non-zero tolerances are normally determined through arrangements between private sector actors in international supply chains—who are usually interested in ensuring that the tolerances are sufficiently lax to ensure the continued commercial viability of trade in the product that has been inadvertently comingled with a non-traded product—and governments that wish to protect their consumers from either health hazards or purchasing low quality products (i.e. the presence of insect fragments in processed meat lowers the quality of the meat). These tolerances are generally developed with commercial viability being an important part of the rule-setting process. They are not included in trade agreements. For example, these types of tolerances exist for weed seeds, livestock antibiotics, insect fragments, chemical residues, manure and stones in various agri-food products (for a discussion of tolerance levels for common unwanted presences in agri-food trade see Olsen et al. 2001). Taking commercial considerations into account means that the tolerances are not considered trade barriers.

If tolerances are imposed at low levels—as opposed to negotiated—they can be considered a trade barrier and recourse to trade agreements may be sought. Launching a trade dispute at the WTO, however, is an expensive and time-consuming effort that requires support from government and seldom provides a quick 'fix' to a commercial disruption. The interest in LLP as a potential issue in international trade in recent years has arisen in response to the European Union's policy of zero tolerance for imports of products comingled with genetically modified (GM) products that have not been approved for import into the European Union. Taken literally, zero tolerance means a rejection of any import shipment where comingling is detected.

Under its zero tolerance policy, the EU has been routinely rejecting imports where comingling of unapproved GM material is detected. To date, however, these rejections have been one-offs rather than a continuing commercial problem. Examples include: Starlink corn (processed corn approved for animal feed but not for food in the US, 2000); Bt10 corn (comingled in corn gluten for feed, 2005); Liberty Link rice 601 and 604 (2006); Herculex maize (2006/2007); Roundup Ready II and Liberty Link soybeans (2008); BT 63 rice (2008); and MON88017, MON89034 and MIR 604 corn in soybeans (2009). While individual comingled shipments were rejected, the problems were fixed at source relatively easily so that future imports were not comingled and, hence, not threatened with rejection. In these one-off cases, the launching of a dispute at the WTO could not be justified. Knowledgeable insiders report that adventitious comingled products (both GM and non-GM) are endemic in the global grain trade, but that they are handled expeditiously and without conflict when left for the market to resolve.

There have not been many case where outright rejections of comingled product suspended international trade, but once such example was the detection of GM flax by the EU. The EU banned flax from Canada after it was found to be comingled with CDC Triffid, a GM flax variety approved in North America but not approved in the EU. Comingling was found in multiple shipments and subsequently throughout the entire flax supply chain (see Viju et al. 2011, for discussion of the Triffid flax case). Given the widespread nature of the comingling, eliminating it was not technically feasible. The EU imposed a short-term import ban on flax shipments from Canada, resulting in considerable costs borne by both the Canadian flax sector and those EU importers dependent on Canadian supplies (Dayananda 2011). This disruption to trade could have justified the launching of a trade action at the WTO but, in the end, a commercially viable system for establishing tolerances was negotiated between the EU and representatives of the flax industry in Canada. Flax shipments to the EU resumed within a year, albeit at reduced levels due to the cost of the ongoing testing and monitoring system that was put in place to ensure shipments were within the agreed tolerances.

The Canadian response is worth exploring. The industry responded quickly to the closure of the EU flaxseed market by developing a Sample and Testing Protocol. The Protocol was developed by the Canadian Grain Commission (CGC)[1] and the Flax Council of Canada (the private sector representative of the Canadian flax industry) together with the EU Directorate-General for Health and Consumer Affairs (DG SANCO) and other EU stakeholders. Agriculture and Agri-Food Canada (AAFC), Foreign Affairs and International Trade Canada (DFAIT) and the Canadian Food Inspection Agency (CFIA) were also involved in the development of the Protocol. Strictly speaking, however, the Protocol represents a private sector arrangement. The main goal of the Protocol is to meet the EU's zero tolerance for unauthorized GM flax and to assure a secure flaxseed supply. The Protocol established the sampling, testing and documentation measures that must be followed along the entire supply chain of Canadian flaxseed destined for the EU (CGC 2010). Two things are important to note. First, while the testing and monitoring regime established in the Protocol is both onerous and costly, it was accepted by the Canadian industry as a necessary investment to secure EU access. Second, the tolerance level established through the Protocol is not zero, but rather based on the results of a testing regime with non-zero parameters (Viju et al. 2011; Dayananda 2011).

[1]This quasi-government regulator of grain quality in Canada is not officially part of the Canadian government.

The question arises, however, what might happen if a commercially acceptable (and hence viable) import regime could not be negotiated? What would the recourse be to the WTO? This question is important as more and more GM crops are approved in some parts of the world while approvals are denied or delayed in other parts of the world; as discussed in Chap. 3, asynchronous approvals are rising. As a result, the potential for unintentional comingling and an increasing number of import bans under a strict policy of zero tolerance will increase.

While there are other examples of zero tolerance in official policies,[2] the European Union's policy of zero tolerance for unapproved GM products is currently creating the most concern. Thus, the remainder of this chapter will be framed within the context of the EU policy pertaining to imports of GM agricultural products. Other examples of zero tolerance would have to be examined on a case-by-case basis.

The basis for the EU policy of rejecting imports of genetically modified products is that they pose a threat to human, animal or plant health or a risk to the environment.[3] Trade barriers justified on this basis come under the WTO's Agreement on the Application of Sanitary and Phytosanitary (SPS) Barriers. But the SPS rules include nothing specific regarding LLP. The SPS affirms it is a country's fundamental right to protect its citizens, animals, plants and environment from threats arising from imports and countries are allowed to erect trade barriers to achieve that end. There has been a long history, however, of countries using trade barriers justified on the basis of SPS concerns when the actual objective was to provide economic protection.

The Uruguay Round of the General Agreement on Tariffs and Trade (GATT) negotiations that led to the establishment of the WTO was the first serious attempt to bring trade in agricultural products under general GATT disciplines (Kerr 2000). There was considerable concern among those negotiating that, faced for the first time with serious constraints on the ability to extend farmers economic protection from foreign competition, countries would begin to impose trade barriers on the basis of spurious SPS justifications (Kerr 2003). In an attempt to prevent the use of SPS measures for nefarious purposes, the best-efforts provisions regarding SPS measures in the GATT were replaced by the much more specific SPS Agreement (Isaac and Kerr 2003).

The SPS follows the two general WTO principles: non-discrimination in that 'like' products from all countries should be treated the same (the Most Favoured Nation principle); and foreign goods should not be regulated differently than 'like'

---

[2]Examples are various Californian legislation banning the use of certain pesticides in agricultural production (Cash et al. 2003).

[3]It has been argued that the actual reason for the reticence of the EU in approving GM products is resistance from groups in civil society—both consumer groups and environmental groups. As the WTO has no direct mechanism to deal with governments wishing to put trade barriers in place due to consumer or environmentalist requests for protection, they have had to attempt to justify their trade restrictions on the basis of existing WTO rules—and in particular the SPS (see Chap. 6 for a discussion of this issue).

domestic products (the National Treatment principle). Furthermore, any trade barrier put in place should not be more onerous than is necessary to achieve its ends. If countries wish to use trade barriers justified on the basis of SPS concerns, they must provide a scientific justification for the import ban (or other trade restriction) and they must undertake a risk assessment that shows a risk posed by imports that would be moderated by the proposed measure.

The EU's system for approving new GM products allows for a scientific assessment of potential hazards and risks associated with new GM products by the European Food Safety Authority (EFSA). In addition, however, there is a layer of political approval that can override the scientific assessment. Thus, the EU's approval regime for GM products, including approvals for imports, may not be WTO compliant—but that will have to await a disputes challenge at the WTO.[4] Table 10.2 highlights the delays in the variety approval process and the delays indicated continue to this day, as none of the varieties listed have been approved.

The reason given by the EU for the ban on imports of products that are comingled is that the product unintentionally included in import shipments has not been approved in the EU. The threshold for comingled unapproved products is 0.1%. As there is no specific provision for LLP in the SPS, general SPS provisions apply. According to the SPS, however, to put such a ban in place, a Member State must provide a scientific justification and undertake a risk assessment. In the case of the import bans put in place for LLP comingling, the EU provides neither. Instead, the unapproved GM products are assumed to pose a scientific threat and an unacceptable level of risk. This would seem to run directly contrary to the requirement of the SPS. In the case of the EU ban on imports of beef produced using growth hormones, the WTO Disputes panel ruled that the EU was in violation of its SPS commitments because, in part, it had not done a proper risk assessment (Roberts 1998; Kerr and Hobbs 2002).[5] Without a risk assessment, the EU system for new GM products might be considered a trade barrier that was more onerous than required to achieve its objective. According to Article 5 (6) of the SPS:

> when establishing or maintaining sanitary or phytosanitary measures to achieve the appropriate level of sanitary or phytosanitary protection, Members shall ensure that such measures are not more trade-restrictive than required to achieve their appropriate level of sanitary or phytosanitary protection, taking into account technical and economic feasibility.

The EU approval process is very long and costly (Viju et al. 2012). In some cases, approval for the unintentionally comingled product would never be sought in the EU—the product might not be agronomically suitable for cultivation in the EU,

[4]See Viju et al. (2012) for a discussion of the EU regime for approval of GM products in the context of compliance with the WTO.

[5]In the case of the import ban on beef produced using growth hormones the EU chose not the comply with the Panel's ruling and, instead opted to accept retaliation, as is its right under the WTO. Failing to comply with a WTO Panel's ruling was an unprecedented decision by the EU and calls into question the credibility of international commitments. See Kerr and Hobbs (2005) for a discussion of the wider ramifications of the EU's decision to ignore the Panel's ruling.

**Table 10.2** Timeline for GM products in Europe

| Product | EFSA opinion | Number of days for the Commission to vote at the committee level | Number of days for the commission to vote in the appeal committee | Days after council/appeal vote—waiting for approval |
|---|---|---|---|---|
| 1507 maize | 03/03/2005 | Voted after 1462 days | 1131 days and counting | |
| Bt11 maize | 19/05/2005 | Voted after 1385 days | 1131 days and counting | |
| LL rice 62 | 30/10/2007 | 1613 days and counting | | |
| NK603 maize | 11/06/2009 | 1007 days and counting | | |
| Mon810 maize | 30/06/2009 | 1006 days and counting | | |
| MS8xRF3 rapeseed | 22/09/2009 | 922 days and counting | | |
| Gt73 oilseed rape | 15/12/2009 | 838 days and counting | | |
| MON863 maize | 30/03/2010 | 733 days and counting | | |
| MON89034x1507x MON88017x59122 maize | 27/09/2010 | 552 days and counting | | |
| MON89034x1507xNK603 maize | 27/09/2010 | 552 days and counting | | |
| MON531 cotton | 16/09/2011 | 198 days and counting | | |
| MON88017 maize | 10/11/2011 | 143 days and counting | | |
| MON1445 cotton | 16/12/2011 | 107 days and counting | | |
| GA21 maize | 16/12/2011 | 107 days and counting | | |
| MON87701xMON89788 soybean | 15/02/2012 | 46 days and counting | | |
| MON531xMON1445 cotton | 28/03/2012 | 5 days and counting | | |

*Source* Bobo (2012)

the market might be perceived as too small to justify the expense of seeking approval or the GM product may be obsolete. Remember, it is not the primary export that needs approval, only the unintentionally comingled product that cannot be reduced to zero in the supply chain of imports. The case of Triffid flax is illustrative. GM Triffid flax was developed in Canada at the end of the 1990s and sought and received regulatory approval for cultivation in Canada and the US. The major market for Canadian flax is the EU, but Triffid was reaching the stage of commercialization at the point in time when biotechnology was becoming a major political issue in the EU. Concerned about losing the important EU market for flax, the developers of Triffid flax voluntarily withdrew the product and it was subsequently de-registered in North America. Triffid had never been grown commercially but seed companies were growing the variety in preparation for its commercialization. The stocks of Triffid flax were destroyed and, as far as anyone knew, it was no longer in the supply chain (Dayananada 2011). A decade later traces of the Triffid variety were found initially in the EU and subsequently throughout the Canadian supply chain. In the intervening years, the Triffid variety had become agronomically obsolete having been replaced by superior, but non-GM varieties. Hence, approval for Triffid had never been sought in the EU. It was the market for Canadian non-GM flax that suffered from the import ban on comingled GM flax. Hence, in the absence of a risk assessment, the alternative to the import ban provided by the EU—seeking approval to import—could be argued as being an overly onerous barrier to trade. Of course, clarity on this issue will have to await a WTO disputes challenge and a Panel ruling.

The SPS does, however, allow the imposition of import barriers if there is insufficient scientific information—as a precaution. According to Article 5 (7) of the SPS:

> In cases where relevant scientific evidence is insufficient, a Member may provisionally adopt sanitary or phytosanitary measures on the basis of available pertinent information, including that from the relevant international organizations as well as from sanitary or phytosanitary measures applied by other Members. In such circumstances, Members shall seek to obtain the additional information necessary for a more objective assessment of risk and review the sanitary or phytosanitary measure accordingly within a reasonable period of time.

The import bans on LLP under the policy of zero tolerance put in place by the EU do not explicitly state that they are justified on the basis of precaution—insufficient scientific information exists. If challenged, however, this would seem to be an avenue that could be used to support the import bans. It is the other aspects of Article 5 (7) where it could be argued that the EU system fails to comply. When imposing its import bans, the EU does not examine the sanitary and phytosanitary measures that other Members apply or the scientific evidence that underlies those measures. The EU import bans are automatic for unapproved GM products. There is no process for examining the scientific evidence of other countries and to consider such evidence sufficient to allow imports—the only recourse is to seek formal approval of the GM in the EU.

The EU does make a distinction between two types of unapproved GM products. The first is commonly known as a low level presence where the GM product is approved in the export market but not in the importing market. The second form of GM event is known as 'adventitious presence', occurring when the GM product is not approved in any market (i.e. is an experimental product or is cultivated under confined field trials) (Viju et al. 2011). In the latter case, where sufficient scientific information does not yet exist, a precautionary ban may be accepted as a justification. In the first case, however, scientific evidence does exist because the products have received approvals by foreign authorities based on scientific evidence. For example, in the case of Triffid flax, it had received regulatory approval for cultivation in both Canada and the US. While an importing country may chose not to accept such evidence as being sufficient to approve imports, as in the EU where import bans are automatic and such evidence is not taken into account, it could be argued as not being in compliance with SPS Article 5 (7).

Further, if the scientific evidence is not considered sufficient and an import ban is imposed as a precaution, then it is incumbent upon the Member imposing the ban to proactively seek the additional information required to undertake a full scientific examination and a risk assessment. There is no evidence of the EU seeking such information. The EU regulatory approach is entirely passive, only allowing those wishing to import to use the processes established for formal approval of new GM products. The case of a Member failing to "seek to obtain the additional information necessary for a more objective assessment of risk and review the sanitary or phytosanitary measure accordingly within a reasonable period of time (SPS Article 5 (7))" has never been brought to a disputes panel and clarity would require a ruling from a panel.

## Conclusions

The global agri-food markets, and especially markets for genetically modified foods, are at a crossroads. It is probably fair to assume that consumers will not get less demanding over coming years or that technology will stop advancing. The real uncertainties relate to how regulators and industry will manage the two trends.

There would appear to be three possible market responses. The pessimistic future might be that technologies swamp the capacities of consumers and regulators to assess, test and adopt new products and of producers to confidently and appropriately quality assure their products. In that case, we might face higher or more impermeable barriers in the international marketplace, and food security would suffer. An alternate, optimistic outcome might be for regulators, industry and interest groups to come together through negotiation or litigation and adopt common standards, testing protocols and regulatory processes that can effectively and efficiently deliver the right quantities and qualities to meet the demand. This is not impossible to achieve but is unlikely given the trajectories of governments and industry. The most likely outcome would involve us muddling through, with firms, sectors, standards organizations and governments just about keeping ahead of the pressures for product differentiation coming from consumers, the technology and regulators. To some extent, this would be like moving along the hogs-back of a steep hill—one small deviation either side would tip the sector into a new direction.

## References


Ackerlof, G. 1970. The market for lemons: Quality, uncertainty and the market mechanism. *Quarterly Journal of Economics* 84: 488–500.

Bakhshi, S., and W.A. Ker. 2008. Incorporating labour standards in trade agreements: Protectionist ploy or legitimate trade policy issue? *International Journal of Trade and Global Markets* 1 (4): 373–391.

Belcher, K., J. Nolan, and P.W.B. Phillips. 2005. Genetically modified crops and agricultural landscapes: Spatial patterns of contamination. *Ecological Economics* 53: 387–401.

Benacek, V. 2005. Three dimensions of modern social governance: Markets, hierarchies and kinships. *Pracovni Sesity Working Papers*, No. 1/2005.

Bobo, J.A. 2012. *Senior advisor for biotechnology*. United States Department of Agriculture, Personal Communications.

Boulding, K. 1970. Organizers of social evolution. In *A primer on social dynamics: History as dialectics and development*, ed. K. Boulding, 19–36. London: Collier MacMillan Ltd.

Buckwell, A., G. Brookes, and D. Bradley. 1999. Economics of identity preservation for genetically modified crops. *Food Biotechnology Communications Initiative*.

Canadian General Standards Board. 2004. *National Standard of Canada: Voluntary Labelling And Advertising of Foods That Are and Are Not Products of Genetic Engineering*. CAN/CGSB-32.315-2004. Online at: http://www.tpsgc-pwgsc.gc.ca/cgsb/on_the_net/032_0315/standard-e.html.

Canadian Grain Commission. 2010. *Background Information on Genetically Modified Material Found in Canadian Flaxseed: Low Levels of Genetically Modified Material Found in Canadian Flaxseed*. Winnipeg: Canadian Grain Commission, February 1. Online at: http://www.grainscanada.gc.ca/gmflax-lingm/pfsb-plcc-eng.htm.

Cash, S., D.L. Sunding, A. Swoboda, and D. Zilberman. 2003. Indirect effects of pesticide regulation and the Food Quality Protection Act. *Current Agriculture, Food and Resource Issues* 4: 73–79.

Coase, R. 1937. The nature of the firm. *Economica*, n.s.4 (16): 386–405.

Coase, R. 1960. The problem of social cost. *The Journal of Law and Economics* 3 (1): 1–44.

Codex Alimentarius Commission. 2001. *Matters arising from Codex committees and task forces: Traceability*. Rome: Food and Agriculture Organization of the United Nations. Online at: ftp://ftp.fao.org/codex/ccexec49/al0121ee.pdf.

Dayananda, B. 2011. The European Union Policy of Zero Tolerance: Insights from the Discovery of CDC Triffid. M.Sc. thesis, University of Saskatchewan.

European Commission. 2003. *Commission Recommendation of 23 July 2003 on guidelines for the development of national strategies and best practices to ensure the coexistence of genetically modified crops with conventional and organic farming*. 2003/556/EC. http://eur-lex.europa.eu/LexUriServ/LexUriServ.do?uri=OJ:L:2003:189:0036:0047:EN:PDF.

European Commission. 2009. *Report from the Commission to the Council and the European Parliament on the coexistence of genetically modified crops with conventional and organic farming, COM(2009) 153 final*. Brussels, April 2.

European Union. 2001. *Regulation of the European Parliament and of the Council concerning traceability and labelling of genetically modified organisms and traceability of food and feed products produced from genetically modified organisms and amending Directive 2001/18/EC*. Accessed at: http://europa.eu.int/comm/food/fs/biotech/biotech09_en.pdf.

Furtan, W., R. Gray and J. Holzman. 2005. Regulatory approval decisions in the presence of market externalities: The case of genetically modified wheat. *Journal of Agricultural and Resource Economics* 30 (1):12–28.

Furtan, W., A. Guzel, and S. Weseen. 2007. Landscape clubs: Co-existence of genetically modified and organic crops. *Canadian Journal of Agricultural Economics* 55 (2): 185–195.

GM Coexistence Conference (GMCC). 2007. *Third International Conference on Coexistence between Genetically-Modified (GM) and non-GM based Agricultural Supply Chains.* Conference Website, Seville, Spain. Accessed at: http://www.gate2biotech.cz/third-international-conference-on-coexistence-between-genetically-modified-gm-and-non-gm-based-agricultural-supply-chains-gmcc/.

Gosnell, D. 2001. Non-GM Wheat Segregation Strategies: Comparing the Costs. M.Sc. thesis, University of Saskatchewan.

Hobbs, J.E., and W.A. Kerr. 1999. Costs/benefits of microbial origin. In *Encyclopedia of food microbiology*, ed. R.K. Robinson, C.A. Batt, and P. Patel, 480–486. London: Academic Press.

Huso, S., and W. Wilson. 2005. *Impacts of genetically modified (GM) traits on conventional technologies.* ND State University Agribusiness and Applied Economics Report No. 560-S, March.

Isaac, G., and W.A. Kerr. 2003. GMOs at the WTO: A harvest of trouble. *Journal of World Trade* 37 (6): 1083–1095.

Isaac, G., M. Phillipson, and W.A. Kerr. 2002. *International Regulation of Trade in the Products of Biotechnology, Estey Centre Research Papers No. 2.* Saskatoon: Estey Centre for Law and Economics in International Trade.

James, C. 2015. Global status of commercialized biotech/GM crops: 2015. ISAAA Brief 51.

Kerr, W.A. 2000. The next step will be harder: Issues for the new round of agricultural negotiations at the World Trade Organization. *Journal of World Trade* 34 (1): 123–140.

Kerr, W.A. 2003. Science-based rules of trade—A mantra for some, an anathema for others. *Journal of International Law and Trade Policy* 4 (2): 86–97.

Kerr, W.A., and J.E. Hobbs. 2002. The North American-European Union dispute over beef produced using growth hormones: A major test for the new international trade regime. *The World Economy* 25 (2): 283–296.

Kerr, W.A., and J.E. Hobbs. 2005. Consumers, cows and carousels: Why the dispute over beef hormones is far more important than its commercial value. In *The WTO and the regulation of international trade*, ed. N. Perdikis, and R. Read, 191–214. Cheltenham: Edward Elgar.

Khaminwa, A. 2003. Co-existence. In *Beyond intractability: A free knowledge base on more constructive approaches to destructive conflict.* Online at: http://www.beyondintractability.org/essay/coexistence/.

Kuntz, G. 2001. Transgenic Wheat: Potential Price Impacts for Canada's Wheat Export Market. Master's thesis, University of Saskatchewan Department of Agricultural Economics, Saskatoon, Sk.

Lancaster, K.J. 1966. A new approach to consumer theory. *Journal of Political Economy* 74 (2): 132–157.

Lin, W. 2002. Estimating the costs of segregation for non-biotech maize and soybeans. In *Market development for genetically modified foods*, ed. V. Santaniello, R.E. Evenson, and D. Zilberman. Wallingford, UK: CABI Publishing.

Mahoney, J. 1992. The choice of organizational form: Vertical financial ownership versus other methods of vertical integration. *Strategic Management Journal* 13: 559–584.

Maltzberger, R., and N. Kalaitzandonakes. 2000. The hidden costs of identity preservation. *Feedstuffs* 72 (36): 1–2.

Mokyr, J. 1997. *The political economy of technological change: resistance and innovation in economic history.* Online at: www.faculty.econ.northwestern.edu/faculty/mokyr/berg.pdf.

North, D. 1991. Institutions. *Journal of Economic Perspectives* 5: 97–112.

Office International Des Epizootics (OIE). 2000. *A short history of the office international des epizootics.* Paris: OIE Press. Online at: www.oie.int/eng/OIE/en_histoire.htm.

Olsen, A.R., J.S. Gecan, G.C. Ziobro, and J.R. Bryce. 2001. Regulatory action criteria for filth and other extraneous materials v. strategy for evaluating hazardous and nonhazardous filth. *Regulatory Toxicology and Pharmacology* 33: 363–392.

Olson, M. 1965. *The logic of collective action: Public goods and the theory of groups*. Harvard: Harvard University Press.

Paquet, G. 2001. The new governance, subsidiarity and the strategic state. In *OECD, governance in the 21st century*. Paris: OECD. Online at: http://www.oecd.org/dataoecd/15/0/17394484.pdf.

Phillips, P.W.B. 2007. *Governing transformative technological innovation: Who's in charge?*. Oxford: Edward Elgar.

Phillips, P.W.B., and H. McNeill. 2001. Labelling for GM foods: Theory and practice. *AgBioForum* 3 (4): 219–224.

Phillips, P.W.B., and S.J. Smyth. 2004. Managing the value of new-trait varieties in the canola supply chain in Canada. *Supply Chain Management* 9 (4): 313–322.

Picciotti, R. 1995. *Putting Institutional Economics to Work: From Participation to Governance, World Bank Discussion Paper 304*. Online at: http://www-wds.worldbank.org/servlet/WDSContentServer/WDSP/IB/1995/09/01/000009265_3961219115337/Rendered/PDF/multi_page.pdf.

Roberts, D. 1998. Preliminary assessment of the effects of the WTO agreement on sanitary and phytosanitary trade regulations. *Journal of International Economic Law* 1 (3): 377–405.

Schumpeter, J. 1954. *Capitalism, socialism, and democracy*. London: George Allen and Unwin.

Seabright, P. 2005. *The company of strangers: a natural history of economic life*. Princeton: Princton University Press.

Smyth, S.J., and P.W.B. Phillips. 2003. Product differentiation alternatives: Identity preservation, segregation and traceability. *AgBioForum* 5 (2): 30–42.

Smyth, S.J., P.W.B. Phillips, W.A. Kerr, and G.G. Khatchatourians. 2004a. *Regulating the liabilities of agricultural biotechnology*. Wallingford: CABI Publishing.

Smyth, S.J., G.G. Khachatourians, and P.W.B. Phillips. 2004b. The liability from regulating gene flow in plant-made pharmaceuticals. *Biotechnology and Genetic Engineering Reviews* 21: 277–297.

Thompson, P. 2006. Should we have GM crops? *Santa Clara Journal of International Law* 4: 75–95.

Viju, C., M.T. Yeung, and W.A. Kerr. 2011. *Post Moratorium EU Regulation of Genetically Modified Products—Triffid Flax, CATPRN Commissioned Paper, No. 2011-03*. Canadian Agricultural Trade Policy and Competitiveness Research Network. Online at: http://www.uoguelph.ca/catprn/PDF-CP/CP2011-03-viju-yeung-kerr.pdf.

Viju, C., M.T. Yeung, and W.A. Kerr. 2012. The trade implications of the post-moratorium European Union approval system for genetically modified organisms. *Journal of World Trade* 46 (5): 1207–1238.

Williamson, O. 1979. Transaction-cost economics: The governance of contractual relations. *Journal of Law and Economics* 22: 233–261.

Wilson, W., E. DeVuyst, R. Taylor, W. Koo, and B. Dahl. 2008. Implications of biotech traits with segregation costs and market segments: The case of Roundup Ready® Wheat. *European Review of Agricultural Economics* 35 (1): 51–73.

# Chapter 11
# Multilateral Trade Negotiation Options

**Abstract** The advent of biotechnology products has triggered a spirited debate about how we should assess risks, who should undertake the task of defining what is acceptable, what rules we should draw upon and where we should vest the authority to decide. In recent decades there has been a strong move to normalize and institutionalize a 'science-informed' system in international science and international treaties. Recently there has been a substantive pushback against the privileged role scientists and science institutions play in decision-making, especially in the introduction and use of genetically modified (GM) crops. We assess the underlying information, valuation and selecting rules involved in the battle between 'science-informed' decision-making and rules incorporating socio-economic considerations. This chapter examines the underlying rules, processes, structures, efforts and outcomes to normalize innovative technologies in global agri-food trade.

## Introduction

The development of genetically modified (GM) crops began in the 1980s and the first GM products rolled out into the market in the mid-1990s with little or no consumer backlash. As the products of science, they simply slipstreamed into the science-based international regulatory regime, with scientists largely filling in the missing pieces in the system. Even in the EU, products labeled as GM were leading sellers (Krebs 2000). As the decade progressed, new voices with new concerns were heard, but, as suggested in previous chapters, they found little or no room in the science-based system for their views to be represented (Kerr 2010). In response, some opponents and their governments attempted to find ways to broaden the scope of the scientific base that would be considered in product evaluations, using a variety of clauses in national legislation and international treaties, such as the World Trade Organization (WTO), but often found that those systems could not fully incorporate their concerns (as discussed in Chap. 7). In response, there was an international effort, led by the European Union (EU), to develop complementary regulatory and treaty systems that would add an additional socio-economic filter to

 
S.J. Smyth et al., *Biotechnology Regulation and Trade*, Natural Resource Management and Policy 51, DOI 10.1007/978-3-319-53295-0_11

new product assessment. While some of these processes were incorporated into domestic law, there was a strong push throughout the 1990s and into the 21st century to develop new international treaties that could complement or, at times, supplant the science-based approach with a broader set of perspectives and actors. As a result, two parallel systems currently exist that acknowledge their differences and have at least accepted that dueling systems may not achieve either the efficiency of science-based review or the reflexivity of socio-economic considerations (SECs).

This chapter examines the task of defining what is acceptable, what rules we should draw and where we should vest the authority to decide. Over the past few generations there has been a strong move to normalize and institutionalize a 'science-based' (or more accurately 'science-informed') system in international treaties, which offers a set of processes and experiences one can assess and consider. Meanwhile, the effort to construct a complementary or competing system based on SECs is less developed. The authority to undertake such efforts is now embedded in a range of institutions, but the norms, practices, principles and institutions that will deliver authoritative and repeatable evidence are not yet in place. This chapter applies Ostrom's (2005) institutional analysis and development (IAD) framework that was outlined in Chap. 7 to animate our institutional analysis. This approach begins with the assumption that decision systems define the scope of review, the ways relevant information is agglomerated, the normative value placed on different types of knowledge and the process(es) of selecting the pathway(s) of management, all in the context of constraints (e.g. a budget constraint or bounded rationality). Ultimately, this framework offers an approach to unpack the multiple layers of processes. The framework is applied to assess the underlying information, valuation and selecting rules involved in the battle between 'science-informed' decision-making and decisions informed by SECs.

## Background

In many ways, the agricultural biotechnology industry is at the leading edge of the battle between our largely modernist economy and society and the post-modernist critique. As a product of deductive, reductionist, corporate science, GM crops are in many ways the natural conclusion of more than 150 years of increasingly programmed, organized and managed scientific enterprise, where the universities, industry and governments have collaborated to generate 'progress' in the name of and for their citizens and consumers. While there is much one might criticize about our modern economy and society, proponents of modernism can fairly assert that this power system has, for the majority of the world's population, created unprecedented wealth and quality of life.

Most of the regulatory and governance systems and subsystems erected in OECD countries accept and build upon the notion of economic and social progress as the ultimate goal of society. In the extreme, critics argue that the modernist enterprise represents a meta-narrative which asserts that humankind can and should discover 'objective truths' through scientific discovery and apply those advances to increase the aggregate wealth of mankind.

As early as 1870, at the height of the second industrial revolution, one can see the stirrings of a counterrevolution that challenges the notion that there is any objective truth. A wide range of critics of modernism assert that all apparent realities are simply social constructs, reflecting the power relations of their time. What started as a critique of art, culture and social structures has now evolved into an all-out critique of the structures and artefacts of modernism, including the systems that generate and govern the products of agricultural biotechnology.

While the modernist, scientifically-based system has continued to add to and refine its understanding of science and its uses, the post-modernist counterrevolution has worked to define and develop an alternate power system for adjudicating what should be done and by whom. The modernist science-based industrial and regulatory system accepts as given the power system and the norm of progress, and then focuses on further modularizing the knowledge they believe is needed to optimize progress. Most of our existing national and international scientific and regulatory structures—such as the risk analysis framework (RAF) that all Organisation for Economic Cooperation and Development (OECD) countries have embedded in their national regulatory systems, the specialty technical agencies of the United Nations (e.g. the International Plant Protection Convention, (IPPC); World Organization for Animal Health, (OIE); World Health Organization, (WHO); Food and Agriculture Organization (FAO); and the *Codex Alimentarius Commission* (Codex), (Codex)), the international trade regime (e.g. the WTO and its sub-agreements, World Intellectual Property Organization (WIPO) and regional trade agreements such as the European Union and the North American Free Trade Agreement (NAFTA)) and the OECD itself—accept these norms and principles and, for the most part, operate as it they are unaware or at least uninterested in their critics' desire for a more reflexive, open and flexible range of decision-making systems and outcomes.

Post-modernists have promoted change from both within the existing system and by working to create new alternative decision-making systems. In the context of existing systems, there have been efforts to exercise the powers of derogation to make decisions that deviate from the central tendency of scientific justification and economic efficiency. In some cases this works; most often it is not effective.

As noted, governance and regulation of GM crops is a key battle ground. Both modernists and post-modernists acknowledge that transformative technological innovations like GM crops exhibit a range of the attributes that challenge our existing systems' ability to fully address all of the possible questions. As noted in Chap. 6, a few elements of transformative technologies stand out. Transformative technologies involve disjointed, step adjustments in our productive and institutional capacity, displacing, destabilizing or overturning precursor systems. In technical terms, they often are driven by new epistemologies, they offer significant complementarities (within and across sectors), and they tend to involve recombined or hybrid technologies (one might actually label them convergent technologies). Furthermore, although they often emerge wrapped in an optimistic fervour, they are hard to anticipate because they infiltrate and influence multiple sectors, markets and domains over a long and variable timeline—in spite of the rhetoric, none of the identified technologies overwhelmed society quickly. As mentioned above, the changes brought by the automobile have yet to complete the move to a new equilibrium (Kerr 2016). This makes it difficult to manage or govern them through established markets or authorities. The nebulous nature of transformative technology creates uncertainty, which tends to generate debate and controversy (Phillips 2007).

These types of technologies not only challenge the modernist architecture for science, regulation and intellectual property rights, they open the door to new actors with new questions and new values, interests and beliefs. In short, there is both criticism and competition within and between the modernist and post-modern camps.

The modernist enterprise is well advanced in its efforts. The emergence of the WTO in 1995 provides a clear and unambiguous venue for advancing the science-based system. The WTO, and the earlier GATT, is explicitly focused on optimizing economic welfare through trade, by implication and design limiting national efforts to restrict international trade in products of science and technology (see Chap. 6). The principles-based GATT has always allowed limited derogations based on a relatively wide range of 'legitimate objectives' (such as market making [Art. XI] and health, safety, morality and protection of the environment [Art. XXb]), provided they accord with the principles of non-discrimination (Art. I Most Favored Nation (MFN) clause) and national treatment (Art. III). Formal incorporation of the Sanitary and Phytosanitary (SPS) and Technical Barriers to Trade (TBT) sub-agreements into the WTO in 1995 was designed to close many most of the more aggressive derogations. Three key panel decisions related to SPS measures confirmed the limited scope for derogations: the EU hormones case (Kerr and Hobbs 2005); the Australian salmon case; and the Japanese agricultural products case. In all three cases a contested domestic SPS measure was struck down on the basis that there was no risk assessment completed to support the SPS measure, or the risk assessment was improperly done. This recognition of the pivotal importance of a proper science-based assessment is noteworthy. The salmon case decision, in particular, set out three key criteria for a 'proper' risk assessment: (1) it must identify the diseases whose entry, establishment or spread a Member wants to prevent within its territory, as well as the potential biological and economic consequences associated with the entry or spread of these diseases; (2) it must evaluate the likelihood of the entry, establishment or spread of these diseases, as well as the potential biological and economic consequences of failing to prevent introduction; and (3) it must evaluate the likelihood of the entry or spread of these diseases according to the SPS measures which might be applied (WTO 1998, para 121). As suggested above, if a risk assessment does not even refer to the SPS measure, it is doomed to failure before a WTO Panel. In short, nations that tested the new rules have discovered that while dispute settlement rulings accept derogations for legitimate objectives, they are very narrow in terms of their scale, scope, duration and flexibility. Even where derogations are allowed, the strict rules of science and economics govern their application, which simply confirms the post-modernist assertion that these systems are more about power than universal truth.

Overall the scientifically-informed, economic model underpinning the WTO—which predicts and offers remedies only for cases where producers seek protection in search of economic rents—is unable to accommodate the demand of consumers, citizens and belief-based environmental non-governmental organizations (eNGOs) who are seeking protection from, or control over, the decision-making system.

To many, 'science' is simply not working. In the absence of any agreement on what makes a 'scientific consensus' (Kuhn 1970, called consensus a 'paradigm' which incorporates known theories and known evidence, aka his known-knowns), it is not clear when there is enough science. In that context, simply addressing Kuhn's unknown-knowns (i.e. where we either have known theory or evidence, but not both) will not satisfy post-modernists, as they assert paradigms are simply reflections of the prevailing power systems and not ultimate truths.

Those unwilling to simply continue the slow, patient work of filling in the unknowns fall into two camps. Some want a whole new set of rules that reflect new norms while others simply want a way to pause the process for a while to allow the science and people to catch up to the new realities. Both camps have used the emerging norm of 'precaution' as a way to achieve their, albeit different, outcomes. All countries have some form of precautionary approach, either formally articulated or informally used, that enables them to delay or suspend judgment on a product that is suspected of posing unacceptable risks. The first formal articulation of precaution was in the World Charter for Nature in 1982. Since that time the policy has been expressed in numerous national and international regulatory systems. Principle 15 of the 1992 Rio Declaration on the Environment and Development states:

> [i]n order to protect the environment, the precautionary approach shall be widely applied by States according to their capabilities. Where there are threats of serious or irreversible damage, lack of full scientific certainty shall not be used as a reason for postponing cost-effective measures to prevent environmental degradation.

The Commission of European Communities offered guidelines for using its version in a politically transparent manner:

> measures … must not be disproportionate to the desired level of protection and must not aim at zero risk … comparable situations should not be treated differently and … different situations should not be treated in the same way, unless there are objective grounds for doing so … measures … should be comparable in nature and scope with measures already taken in equivalent areas in which all the scientific data are available … the measures must be of a provisional nature pending the availability of more reliable scientific data … scientific research shall be continued with a view to obtaining more complete data (EC 2000, pp. 19–21).

While in the US and Canada the concept of precaution advocated by the Convention on Biological Diversity (CBD) and EU (which they argue has been used to delay decisions excessively), they both have precaution as a guiding principle in their assessment systems. The difference would appear to be more in tenor and intent rather than general principles.

Those who want a more permanent change in the power system are seeking to place socio-economic considerations (SECs) at the centre of decision making, by working through other venues. As noted, WTO Articles XI and XXb and the SPS and TBT Agreements offer venues for further, albeit limited, elaboration. Some assert that the scope for SEC is not fully tested. For example, the WTO is not about 'trade at any cost'—the preamble of the Marrakesh Agreement Establishing the World Trade Organization (1994) affirms "…the objective of sustainable development, seeking both to protect and preserve the environment…". How this might play out in the case of a dispute explicitly involving SECs for GM crops has yet to be tested. It is not clear what flexibilities and policy space is available in the sub-agreements, in particular. Risk assessment under the SPS Agreement, for example, already involves a mix of scientific and economic considerations. Article 5.3 of the SPS Agreement provides that in assessing risks to animals and plants:

> Members shall take into account as relevant economic factors: the potential damage in terms of loss of production or sales in the event of the entry, establishment or spread of a pest or disease; the costs of control or eradication in the territory of the importing Member; and the relative cost-effectiveness of alternative approaches to limiting risks (Article 5.3).

In the meantime, other venues have emerged where SECs are accepted and may become more influential than science in determining decisions. Article 8j of the CBD directs Parties to consider traditional knowledge from a range of perspectives, which is then embodied and amplified in Articles 9 (farmers rights) and 13 (benefits sharing) in the International Treaty on Plant Genetic Resources for Food and Agriculture. More recently Article 26(1) of the Cartagena Protocol on Biosafety (CPB) directs that member states may assess the impact on the conservation and sustainable use of biological diversity arising from the first transboundary movement, handling and use of a living modified organism (LMO); the impact analysis may include the potential effects on biological diversity, which may include social or economic considerations. Article 26 also specifically mentions that the 'value' of biological diversity to indigenous and local communities may be considered, which links to Article 8(j) of the CBD and the Treaty and could involve intrinsic valuation of knowledge, innovations and practices of indigenous and local communities embodying traditional lifestyles. These venues have all created space for SECs to be addressed but have yet to specifically discuss how they might create some norms and rules to reduce to practice and use what is a quite diverse set of socio-economic methods (Ludlow et al. 2014). The CPB has sponsored on-line dialogues that highlighted the gap between the aspirations and realities of those seeking more formal SEC consideration.

One feature so far not fully explored is the potential impact of poorly designed and structured rules based on SECs. From past experience in the international trade realm, when trade rules are poorly structured and adjudicated private firms often find a way to extract an economic rent from the resulting market imperfections. There is a real risk that firms or other rent-seeking actors will similarly use poorly operating systems based on SECs to extract a rent, with the likely outcome that both the economy and society will be poorer as a result.

## Institutional Analysis and Development Framework

Ostrom's Institutional Analysis and Development (IAD) framework has been used recently in an attempt to explain the institutional underpinnings of many complex problems faced in today's world (Phillips et al. 2015). The intention behind developing such a framework was to coherently and comprehensively deconstruct the complexly nested layers of organizational environment, actors, rules and the anticipated outcomes. The focus of this framework is not just on the organizational structures; it also reveals the complex layers of interactions between actors and the organizational codes as well as interactions among the actors themselves. It highlights the behavioural underpinnings of many of the problems and the complex solutions that emerge in institutional contexts. In doing so, it sheds light on the feedback loops between the exogenous environment, the community, the rules, the action situation, the actors and their interactions with the exogenous and endogenous variables and, finally, the outcomes. The IAD takes into account that the connections between any two or more of these variables are not linear; rather they are complexly nested. This, as many critics argue, adds to the already complex and dynamic picture of institutional analysis. However, understood properly, the IAD framework puts forward a coherent vocabulary for the study of institutions. Applied correctly, it can help resolve much of the institutional complexity surrounding any given problem without taking away any of the pertinent details.

The IAD framework takes a systems approach to policy processes, whereby inputs, policy-makers, outputs, outcomes, evaluative criteria and feedback effects are connected to each other in a dynamic manner. As outlined above, Ostrom argues that any complex system can be viewed as being composed of 'holons', which constitute complete sub-systems within an overarching system. These sub-systems, in a complex adaptive system, "can be 'dissected' into its constituent branches on which the holons represent the nodes of the tree, and the lines connecting them to the channels of communication, control or transportation" (Ostrom 2005, p. 12). While the notion of complex systems has been in use since the seminal work by Simon (1955), the IAD framework formalizes the analysis by clearly spelling out the constituent parts and providing a coherent understanding of what these components mean for the entire system. It also highlights the fact that many of the interactions within and across the sub-systems occur simultaneously at multiple levels. The IAD framework therefore provides the institutional analyst with the luxury either to analyze the system as a whole or to focus on a set of sub-systems independently or jointly.

Another key analytical feature of the IAD is that it allows the analyst to account for possibilities of polycentric governance—"a system of governance in which authorities from overlapping jurisdictions (or centers of authority) interact to determine the conditions under which these authorities, as well as the citizens subject to these jurisdictional units, are authorized to act as well as the constraints put upon their activities for public purposes" (McGinnis 2011, p. 171). Given that most knowledge-based economies rely on polycentric systems of governance, it is

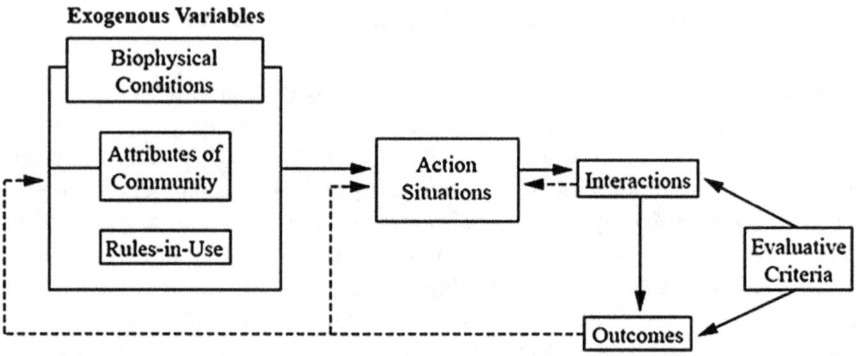


**Fig. 11.1** Components of the IAD framework. *Source* Ostrom (2005, p. 15)

imperative to be able to dissect the underlying components of a system to determine the governing forces which exert influence on that sub-system and in turn on the overall system.

At the simplest analytical level, the IAD framework can be viewed as consisting of exogenous variables, an action arena, and the interaction between these, the outcomes, and the evaluative criteria. All these variables have feedback effects on each other, which transforms the system as a whole as well as the variables themselves. The elements defining the exogenous variables include biophysical or material conditions, attributes of community and rules. The action arena is composed of the action situations and participants. Interactions between the action arena and the exogenous variables determine the outcomes, which are then evaluated using the criteria appropriate for the given system. Outcomes are fed back into the action arena and the exogenous variables, which results in transformations within the holons or in the entire system or both (Fig. 11.1).

## A New Intellectual Architecture?

Risk management is fundamentally the result of a nested set of subsystems. While one can find an almost infinite number of subsystems or holons, for the sake of this argument we start with the attribution and assignment of ownership—through both formal, codified and through informal, community-based intellectual property rights —and then examine the engaged and latent subsystems that can or could be part of delivering effective, efficient and reflexive decisions on the introduction and use of new products of biotechnology.

As suggested above, each holon can be framed by asking the basic journalistic questions—Who? What? Where? When? Why? and How?

As for who, one basic assumption accepted by many is that governments are the only legitimate, authoritative actors in the field of risk analysis. While that may be true at times, there is ample evidence that sub-systems operating in the RAF are sometimes led or dominated by firms and a range of civil authorities, some strongly values driven, some fundamentally interests-based and some primarily focused on enforcing epistemological norms (Phillips 2007). In the context of GM foods, the epistemic civil authorities command key parts of the system. These epistemic communities—e.g. scientists, lawyers and economists—have recognized expertise and competence in a particular domain and an authoritative claim to specific knowledge within that domain or issue area. These communities and their members have a shared set of normative and principled assumptions, shared causal beliefs, shared notions of weighing and validating knowledge in the domain, a shared technical vocabulary and a common set of practices associated with a specific set of problems (Haas 1992). Most epistemic communities in this area span international boundaries and have representatives in industry, government and many other civil authorities. Unlike values-based social movements and interest-based groups (both of which are unlikely to share causal beliefs), epistemic communities organized around disciplines or in professions share causal beliefs and a consensual knowledge base, but do not share principled beliefs or interests (which are the unifying feature of social movements and interest groups). For the same reason, epistemic communities do not cohabitate easily with markets or the state—their foundational commitment to a set of causal beliefs and consensual knowledge is inconsistent with the practical politics of legislators and bureaucratic agencies or the self-interested nature of market actors. Nevertheless, epistemic communities are vital actors in the identification of what is known and unknown about new technologies, how to evaluate the impact of a new technology and the technical options for filling in our knowledge gaps. They are not, however, able to authoritatively define the boundaries of what is acceptable risk—that is inherently a political process, involving balancing of goods and harms.

As for what gets normalized, the natural scientists and risk analysis experts have developed a language and a common set of concepts, methods and processes to address the science. With the advent of genomics, molecular biology and an ever expanding list of 'omics', scientists have gone a long way to defining the structure or biological matter (using the four base pairs of AC and GT,[1] chemistry and proteomics to almost uniquely identify sources and functions of genes). The regulatory community, involving many leading scientists, have worked within nations and at the international level to scope out the structure and function of the RAF (e.g. the Red, Blue and Orange Books, NRC 1983, 1994 and 1996) and then harmonize it through international standards (e.g. OECD 2005). The scientific and regulatory community has also developed a range of methods and mechanical tests to validate the presence and effect of specific traits—e.g. the western immunoblot analysis, mass spectral analysis, segregation analyses, Southern blot analysis, in vitro fate studies, glycosylation and enzyme-linked immunosorbent assays (ELISAs)—which have then been standardized in the academic literature, in international standards

[1]A, C, G and T represent the four nucleotide bases of adenine, cytosine, guanine and thymine.

and normalized through multi-party ring-tests. The language, concepts, methods and processes related to SECs, in contrast, has not undergone such extensive normalization. While different fields will assert that there are strong norms with central tendencies, the simple diversity of approaches—ranging from deductive, experimental, highly stylized social sciences, such as empirical economics, to inductive, discourse driven humanist analysis—ensures that across the SEC space, there is no central tendency. Who you ask and what you ask will determine the answer more often than not.

We have already discussed the 'where' in some detail above. To some extent one is faced with a 'forum shopping' problem, with disputants choosing the venue based on what members it has, what evidence it accepts and what it is attempting to optimize. Firms and scientists focused on commercialization generally look first to the science-based risk analysis systems operating in many OECD countries as the primary source of evaluation, and then rely on the science-informed decision-making in the international trade regime to backstop those decisions (e.g. in the NAFTA and WTO, and related authorities in the IPPC, OIE and Codex). Those seeking to optimize reflexivity, as discussed, have attempted to use the precautionary provisions in the RAF and derogations for legitimate objectives in international trade systems but have for the most part found more positive outcomes by working to frame new international treaties to incorporate SECs at their core (e.g. CBD, CPB, International Treaty on Plant Genetic Resources (ITPGR) and Kyoto Protocol).

The 'when' is complicated as transformative technologies deliver unpredictable results in often unexpected ways. Given that the emergence of new technologies is driven almost entirely by market proponents, even highly anticipated products may often be delayed due to market volatility or incomprehensible in-firm dynamics. Most of the time, regulators and the broader community are unaware of the basic and applied research being done inside firms until a candidate product is almost ready. Thus, it is very difficult to anticipate the issues and position the regulatory effort and social debate to address these before products emerge. This poses significant problems as firms by then have sunk significant financial and strategic resources in the development of a product. The science-based, modernist compromise is that as long as something is 'safe enough', regulators will allow new products to enter the market. Firms are responsible for handling any SECs. While some firms have attempted to anticipate the social response to their offerings by operating ethics boards and industry advisory groups, these efforts have only had mixed success. Monsanto, for example, as early as the mid-1980s worked with key eNGOs and farmer groups to solicit their views and concerns about different GM crops, but still was roundly vilified when it introduced Roundup-tolerant soybeans and canola and Bt-resistant corn in 1995.

One thorny question that is not often asked is 'why' does one care? Some are clearly modernists, simply seeking to ensure new products do not harm people or the environment. Others have broader goals, seeking to direct technological innovation to avoid (or achieve) general or specific socio-economic outcomes, such as enriching modern agriculture. Some engaged in the dialogue about regulation of GM foods simply see this as a good battleground for advancing their post-modernist agenda. The logic is that it is much easier to forestall the introduction of something that has not yet been accommodated into our economy and lifestyle; in contrast, it is very difficult to get people to give up something they already have and value. Given the diversity of goals, one can see that some individuals are likely to be dissatisfied, regardless of the outcome—the interests do not converge.

The 'how' warrants some further discussion. In one sense 'how' helps to unpack a range of holons or sub-systems, where different participants act purposefully in narrow, specific action arenas, facing different exogenous variables, developing and using different evaluative criteria, and delivering different discrete outcomes, that either support or undercut the RAF and related governance system. Six specific holons or sub-systems present opportunities to refine or develop a new intellectual architecture for GM foods: nationalized regulation and governance; development of international specialized expertise; industry-based management; intergovernmental negotiation; references to 'wise men'; and the development of case law through litigation and dispute settlement. One could imagine some of, or all of, these holons engaging at various times in an effort to complete or renew the overall system.

## *Nationalized Regulation and Governance*

In absence of evidence of credible efforts by others to develop norms, rules and processes to handle scientific and SECs, the state will need to respond. This has already happened in most nations developing, producing and trading the products of GMOs. According to Clive James at ISAAA, in 2015, the twentieth anniversary of the commercialization of biotech crops, total accumulated area from 1996 to 2015 exceeded 2 billion hectares. Twenty-eight countries planted biotech crops in 2015—the top 10 countries each grew more than million hectares. Of the 28 biotech crop countries in 2015, 20 were developing and 8 were industrial countries; in addition, another 30 imported biotech crop products (James 2015). In total, for 2016, 39 countries plus the EU have approved the use of at least one biotech crops, either for planting or importing (James 2015).

In one way or another, each of these states has developed some form of regulation and governance related to the introduction, cultivation, export, import or use of the products. Given the close links between the scientific and regulatory epistemic communities, even radically different governments and countries are likely to adapt and adopt many elements of the language, methods and processes that have been internationalized through the efforts of the major players (e.g. the US and EU), the international governmental organizations (e.g. the WTO, OECD and Codex) and the science community itself (e.g. through scientific consensus reports commissioned by the National Academy of Science, FAO, WHO, OECD, Royal Society and Third World Academy of Science and through hands-on extension by the Consultative Group on International Agricultural Research system and its centres).

In one sense this is not a bad place to start. Many regulatory and governance innovations in the past have usually come from national and sub-national action arenas, where many variables that hold up international negotiations are for all intents and purposes exogenous, allowing actors to interact relatively easily, often in close knit communities, to develop rules that deliver appropriate outcomes in the local system. In the absence of trade, this could be optimal.

Problems can and will emerge where trade can and should exist (based on comparative advantage). Different rules, which might be locally optimal, often erect barriers to trade, which all other things being equal, generate economic rents that can be exploited. Rent seeking has been demonstrated, by Bhagwati (1982) and Laband and Sophocleus (1988) and others, to create major distortions in an economy and dissipate non-trivial parts of an economy in wasteful activities or can contribute to a culture of cronyism and corruption.

## *Development of Expertise in Specialized Institutions*

One alternative to having every nation state work independently to develop its own rules and systems is to do that work collectively in the context of one of the many specialized international science or regulatory agencies. This effort could come without any reference to national policies and rules, but in most cases there is an effort to build upon national competence to develop new, purpose-built international rules and structures to manage safety and trade.

This work tends to be most successful when the action arena is narrowly controlled, with participants drawn from a narrow range of epistemic communities. In that way the process is able to effectively make exogenous many of the views and perspectives that would make it difficult to reach consensus.

There are a wide range of efforts that have done just this. The OECD, for example, has taken the initiative to develop ConsensusDocuments (CDs) that comprise technical information for use during the regulatory assessment of products of biotechnology and are mutually recognized among OECD Member countries. As of 2016, more than 50 CDs have been developed; the great majority of them remain current and operational. They focus on the biology of organisms (including crops, for trees and micro-organisms), introduced novel traits (in 2016, they related to the five most common introduced traits) and various facilitating norms (11 CDs). Meanwhile, industry standards (either brands or collective criteria) as encompassed in the International Organization for Standardization (ISO) or hazard analysis and critical control point (HACCP) systems can also provide a starting point for international standards. Standards from Codex, the OIE and the IPPC have at times been used to build up a basic text of agreed acceptable measures. The usual process is for bilateral talks between major producer and consumer nations to arrive at a common standard, which is then put forward as the base for an international standard for harmonization.

While practical, this 'patchwork' approach defers to specialization and, hence, presents certain problems. Some issues are not immediately taken up by any institution (take for example SECs). Alternatively, it is never clear what to do when there is inadequate data; one of the biggest challenges of current science-based processes is that there is often a lack of scientific data upon which to base new standards. Moreover, there is simply no objective way to decide how to deal with scientific uncertainty about the long-term effects on human health or the environment.

## *Industry-Based Management*

One response to slow development of national and international rules has been for the companies or parts of the biotechnology industry to implement self-regulation to maintain market access. There are a number of cases where parts of the agri-food industry have developed systems to deliver products with higher standards than domestic or even international minimum standards. The red meats industry in Australia (Spriggs and Isaac 2001), the canola industry in Canada (Gray et al. 2006; Phillips and Smyth 2004), retailers and processors in the EU (Hobbs 2010), North America and Asia (Phillips and McNeill 2002) and the corn industry in the US, have all adopted private standards at one time or another in the recent past. Over time, private standards, supplemented by HACCP protocols or ISO ratings (particularly 9000 and 14000 series) could supplement or replace public regulation. In order to address market demands for traceability and separability, new physical and organizational infrastructure may be required. Already ISO has developed new eco-labeling standards (ISO 14020 and ISO 14024) and has presented industry with the opportunity to use the standards as a way to avoid environmental challenges to their products introduced into domestic or foreign markets.

This approach is largely underdeveloped at present but represents an interesting possibility for industry to develop its own standards that will ensure regulatory and market acceptance of their products. A major advantage is that such an approach would not require industry to bite off large chunks of the regulatory apple. Instead, problems can be resolved in bite-sized pieces as problems and opportunities are identified.

## *Intergovernmental Negotiations*

In the past, once nations have structured their own internal systems, as discussed above, and have begun to trade, there has been a tendency for either industry or governments to search out ways to 'level the playing field' in the international marketplace. Perhaps it is time to go back into the past and adopt pre-existing approaches to address issues around this new technology. A number of approaches to negotiation have been used successfully in the past. The main action arena in the past has been the GATT/WTO, which at its root has a few core principles that drive all of the activities. The most-favored-nation (MFN) principle (Art. 1 in GATT, Art 2 in GATS and Art 4 in the TRIPS) establishes that countries cannot normally discriminate among their trading partners—any concessions granted to one Member (such as a lower customs duty rate for one of their products) must be provided to all WTO members (Article 3 of GATT, Article 17 of GATS and Article 3 of TRIPS). Other important principles include the *like products* provision, whereby similar products should be treated the same, and the pervasive commitment to transparency in all trade related measures.

In the early rounds of GATT negotiations the process was predominantly one of reciprocal negotiating related to key issues and key markets. A country made bids and offers with key traders to liberalize specific areas; once bilateral agreements were set, they were multilateralized through the most-favoured-nation principle. In this way, the negotiations focused engaging key partners on those trade issues that had the greatest commercial importance. This strategy would entail those countries producing and exporting the bulk of the GM crops—the US, Brazil, Argentina and Canada—engaging in narrowly based negotiations with the key importers—the EU and Japan—related to a handful of GM crops—soybeans, corn, cotton and canola. The strong reciprocity of interests in continued international trade in those products among those countries should improve the likelihood of success.

When the number of GATT Members expanded with decolonization in the 1960s, a second negotiating approach was added. Instead of simply negotiating for each issue, tariff or product, negotiating groups would adopt formula for dealing with a wide range of measures. One approach, first adopted in the Kennedy Round (1963–67) was to negotiate a formula for cutting tariffs, called the Swiss formula, whereby tariffs were bundled into four bands, and higher cuts were assess for higher tariffs (Reinert et al. 2009). The Uruguay Round of negotiations introduced and implemented two new approaches. First, they created a practical system to convert non-tariff measures (such as quotas) into tariffs. Then they adopted a traffic light analogy to rank policies in the agri-food talks, whereby Green box domestic or trade policies were deemed to be minimally trade distorting and excluded from domestic support reduction commitments (e.g. policies dealing with research, extension, inspection and grading, environmental and conservation programs, disaster relief, crop insurance, domestic food assistance, food security stocks, structural adjustment programs, direct payments not linked to production and export market promotion), Amber box policies such as market price support, payments related to current production or prices, and input subsidies were subject to careful review and binding reductions over time and Blue box policies (e.g. acceptable, but temporary, or transition policies such as payments under production-limiting programs that with further reforms could turn Green) were bound and targeted for transition. Red box policies were prohibited and must be immediately withdrawn.[2]

[2]With the exception of agriculture where export subsidies were capped (Hobbs and Kerr 2000).

Each of these previous attempts to innovate new negotiating processes has been the result of shifting action arenas, adding new actors with new interests and occasionally by forcefully interjecting new norms and concepts. The shift from the bid and ask system to formulas and analogies came when membership in the GATT swelled from a dozen or so members to more than 60 in 1963 and then to more than 160 today. The relatively closed shop of lawyers and trade diplomats that traditionally supported the system was not the source of inspiration—rather economists, versed in mathematics and algebra, showed how the insurmountable details could be agglomerated into formulae and then negotiated without explicit reference to specific measures, except as they involved exceptions. Similarly, the stop light analogy (i.e. Red, Amber and Green box subsidies) came from a request of the scholarly academic community to offer suggestions on how to deal with the perennially thorny question of agri-food policy. The increasing globalization of the scholarly community, especially the introgression into economics of mathematics as the common language and conceptual reasoning, provided the ultimate basis for this breakthrough.

One risk of relying on issues-based negotiations is that they often focus on older issues and not on newly arisen concerns. Furthermore, this approach can often isolate many of the recently mobilized developing nations and actors, which simply puts pressure on other forums.

## *Reference to 'Eminent Persons'*

In some cases there is a strong tendency to punt the ball to individuals or groups of 'eminent persons' to see if their knowledge and authority can develop something new and interject it into the policy dialogue. In the context of GM foods, the policy dialogue has been mostly a punting game, with few tries or even field goals. There have been two somewhat distinct approaches. Most governments have created or commissioned experts or epistemic groups to offer advice while an increasing number have engaged in a range of 'democratic engagement' processes that seek to democratize the dialogue and engage a broader range of values, interests and beliefs.

From the beginnings at the Asilomar Conference in 1975, where an informal group of industry, academic and government scientists established a self-defined moratorium for genetic engineering, governments have looked to eminent persons or groups to help them clarify the knowledge-base, frame the issues, scope out options and provide advice on choices. Most governments have used formal or informal advisory councils, royal commissions, royal societies, academies and other review bodies to examine and advise on some aspect of GM foods. The shelves are full of these reports. In most cases expert groups have provided expert advice, which is then often adapted and adopted into the regulatory and governance system. While quite valuable at times, few if any of these reports have offered much more than incremental, iterative advice on how to sustain, enhance or extend the largely modernist system.

The second approach has been to expand the scope of dialogue to include those without epistemic expertise. This has been done in two ways. First, some governments have reformulated their advisory structures to engage new actors. In Canada, for instance, the Government of Canada replaced its expert National Biotechnology Advisory Committee in 1999 with the more pluralist Canadian Biotechnology Advisory Committee while in New Zealand a Royal Commission of Inquiry on Genetic Modification in 2001 engaged a mix of participants. The other prong has been to use a range of purpose built processes to engage new actors. This effort was pioneered in the US where the National Institute of Health (NIH) brought together medical experts to assess the safety and efficacy of medical technologies and then was transformed by the Danish Board of Technology from an experts-only model into a mediated process that engages both experts and citizens (Einsiedel and Eastlick 2000). A range of new models has emerged and been tried, including consensus conferences (e.g. in Canada and Denmark), citizens' juries (e.g. in France), focus groups and public consultations (such as the UK GM Nation process in 2003) to offer SEC perspectives. In most cases, the advice from both the advisory groups and the processes are acknowledged but rarely adopted and adapted. Most of the advice from these processes tends to be more nebulous and less specific, so it is harder to translate into policy and practice. In those instances of clear concordance between the advice and government action, such as France's proposed moratorium on GM food in 1999 following the report of a citizens' jury that concluded GM foods were not desired, are usually examples of mutual interests rather than true influence. The challenge is that these pluralist action arenas are seldom, if ever, defined as authoritative venues for decision-making—instead they are side shows that are used to amplify prior positions or ignore if incompatible.

## *Disputes and Case Law*

In absence of action in any of the other arenas, the binding dispute settlement body (DSB) in the WTO offers one accessible venue for states to engage in refining and clarifying the rules. The processes for handling trade disputes are, for the most part, in place at the WTO. The benefit of this approach for countries exporting GM crops is that it does not require further negotiations and is most likely to deliver pro trade-, science- and rules-based decisions. The decisions related to the SPS and TBT Agreements noted earlier, combined with the DSB opinion related to the EU moratorium, have affirmed that the WTO will respond to cases when they are engaged, but that their decisions will be relatively narrow and focused on resolving the case at hand rather than broadly interpret the treaty and its principles.

The case-by-case approach has some very serious drawbacks. Coverage of issues affecting biotechnology depends on cases actually being brought before the tribunals in question. The IPPC tribunal has heard no cases. WTO panels have heard cases on health and food safety issues but so far have not shown that they are equipped to deal with other thorny questions, such as how to balance environmental and socio-economic issues with trade concerns.

A number of other features, some practical and some systemic, make this approach problematic. First, there is the long-standing problem of unequal resources between developed and developing countries that determine if and when a case (or several of them) might be initiated at the WTO. Second, depending on the issues that member states would want to take forward to the dispute settlement system, it usually takes years to fully develop a body of law sufficient to make judgements. Finally, even when decisions are clear and complete, states do not always comply with the WTO's binding decisions (e.g. the EU has not complied with the panel ruling on the beef hormones case). At the end of the day, a losing party can side-step the consequences of a negative decision by paying compensation to the successful party. If such compensation is not forthcoming, the successful party can obtain an authorized suspension of proportionate trade benefits. In the end, however, the importing country's trade is not enhanced. Given the complexity and strongly held beliefs related to these cases, non-compliance is very possible, which could over time undermine the entire WTO system.

More fundamentally, the case-by-case approach can be attacked in that it is science-based and there is currently no institution that can develop a concurrent case law approach to deal with outstanding socio-economic issues.

## Parallel or Convergent Paths

At this point in the discourse there are two parallel, or possibly divergent, regulatory paths. The science-based or modernist path is led by the US and to a lesser extent the Austral-Asian region. The SEC or post-modern path is led by the EU and, through post-colonial connections, most African nations. The post-modern path is not about ensuring the personal safety of GM food consumption. If it were, no European would travel to North America as this would require them to eat the food in North America. This is confirmed by a report to the European Commission (2011, p. 8) on socio-economic implications of GM crop production. This report states that: "[r]esponses given by consumers when prompted by questionnaires about GM-food are not a reliable guide to what they do when shopping in grocery stores. Europeans buy GM-foods when they are physically present on the shelves."

Given that food and health safety are not the major concerns of post-modernists, the driver then has to be politics. It is not the political agenda of most average citizens/consumers, but the political agenda of narrow, focused special interest groups—with strongly held preferences. These interest groups have entrenched interests with strong political agendas. On might rightly ask: can the world continue with two competing, parallel regulatory and trade paths, or will convergence occur?

It is conceivable that a duel regulatory approach could function for the next 10–20 years without creating substantial problems. The Americas and the Austral-Asian countries could continue to move forward with the commercialization of innovative agricultural technologies and the European-African alliance could continue to fall further behind in terms of technological advancement and increased production. However, at some point the loss of one-half of the potential market for new products will begin to have an impact on investment decisions and pressure will mount to find a resolution. That is, states and firms will seek to reduce the investment uncertainty and increase the regulatory transparency of SECs, so that the polarized world can be replaced by an international regulatory system that provides consistent decisions (see Chap. 7).

The challenge of embedding SECs in risk regulation is their potential for nefarious use. All too often, SECs are used by governments to mask third party protection or to prevent the import of competing products from other countries. Examples of SECs can include, but are not limited to: food security; benefits to society; indigenous knowledge; impacts on biodiversity; health impacts; environmental impacts; market access; consumer choice; labour impacts; intellectual property rights; ethical implications; and religious/cultural implications (Ludlow et al. 2014; Isaac and Kerr 2003).

It is theoretically conceivable that some of the above list could be methodologically incorporated into a science-based regulatory framework. Of course, this would add cost to the regulatory process in terms of research investment and time required for regulatory approval. However, the fundamental challenge that will still remain is how to deal with those SECs that cannot be incorporated into a science-based framework. Even if a mechanism could be developed to incorporate SECs, the practical problem of how to prevent them being captured by vested economic interests remains.

The central tendencies of states and markets are well known and understood, but the incorporation of 'social voice' is problematic. The problem is that social-based regulations have no process for reconciling differences among the three classes of civil authorities in the way science-based regulation reconciles disputes among scientists and economists.

Those authorities that are strongly values-driven (with strong preferences) have attracted much of the consideration. Values ultimately reflect what people want out of life, and what they are willing to accept. They may be defined in the context of personal attributes (e.g. honor, integrity, honesty and spirituality), cultural traits (e.g. ethnic, linguistic or national roots), social aspects (e.g. equality or altruism) or workplace related qualities (e.g. loyalty, contentiousness or professionalism). Generally values are not easily mutable—they are grounded in us by our upbringing, our communities and our experiences. The conventional view is that collective action is and should be all about reflecting values, beliefs and belonging (Wilsden and Willis 2004). While collective organizations are often very effective in mobilizing voice, they frequently complicate the governing process. Organizations often reflect a narrow range of values—the more homogenous their membership, the more effectively they can define their message. Challenges arise when competing civil authorities articulate values that are incompatible or difficult to pursue jointly. In those situations, some, if not all, of the civil authorities are bound to be disappointed and frustrated at any compromise, because the nature of their claim is not subject to compromise due to their strong individual preferences.

A second group of civil authorities are driven by the interests of their members. While these institutions may articulate some common values, the fundamental driving force for most of these organizations is to derive some type of benefit for their membership. These interests may range widely, often involving structuring rules to deliver desired outcomes or to redistribute impacts of specific market or public policy outcomes. Furthermore, many of these authorities have an evolving set of interests that will be played out in a succession of repeated processes. The absence of any fundamental values at stake and the nature of the repeated games these groups are engaged in tends to dampen the rhetorical excesses and lead to more cooperative engagement.

Third, there are many authorities engaged in developing, sustaining or protecting some epistemology or process. These epistemic communities have recognized expertise and competence in a particular domain and an authoritative claim to specific knowledge within that domain or issue area (Haas 1992; Mokyr 2002). These communities and their members have a shared set of normative and principled ideas, shared causal beliefs, shared notions of weighing and validating knowledge in the domain, a shared technical vocabulary and a common set of practices associated with a specific set of problems. Many epistemic communities span international boundaries. Unlike values-based social movements and interest-based groups, epistemic communities share causal beliefs and a consensual knowledge base, but do not share principled beliefs or interests (which are the unifying feature of social movements and interest groups). For the same reason, epistemic communities do not cohabitate easily with markets or the state—their foundational commitment to a set of causal beliefs and consensual knowledge is inconsistent with the practical politics of legislators and bureaucratic agencies or the self-interested nature of market actors.

Applying Kuhn's paradigm, one could interpret the challenge of SECs is a classic case of unknown theory and unknown evidence (Kuhn 1970). In the presence of profound uncertainty, there is a tendency to delay making a decision, which can result in a type II error. The difficulty is that there is seldom any value attributed to the foregone decision. As many SEC factors cannot be quantified in a scientific process, regulators become stymied. As a result, decisions often get bumped upstairs to the political level, which opens the regulatory process to all kinds of manipulation, depending on which party is in power at any given point in time. Removing politics from international regulation and trade was the driving reason behind the establishment of the SPS (Kerr 2000). Without science-based decision-making as the foundation for a regulatory system, the politicization of the process could naturally lead to manipulation of the system and the inhibiting of international trade of agriculture products.

## Conclusions

A number of observations can be made about the present duel nature of the regulatory framework for biotechnology derived products.

Biotechnology, especially the area of GM foods, offers a real-time experimental space where the modernist and post-modernist systems are seeking accord. The battle is engaged, the action arenas mostly defined and the actors are mobilized. New norms, processes and institutions are emerging while at the same time industry is pushing ahead with adoption. Twenty years into the experiment, GM foods are widely produced on every continent and widely traded in more than half of international commerce. James (2015) concludes that GM technologies have been the fastest and most extensively adopted new agri-food innovation in the history of agriculture. Nevertheless, tensions remain. Some important actors remain disconnected from any authoritative action arena, there are real and significant gaps in our knowledge systems, in many areas arguments continue over the norms that should drive our system and the connections that can and should drive the system to decision-making are in many cases tenuous or ineffective, in that the US and some other key countries are not members of all of the institutions.

In conclusion, the complexity of the various issues makes it very unlikely for any comprehensive approach to the regulation of biotechnology. Such an approach is also probably undesirable as comprehensive negotiations often lead to 'horse-trading' that produces anomalies and inconsistencies (Kerr 2003 notes, for example, the inconsistency of having intellectual property protection within the WTO). Furthermore, it is not clear whether there is any institution ready to accept such a gargantuan task. The OECD is facilitating dialogue and collecting data but suffers from a limited membership. The WTO has its hands full with other trade issues. The CPB has focused on environmental issues and will be fully occupied for the foreseeable future establishing a framework for balancing international environment interests in the international trade system. Consumer concerns and continuing anxiety surrounding GM products coupled with the atmosphere of protest against large organizations suggest that even if comprehensive negotiations could deliver a result, it would be very difficult to sell any outcome to consumer and activist groups.

## References

Bhagwati, J. 1982. Directly unproductive, profit-seeking (DUP) activities. *Journal of Political Economy* 90, October: 988–1002.

Einsiedel, E., and D. Eastlick. 2000. Consensus conferences as deliberative democracy: A communications perspective. *Science Communication* 21: 323–343.

European Commission. 2000. *Communication from the Commission on the Precautionary Principle*. Brussels: European Commission Press.

European Commission. 2011. Report from the Commission to the European Parliament and the Council on socio-economic implications of GMO cultivation on the basis of Member States contributions, as requested by the Conclusions of the Environment Council of December 2008.

Gray, R., S. Malla, and P.W.B. Phillips. 2006. Product innovation in the Canadian canola sector. *Supply Chain Management* 11 (1): 65–74.

Haas, P. 1992. Introduction: Epistemic communities and international policy coordination. *International Organization* 46 (1): 1–35.

Hobbs, J.E. 2010. Public and private standards for food safety and quality: International trade implications. *Journal of International Law and Trade Policy* 11 (1): 136–152.

Hobbs, A.L., and W.A. Kerr. 2000. First salvoes: A guide to the opening proposals at the WTO negotiations on agricultural trade. *Agribusiness Paesaggio and Ambiente* 4 (2): 97–108.

Isaac, G.E., and W.A. Kerr. 2003. International trade, intellectual property rights and traditional knowledge—The case of plant genetic resources. *Journal of Aboriginal Economic Development* 3 (2): 75–85.

James, C. 2015. *20th anniversary (1996 to 2015) of the Global Commercialization of Biotech Crops and Biotech Crop Highlights in 2015*. ISAAA Brief No. 51. ISAAA: Ithaca, NY.

Kerr, W.A. 2000. A new world chaos?—International institutions in the information age. *Journal of International Law and Trade Policy* 1 (1): 1–10.

Kerr, W.A. 2003. The efficacy of the TRIPS: Incentives, capacity and threats. *Journal of International Law and Trade Policy* 4 (1): 1–14.

Kerr, W.A. 2010. What is new in protectionism?—Consumers, cranks and captives. *Canadian Journal of Agricultural Economics* 58 (1): 5–22.

Kerr, W.A. 2016. Disequilibrium, trade and the consequences of adjustment. *Journal of International Law and Trade Policy* 17 (2): 59–75.

Kerr, W.A., and J.E. Hobbs. 2005. Consumers, cows and carousels: Why the dispute over beef hormones is far more important than its commercial value. In *The WTO and the Regulation of International Trade*, ed. N. Perdikis, and R. Read, 191–214. Cheltenham: Edward Elgar.

Krebs, J.R. 2000. GM foods in the UK between 1996 and 1999: Comments on "Genetically modified crops: Risks and promise" by Gordon Conway. *Conservation Ecology* 4 (1): 11.

Kuhn, T. 1970. *The structure of scientific revolutions*, 2nd ed. Chicago: University of Chicago Press.

Laband, D., and J. Sophocleus. 1988. The social cost of rent-seeking: First estimates. *Public Choice* 58 (3): 269–275.

Ludlow, K., S.J. Smyth, and J. Falck-Zepeda (eds.). 2014. *Socio-economic considerations in biotechnology regulation*. New York: Springer Science+Business Media.

McGinnis, M.D. 2011. An introduction to IAD and the language of the Ostrom workshop: A simple guide to a complex framework. *The Policy Studies Journal* 39 (1): 169–183.

Mokyr, J. 2002. Innovation in an historical perspective: Tales of technology and evolution. In *Technological innovation and economic performance*, ed. B. Steil, G. Victor, and R. Nelson, 23–46. Princeton: Princeton University Press.

National Research Council. 1983. *Risk assessment in the federal government: Managing the process*. Washington, D.C.: National Academy Press.

National Research Council. 1994. *Science and judgment in risk assessment*. Washington, D.C.: National Academy Press.

National Research Council. 1996. *Understanding risk: Informing decisions in a Democratic Society*. Washington, D.C.: National Academy Press.

Organisation for Economic Cooperation and Development. 2005. An introduction to the biosafety consensus documents of OECD's working group for harmonization in biotechnology, No. 32 in the Series on Harmonization of Regulatory Oversight in biotechnology. Paris: OECD.

Ostrom, E. 2005. *Understanding institutional diversity*. Princeton: Princeton University Press.

Phillips, P.W.B. 2007. *Governing transformative technological innovation: Who's in charge?*. Oxford: Edward Elgar.

Phillips, P.W.B., and H. McNeill. 2002. Labeling for GM foods: Theory and practice. In *Market development for genetically modified foods*, ed. V. Santaniello, R.E. Evenson, and D. Zilberman, 245–261. Wallingford, UK: CAB International.

Phillips, P.W.B., and S.J. Smyth. 2004. Managing the value of new-trait varieties in the canola supply chain in Canada. *Supply Chain Management* 9 (4): 313–322.

Phillips, P.W.B., D. Castle, and S.J. Smyth (eds.). 2015. *Biotechnology, agriculture and development*. Cheltenham, UK: Edward Elgar.

Reinert, K.A., R.S. Rajan, A.J. Glass, and L.S. Davis (eds.). 2009. *The Princeton encyclopaedia of the world economy*. Princeton: Princeton University Press.

Simon, H. 1955. A behavioural model of rational choice. *Quarterly Journal of Economics* 69: 99–188.

Spriggs, J., and G.E. Isaac. 2001. *International competitiveness and food safety: The case of beef.* Wallingford, UK: CAB International Publishing.

Wilsden, J., and R. Willis. 2004. *See-through science: Why public engagement needs to move upstream*. London: DEMOS, Green Alliance, RSA and Environment Agency.

World Trade Organization. 1994. *Marrakesh Declaration of 15 April 1994*, Geneva, World Trade Organization. Online at: https://www.wto.org/english/docs_e/legal_e/marrakesh_decl_e.htm.

World Trade Organisation. 1998. Appellate Decision, Australia—Measures Affecting Importation of Salmon—Complaint by Canada, AB-1998-5 (Appeal from WT/DS18). Online at: http://wto.org.

# Chapter 12
# Labeling and Preferential Trade Deals

**Abstract** The lack of international institutional capacity to oversee the challenge of international trade in GM products has resulted in a conundrum. If the world of global trade is to avoid falling into the scenario of constant trade blockages and barriers, then practical solutions are required to advance the discussion and dialogue. One strategy is to pursue more clear and operational labeling rules through regional trade agreements in order to help facilitate trade.

## Introduction

The labeling of products of modern agricultural biotechnology has become one of the most contentious issues in international trade policy in the wake of the commencement of commercial production of crops that have been genetically modified (GM) in the mid-1990s. The issue remains unresolved to this day, a period that now spans 20 years. Part of the reason why the issue remains unresolved is that the broader issue of multilateral rule-making for international trade has been effectively stalled in the World Trade Organization's (WTO) Doha Round. There has been no progress in multilateral rule-making for international trade since the completion of the Uruguay Round in 1994. The Doha Round commenced in 2001 and remains unfinished—there is both a reluctance to abandon it and a lack of will to see it completed among WTO members. The current multilateral rules on labeling were negotiated prior to commercial agricultural production using genetically modified organisms (GMOs). Thus, they take no account of the demands for differential regulation and pro-active labeling that became part of the broader policy controversy over GMOs that has dogged domestic regulatory development and commercialization around the world since the arrival of the products of biotechnology in food marketplaces. Even in countries which are the major proponents of agricultural biotechnology, such as the US, labeling policy remains a controversial issue. Clark et al. (2014) report US$114 million has been spent in the past three years in the argument against mandatory GMO labeling. Mandatory GMO labeling in the state of Vermont came into effect on 1 July 2016, causing a great quandary.



S.J. Smyth et al., *Biotechnology Regulation and Trade*, Natural Resource Management and Policy 51, DOI 10.1007/978-3-319-53295-0_12

Some food processing firms have begun to label products as containing GM ingredient in response to Vermont's legislation in spite of national efforts to countermand the labeling requirements.

The controversy over labeling has been resolved in different countries in different ways. As a result, there is considerable divergence in domestic regulatory regimes pertaining to the labeling of the products of modern agricultural biotechnology. This divergence can and does translate into trade barriers. Given that there seems little prospect for multilateral progress in resolving issues pertaining to labeling, it is probably not surprising that countries have been seeking alternatives whereby they can garner the perceived gains from trade—the current mechanism of choice for many countries is to negotiate preferential trade agreements among subsets of countries. These agreements are allowed by the WTO, but cannot contravene its rules. Hundreds of preferential trade agreements now exist and a number of major initiatives are in process, including the Transatlantic Trade and Investment Partnership (T-TIP) between the European Union (EU) and the United States (US), the Trans Pacific Partnership (TPP) among 12 Pacific Rim countries including the US and Japan and the Comprehensive Economic and Trade Agreement (CETA) between the EU and Canada. While the reasons why countries are negotiating preferential agreements are multi-dimensional, it is interesting that the T-TIP, the TPP and the CETA all have trade in the products of biotechnology as part of their negotiating mandate. Given that labeling is one of the controversial issues for trade in the products of biotechnology, it suggests that it may be time to examine the issue of labeling in an international context—to determine what is possible in the context of a preferential trade agreement.

The topic has not been extensively explored in the international trade literature because the multilateral rules for labeling in international trade were negotiated in the Uruguay Round, coming into force in 1995. As suggested above, no progress has been made multilaterally since that time. In fact, the two agreements where labeling is dealt with in the WTO—the Agreement on the Application of Sanitary and Phytosanitary Measures (SPS) and the Agreement on Technical Barriers to Trade (TBT) were not opened for re-negotiation in the Doha Round. While there has been considerable discussion of labeling in the context of the SPS and TBT, there has been little examination of what alternatives might be available in arrangements such as preferential trade agreements.

We argue here that, in fact, the economic consequences of mandatory labeling are poorly understood by those who are its strong advocates and that moving to mandatory labeling may not be able to achieve the results they desire and may not be in anyone's best interest. In the next section the current international rules on labeling are explored to illustrate why they are controversial in the context of GMOs. The consequences of moving the main facets of the SPS and TBT that proponents of mandatory labeling object to into a preferential trade agreement are then explored. The interface between potential labeling regimes in preferential trade agreements and non-members of the preferential agreements is examined in the following section. Some practical difficulties for the development of a new labeling regime in the T-TIP, TPP and CETA are outlined. We conclude by assessing the likely role of preferential trade agreements in devising a new model for labeling in international trade.

## Labeling in the Multilateral Rules of Trade

The SPS Agreement was negotiated and the TBT Agreement re-negotiated[1] and strengthened in the Uruguay Round due to fears by member countries that given agricultural tariffs, non-tariff barriers and subsidies would, for the first time, come under multilateral disciplines, countries seeking alternative mechanisms to provide economic protection would put in place barriers based on nefarious SPS grounds or would use technical barriers to discriminate against products of developing countries (Gaisford and Kerr 2001). The General Agreement on Tariffs and Trade (GATT) —the precursor to the WTO—was first negotiated after the Second World War and the basis of economic thinking about international trade at that time was that only producers would ask for protection (to raise the prices they received).[2] Consumer interests were hurt by higher prices and, hence, it was assumed they would never ask for protection. The result was that the GATT made very few provisions for governments to respond to consumers (or any group other than producers) asking for protection. This is at the root of the difficulty that governments face with strong lobbying for protection from imports of GMO within the WTO.

The SPS and TBT, and their provisions on labeling, are very much in the GATT tradition of only responding to protection requested by producers. As suggested above, the concern in the SPS was that governments would try to provide economic protection for agricultural producers using the implementation of trade barriers based on nefarious sanitary and phytosanitary justifications. Thus, they required a scientific justification for any barriers that were put in place. Further, a risk assessment was required to show that imports imposed an unacceptable level of risk —although each country was allowed to determine what was acceptable. The SPS Agreement was not, itself, to devise the scientific standards that were to underlie it. Instead, three already existing scientific organizations were to develop standards—the *Codex Alimenrarius* (Codex) for human health, the World Organization for Animal Health (OIE) for animal health and the International Plant Protection Convention (IPPC) for plants and the environment.

The system accepts that labeling can be a trade barrier.[3] Labels can be costly to verify, thereby raising prices and reducing quantities traded relative to those in

[1]Prior to the Uruguay Round renegotiation technical barriers to trade were dealt with in the GATT's Standards Code.

[2]Formally, a neoclassical partial equilibrium comparative statics model underlies trade policy making at the WTO (Kerr 2007).

[3]As such they represent a subset of technical barriers to trade. The reduction of trade, and hence the loss of trade benefits, associated with technical barriers to trade, is considerable (Hobbs 2007).

absence of labels. Labels can send signals that can affect consumers' perspectives that can reduce demand and, thus, lower trade from levels that would otherwise exist. Thus, from the SPS perspective, labels should not be allowed without a justification. The justification would then need to be based on science. A *hazard* is one compelling justification for labeling under the SPS (Isaac et al. 2002). It is permissible to use mandatory labeling standards for products derived from GMOs where standards may violate the WTO principle of non-discrimination, provided scientific evidence of a hazard exists. There are three categories of hazard: (1) *hazard for all* which allows an import ban and, hence, there is no need for labeling; (2) *hazard for none* so no mandatory hazard label would be allowed; and (3) *hazard for some*, such as allergens, which justifies a mandatory hazard label (Isaac et al. 2002). Thus, unless there is a scientific identification of a hazard, labeling is not allowed under the SPS.

Those that wish there to be mandatory labeling of imported (and domestic) GMOs justify their request on the rationale that it is a *consumer's right to know* if the imported product contains GMOs so they can individually decide whether or not they wish to consume the product. But there is no automatic *consumer's right to know* in the SPS. All members of the WTO agreed to the SPS rules, including the European Union (EU). There is no disagreement that products containing GMOs that represent a *hazard for all* should have an import ban imposed. Most GMOs, however, have not been found to represent a hazard. Thus, under the SPS an importing country cannot require them to be labeled under an SPS justification.

The SPS, however, is premised on two assumptions that those advocating mandatory labeling find difficult to accept. The first is that there can be a scientific consensus that no hazard exists (Smyth et al. 2011). The scientific method relies on individual scientists questioning the ruling paradigm. Thus, on any issue one is interested in, one can likely find scientists that question the ruling paradigm—i.e. where most scientists agree. The critics of biotechnology can always point to examples of scientists who do perceive potential hazards and question why their opinions are not given weight in public policy-making—including labeling decisions. Further, it can always be argued that insufficient science has been undertaken to justify a conclusion of a product being a *hazard for none*. There are always additional sub-populations that could be tested. As the number of sub-populations is virtually limitless, decisions regarding the existence of a hazard will necessarily be arbitrary *lines in the sand* determined by those charged with making a decision. The concern pertaining to insufficient science having been undertaken is part of the debate surrounding the use of *precaution* in decision making in relation to new technologies. While a precautionary approach is applied by all countries in their examinations of new technologies, some countries have enshrined the

---

(Footnote 3 continued)

Removal of technical barriers to trade or their reconstitution to be the *least trade distorting* to achieve a particular domestic policy goal is a central tenant of the WTO and preferential trade agreements (Kerr 2010). A discussion of the welfare implications of technical barriers to trade can be found in Kerr (2006).

*precautionary principle* in law, but without defining how it should be used in decision making. As a result, it has been seized upon by some groups in civil society as a means to thwart the approval of new technologies through the promotion of very strict definitions (Van den Belt 2003). Given the lack of precision in how to make use of the *precautionary principle* in decision making, it can become a means to paralyze the process of technological progress (Sunstein 2002–2003). Without input from experts from the scientific community, there is no means to determine when sufficient science has been done to warrant approval of a new technology.

Second, the SPS assumes that individuals will defer to the expertise from which the scientific decision has been derived (Smyth et al. 2015). Those requesting mandatory labeling often state that they do not trust those in the scientific community that are involved in regulatory decisions pertaining to biotechnology because they accept funds from the large biotechnology firms that have a vested interest in having their products approved. They see no reason that they should defer to such tainted experts.[4] From this perspective, requiring labels to have a scientific justification is nonsense. Better to have the product labeled and individuals be allowed to make their own choice as to whether they wish to consume them. For these reasons, those who propose mandatory labeling wish that an alternative international trade arrangement could be agreed, either at the WTO or in regional trade agreements.

The TBT, in contrast, is designed to deal with labeling for traded products when *hazards* are not involved. In other words, countries can insist that imports be labeled even when there is no safety concern. WTO members, ever mindful of the potential trade effects of labeling policies, do not allow unconstrained use of labels for imports. First, labels must not discriminate between domestic and foreign products or among products of foreign origin. It is legitimate to require labels for novel products—to ensure that consumers are informed that they are consuming something that was not previously in the market. In other words, labels are justified if no *like product* already exists in the market (Isaac et al. 2002). In the case of GMOs in the US, for example, GM corn and non-GM corn have been judged to be *substantially equivalent* meaning that they are *like products*. In this situation, no labeling can be required. Other countries have determined that all GM-products are novel and argue that labeling of imports can be required. However, this line of reasoning has run into a constraint imposed by further TBT stipulations.

In most cases, GM products are not physically distinguishable from their conventional counterparts—one cannot distinguish between GM canola and non-GM canola by inspection (or for that matter consumption)—which makes genetic modification a *credence* attribute (Hobbs 1996). The feature upon which GM products can be distinguished from non-GM products is the technology used in

[4]Although it is somewhat inconsistent that critics of biotechnology who are not willing to accept a judgement of h*azard for none* are willing to accept a judgement of *hazard for all* from the same science-based system.

their production. But the TBT rules do not allow labeling based on production and processing methods (PPMs) . There is good reason for this. Developing countries are particularly worried that advanced developed countries would use differences in production and processing methods as barriers to trade—for example, to impose trade barriers against textiles manufactured using hand looms when domestic production is based on an advanced mechanized technology, even though the end product is not discernably different. Thus, developing countries during the Uruguay Round negotiations were insistent that PPMs should not be used as a basis for labeling. This rule had nothing to do with GMOs, but the underlying principle still applies to them. This has become an increasing problem for developed country policy makers faced with requests from consumers regarding information on how imported goods are produced. Beyond questions about whether biotechnology was used in the production of a food product, consumers and advocacy groups want to know about child labor, appropriate animal welfare regimes, environmentally sustainable methods and many other provenance attributes. As suggested above, the TBT was not *opened* for re-negotiation in the Doha Round so there has been no change in this labeling rule since before GMOs were commercialized. Thus, the TBT and the SPS rules on labeling have led to considerable frustration for those who want GM labeling for imports on the basis of *consumer's right to know* and for governments that favor proactive, mandatory GM labelling.

## Can Preferential Trade Agreements Resolve Gm Labeling Concerns?

The two major facets of the currently agreed multilateral trade rules on labeling that those who wish tighter control of GMO imports find too constraining are: (1) the requirement of a scientific justification for labels to be applied under an SPS justification and; (2) the prohibition of labels on the basis of PPMs—i.e. the use of biotechnology in production. Many are asking whether these could be removed in a preferential trade agreement negotiated among a subset of WTO member states and, if so, what would be the consequences.

While preferential trade agreements are allowed in the WTO, in theory they must comply with the principles and specific provisions of the WTO. They can go further than WTO commitments (e.g. they can lower tariffs below the bound WTO tariffs) and they can make rules where the WTO is silent but they cannot contravene or undercut WTO commitments. The challenge for those wanting labels is that in the case of labeling provisions, the WTO is not silent. Thus, new rules on labeling that go further than those of the WTO could be included in a preferential trade agreement but they would have to be constructed in such a way so as not to contravene the existing WTO commitments. It is hard to conceive how this could be done, given the unambiguous need for a scientific justification for an SPS label and the absolute prohibition on singling out PPMs in the TBT agreement.

Having said that, it is also the case that countries can agree to any trade policy arrangement as long as a successful dispute is not brought to the WTO. In other words, another WTO member country would have to make a formal complaint about the arrangement included in the preferential trade agreement. Given that members of any preferential trade agreement would have agreed to the new arrangement, it is unlikely they would bring a complaint. (We address later the possibility that a complaint could be brought by a WTO member that was not part of the preferential trade agreement.)

Assume for the moment that a new arrangement is agreed whereby both the scientific justification was dropped and PPMs could be used to justify labeling. A government would then be able to impose labeling requirements on imports strictly on the basis of the *consumer's right to know*. One immediate effect is that the trading partners in the preferential agreement would have given up the trade opportunities that the scientific justification protects because it prevents consumers from being given false signals. Consumers might then rightly ask: "Why would the product be labeled if there is no problem? The logical conclusion is that there would not be a label unless there is a problem, hence I will reduce my consumption of labeled products as a precautionary response." The resulting reduction in trade might be willingly accepted by those agreeing to the new rules, but would impose costs on producers and consumers in other WTO member countries not party to the preferential agreement. Such negative economic externalities can be dealt with either through *fencing in*[5] or *fencing out*[6] (or shared fencing responsibilities). It is consistent with the property rights approach to pollution in economics (Coase 1960). The economic solution preferred by the WTO would be to choose the least cost (least trade distorting) alternative. Economic solutions, however, do not always arise from political action. Coexistence legislation often works on the principle of *fencing in*, irrespective of whether it is the least cost solution. In other words, the property rights associated with coexistence are legislatively endowed upon the producers of organic and conventional products. This means that the responsibility for dealing with the potential externality lies entirely with producers wishing to use GMOs in production. The costs associated with buffer zones and liability lie with them. They must *fence in* any externalities or assume the liability if they do not.

The central problem with labeling imports under a *consumer's right to know* arrangement agreed in a preferential trade agreement is the externalities it would generate. Once it is accepted that imported agricultural products produced using GMOs must be labeled, international supply chains would have to adjust to the new

---

[5]*Fencing in* means that the responsibility for containing or mitigating the effects of a negative externality lies with the polluter. The term comes from the regulation of ranching where the responsibility for preventing negative externalities arising from stray cattle lay with the owner of the cattle who was expected to keep them fenced in.

[6]*Fencing out* means that the responsibility for containing or mitigating the effects of negative externalities lies with those receiving the pollution. The term comes from ranching where the responsibility for preventing negative externalities arising from stray cattle lay with the owner of the property where the negative externality would be manifest meaning they were to build a fence to keep out stray cattle.

situation. For those supply chains that wish to deal in GMOs, it is relatively straight forward. They must simply label their products as containing GMOs. This can be accomplished at very low cost (Gaisford et al. 2001). This is because no one cares if products declared to contain GMOs are accidentally comingled with non-GMO products.

On the other hand, those international supply chains that wish to engage in trade in products to be claimed as non-GMO in the importer's market face considerable challenges. This is because they must prove that their products have not been accidently comingled with products produced using GMOs. Consumers looking for GMO-free products care if they contain GMOs. This means that all of the actors along the international supply chain must expend resources to ensure that their non-GMO products are segregated from GMOs. For the most part, this is a very costly process given the extensive blending, pooling and comingling in the existing transport and storage infrastructure—often used sequentially to transport or store GM and then non-GM products. There have been a number of studies that have investigated the costs of segregation which find that at low levels of tolerance for comingling costs rise quickly (Maltsberger and Kalaitzandonakes 2000; Smyth et al. 2014). The level of tolerance for comingling has been a contentious issue in the regulation of GMOs (Hobbs et al. 2014). Those who advocate for a change in labeling rules in preferential trade agreements are likely to also want low tolerances for products identified as not containing GMOs. The tolerances would have to be negotiated in the preferential trade agreement. In the EU for example, there is zero tolerance for non-approved GMO events (Viju et al. 2014). As asynchronous approvals of GMOs among the members of the preferential trade agreement are frequent (Phillips 2016), it will be nearly impossible for non-GMO supply chains to assure that their product is pure without some proactive effort. The stricter the tolerance, the higher the relative cost of non-GMO products when compared to GMO products. This dynamic interaction between GMO and GMO free markets has been investigated by Gaisford et al. (2001), who identified a number of situations where the *non-GMO* market becomes unviable. This is not the outcome those seeking a change to the trade rules pertaining to labeling envision, or likely anticipate.

Unlike coexistence with the externality controlled by *fencing in*, the mandatory labeling of products in international trade is one where the externality must be handled through *fencing out*. Keeping a non-GM supply chain free from comingling will be the responsibility of the members of the supply chain and they will incur all of the costs of segregation. A *fencing in* system is likely infeasible due to the practical difficulties of identifying the sources of GM products, assigning liability and the substantial legal expenses (Bessel et al. 2006). It would also require similar legal systems among the members of the preferential trade agreement, efficiently functioning courts and an absence of corruption in the legal systems in the countries where the supply chains operate. Further, due to the need to maintain reputation, a non-GMO supply chain would probably have to undertake *fencing out* activities. It would not really matter if one won a liability judgement against a source of comingling if ones customers no longer trust your product to be GM free.

A *fencing in* system would likely be simply too expensive for there to be a viable trade in non-GM products.

A further consequence of the relative cost of GM products versus non-GM products is the type of labeling that firms in the supply chain may seek. Given the cost of *fencing out*, firms might seek a soft option for GM labeling—the vague *may contain* claim. Instead of claiming to be GM free and incurring the costs associated with *fencing out* they may simply turn to the less costly *may contain* route. This has been widely observed in the case of allergens such as peanuts when strict labeling and liability regulatory regimes have been put in place. As a result, the labeling information is of little use to those with allergies as most products are labeled as *may contain* the allergen. *May contain* GMOs is not the information those advocating on the basis of the *consumer's right to know* are interested in.

Hence, agreeing to mandatory labeling of GM products in a preferential trade agreement may not lead to the expected result. Unlike coexistence,[7] it is unlikely to be possible to impose the major costs associated with labeling on those wishing to trade GM products.

## Members of Preferential Trade Agreements and Non-members

Members of preferential trade agreements must still trade with countries that do not belong to the new agreement. Those trade relationships will be governed by the agreed WTO disciplines.[8] Labeling of imports can only apply to products that have been declared a *hazard for none*[9]—i.e. those that have passed the threshold for safety. At that point, GM products and non-GM products from members of the WTO, but not members of the preferential trade agreement, should enter the market of the importer unlabeled and undifferentiated (comingled)—they represent a

[7]Where the costs of controlling any externalities pertaining to production are legislatively imposed on those wishing to grow GM crops.

[8]This will be the case unless the importing country is in a different preferential trade agreement with the potential trading partner. In this case there will be three sets of labeling rules being applied in the importer's market—those negotiated in the new agreement with the subset of trading partners, those negotiated in the previous trade agreement with a different subset of trading partners and the WTO's for the remaining trading partners. This will further complicate the importer's marketplace. Of course, over the last decade and a half many preferential trade agreements have been negotiated—well in excess of one hundred—with a number of countries being members of multiple agreements (Perdikis 2007). This is sometimes referred to as the *spaghetti bowl* of regional trade agreements (Kerr 2011). Of course, not all of the preferential trade agreements to which a country belongs may have specific rules for GMO labeling. It is clear, however, that multiple labeling rules will make the importer's marketplace increasingly complex for consumers. This complication is ignored as the discussion of two labeling rules in this section indicates the effects of such complexity.

[9]Leaving aside the case of *hazard for some*.

*hazard for none*. GM products from members of the preferential trade agreement will be in the importers marketplace and labeled as GM. Non-GM products from members of the preferential trade agreements could enter unlabeled but, to garner a commercial advantage from having a non-GM product, are likely to be labeled as non-GM.

Consumers are thus faced with a choice of three products—those labeled GM, those unlabeled and those labeled GM free—and their *right to know* is partially obscured. Consumers who do not wish to consume GM products will be able to choose not to purchase those products labeled as being GM—reducing their demand, and traded quantities. If they are aware, consumers not wishing to purchase GM products can also chose not to purchase unlabeled product—again decreasing demand and lowering traded quantities. The relative strength of these two effects will depend on a number of factors but crucially on how aware consumers are that unlabeled products may contain GMOs. It is an empirical question. It also depends on the relative prices of the three products. If the prices of non-GM products are relatively high due to segregation costs then intra-marginal low income consumers may chose the cheaper unlabeled product (Gaisford et al. 2001).

Having unlabeled products which may be GM in the market lowers the desired impact of the *consumer's right to know* labeling policy. If consumers had perfect knowledge, they would know that unlabeled products were comingled and could avoid them under a voluntary labeling system whereby non-GM products would be identified due to commercial incentives. The reason for requesting a mandatory labeling scheme for GM products, however, is that all consumers do not have perfect knowledge and, hence, may inadvertently purchase GM products when they did not intend to. Thus, the reason for mandatory labeling of GM products is to prevent such *mistakes*. Having unlabeled, and potentially comingled, products from those trading under WTO rules in the market thus largely negates the objective of having the mandatory labeling policy in the preferential trade agreement.

The importing country might attempt to impose labeling rules similar to those in the preferential trade agreement on all trading partners. This would effectively shut out the imports of *hazard for none* products from countries that do not belong to the preferential trade agreement, which would violate the importing country's SPS commitments and could lead to a challenge through the disputes system of the WTO. The importing country would likely lose the case at the WTO and have to remove the labeling restriction on non-members of the regional trade agreement (Read 2007)—and return its market to the situation described above. Of course, a WTO member can always choose not to comply with a ruling of a WTO disputes panel—as the EU did in the case of its import ban on beef produced using growth hormones (Kerr and Hobbs 2002). If a country chooses to ignore a WTO dispute panel ruling, the country (or countries) whose trade opportunities were harmed, in this case by the labeling regime, can be authorized to retaliate against the country imposing the trade barrier (Read 2007).

Thus, it would appear that preferential trade agreement provisions that would mandate labels based on the *consumer's right to know* are unlikely to be effective in achieving their policy intent given a globalized trading marketplace. It may be that arriving at such a new set of rules is better accomplished at the multilateral level. The frustration for those wishing to promote mandatory labeling of imported GMOs is that both the SPS and the TBT remain *closed*—they are not on the current negotiating agenda of WTO members (Gaisford and Kerr 2001). Even if the SPS and/or the TBT could be *opened* for re-negotiation, arriving at an agreement allowing compulsory labeling is likely to be extremely difficult—as it has proved in the current negotiations surrounding major preferential trading agreements such as the T-TIP, the TPP and the CETA as is discussed below.

## What Can One Expect for Agricultural Biotechnology from the T-TIP, TPP and CETA?

There are a number of major initiatives underway to negotiate preferential trade agreements across the globe. Interest in devising new rules of trade in the products of biotechnology features in some but not all new negotiations; the EU-South Korea agreement, the US-South Korea Agreement and the EU-India negotiations all ignored this topic (Khorana et al. 2010, 2015). Three major preferential trade agreement initiatives include discussions of biotechnology in their agendas: T-TIP between the EU and the US; the TPP among 12 Pacific Rim countries; and the CETA between the EU and Canada. In each case, trade in the products of agricultural biotechnology for the participating countries is a contentious trade issue and there is an interest in the rules of trade for these products. It is, however, difficult to discern the course of the negotiations and, in particular, whether there has been any discussion of labeling rules that differ from those of the SPS and TBT. This is because in all three cases the negotiations have been/are being conducted in strict secrecy.

There would appear to be the potential to negotiate a wide ranging deal on trade issues pertaining to biotechnology, including coexistence and labeling, given the overlapping membership in the current set of regional trade initiatives being pursued. If it could be accomplished, it would cover a substantial portion of global international trade. Figure 12.1 illustrates the current overlap in interest in preferential trade agreements. It also outlines the regulatory regimes pertaining to labeling in the various jurisdictions that are negotiating. The TPP includes both the US and Canada. The US is currently negotiating the T-TIP with the EU. The EU has also recently concluded an agreement with Canada—the Comprehensive Economic and Trade Agreement (CETA). The US and Canada are also attempting to revive regulatory harmonization after nearly 20 years of inaction since the signing of the North American Free Trade Agreement in 1994. The Regulatory Cooperation Council (RCC) was formed early in 2011 under the Perimeter Security and Economic Competitiveness agreement between Canada and the United States (Government of Canada 2011). The objective was to, where possible, simplify and align regulatory approaches. Two specific areas identified in the Joint Action Plan

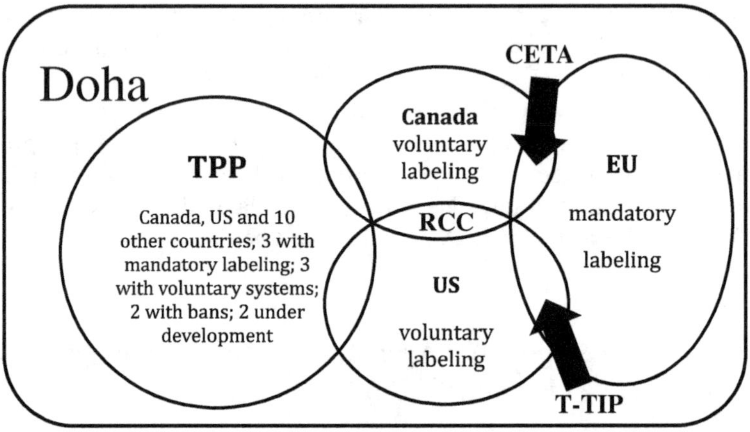


**Fig. 12.1** Overlapping opportunities

were food safety systems and biotechnology. Most of the RCC's work thus far has been related to food safety but the door for cooperation on biotechnology is open.

While there appears to be an opportunity for wider cooperation across a number of venues on issues such as GMO labeling, such an initiative will depend on the willingness of the various parties to see the issue as a priority and their willingness to compromise.

The CETA negotiations were concluded in September 2014 and the agreement was recently ratified by both the EU and Canada (albeit after some last minute high-stakes brokering). Canada certainly had an interest in obtaining better access to the EU market for its GMOs while the EU had an interest in obtaining relief from the SPS and TBT rules for trade in GMOs—and in particular from the threat of disputes being brought forward to the WTO. Labeling policy is one of the areas where the EU is vulnerable because it has a mandatory labeling policy. Most of the Canadian interest in EU policy has focused on the approval process for GMOs and the trade disruptions problems created by low level presence (LLP) comingling (Viju et al. 2014, 2012; Viju and Kerr 2011). Faced with the trade problem caused by these other aspects of the EU's GMO policy, mounting a challenge at the WTO related to EU labeling policy does not appear to have been a priority. However, the EU policy remains vulnerable to a challenge at the WTO.

The text of the agreement is available for scrutiny but the exact details of the negotiations remain confidential. There were indications that the negotiations over biotechnology were difficult (Kerr and Hobbs 2015). This is not surprising given that Canada is a major developer, adopter and exporter of GM products and the EU chafes under its SPS obligations. In the end, the differences proved too difficult to bridge in the negotiations and a classic diplomatic compromise was reached. It was agreed that a formal forum for the discussion of issues pertaining to biotechnology would be established—but with no closure mechanism. This means it is a forum for ongoing discussions but not the resolution when alternative positions on issues arise. A long list of topics which could be discussed in the forum was also included in the formal text of the CETA in Article X.03. These are largely topics that would be of interest to Canada to have discussed. While there is no specific mention of labeling, some of the topics included in the list are sufficiently broad to encompass labeling issues.

The T-TIP negotiations put the two major global advocates of the different approaches to regulating biotechnology across the same negotiating table. The US has been the staunchest defender of the science-based approach to regulation of GM products embedded in the SPS. Its officials continue to stress this commitment in their public statements. EU officials, on the other hand, strongly state that there will be no compromise of food safety in any agreement. There appears, as yet, no room for compromise. Harmonization of regulatory regimes has been floated but what that would mean has not been developed (Kerr 2015). Unless the EU is willing to agree to harmonize to the US regulatory system, or the US is willing to harmonize to the EU system—both outcomes which are extremely remote—any harmonized system would require long, complex and highly technical negotiations which cannot be accomplished by diplomats negotiating umbrella trade agreements. Harmonizing standards requires technical expertise. Thus, what might be expected from the T-TIP is something similar to the CETA compromise—the establishment of a forum to discuss harmonization. Labeling could be a part of those discussions.

The TPP differed from the T-TIP and the CETA in that the EU, with its strong reservations regarding GMOs, was not part of the negotiations. The TPP negotiations took place in strict secrecy. The US and Japan are the major economies in the TPP. The other countries are Australia, Brunei, Canada, Chile, Malaysia, Mexico, New Zealand, Peru, Singapore and Viet Nam. There is, however, wide divergence in GMO policy across the 12 countries (Kerr 2015). Some countries have cultivation bans and some have import bans. In addition to the US, Canada and Australia, three mega adopters, a number of the developing countries in the TPP negotiations are in the process of devising regulatory regimes for GMOs and seriously considering cultivating GM crops. Given this divergence of interests, there is little common ground upon which a direct compromise could have been reached in the negotiations.

In terms of labeling, Australia, Japan and New Zealand each require labeling of GMOs. Canada, Chile, Mexico, Singapore and the US have no mandatory, state-directed labeling requirements, although each has developed measures to facilitate voluntary labeling. Brunei and Peru have import bans on GMOs and thus have no need for a labeling policy. Labeling policies are currently being developed by Malaysia and Viet Nam (Kerr 2015). Given that both the US and Canada are heavily vested in the science-based SPS system, any move away from it on labeling policy would represent a major shift. The issue of GMOs has been a major political issue in Australia, Japan and New Zealand so it is hard to imagine them being able to abandon their labeling policies in favor of a system acceptable to the US and Canada. In short, the issue was too contentious to be dealt with directly in the TPP. In the end, as with the CETA, a forum for ongoing discussions pertaining to biotechnology is to be established. It is sufficiently broadly defined that labeling could be part of those discussions.

Thus, it would appear that there is little likelihood of mutually agreed labeling policies arising from the major preferential trade agreements currently in the process of development. The differences in approach to regulation are simply too strongly held for obvious direct compromises to be identified. If a venue for further discussion were established as a result of a successful conclusion of the T-TIP, it might be useful to have some joint meetings among the forums for discussion of biotechnology issues established in the CETA, the TPP and the T-TIP.

## Conclusions

Differences in regulatory policy pertaining to GMOs have made labeling a contentious issue in international trade. This is because labeling regulations can constitute barriers to trade. Trade agreements can make rules that limit the unfettered use of any potential barriers to trade so that the gains from trade can be garnered by participating countries. Rules of trade allow countries to manage the flow of goods across their borders. While traditionally trade agreements dealt largely with border measures such as tariffs intended to provide economic benefits to those requesting them, latterly the rules of trade have been harnessed to assist in the achievement of domestic social policy. The focus of the WTO, however, is squarely on the trade effects of regulatory policies. The remit of the institution in the area of non-tariff barriers such as labels is to limit their trade effects. In the case of the SPS, the intent is to prevent the imposition of labeling regimes justified on nefarious sanitary or phytosanitary grounds. The TBT compromise on labeling is to prevent developing countries' products having their trade opportunities inhibited due to the PPMs used.

In contrast, the focus of labeling based on *consumer's right to know* is a social policy allowing individual choice regarding the products they purchase. There is no concern with any trade inhibiting effects that such a labeling policy might have. In the international trade context, however, a labeling system based on *consumer's right to know* will likely be trade inhibiting relative to the SPS and TBT rules. Further, the international supply chain effects of a *consumer's right to know* labeling regime may act to raise the relative price of GM free products relative to GM products. This may be contrary to the social objectives of some who advocate *consumer's right to know* as evidenced by coexistence policy which acts to raise the price of GM products relative to organic or conventional products.

In the face of insurmountable obstacles in the WTO, many countries have looked to negotiate labeling rules based on *consumer's rights to know* under a preferential trade agreement. The relative trade benefits would be reduced compared to the existing WTO rules but that may be acceptable to those entering into preferential trade agreements—in the name of achieving social policy objectives. Problems would undoubtedly arise with having multiple sets of labeling rules for products in the importer's marketplace as goods from countries outside the preferential trade agreement would trade under the WTO labeling rules.

Among the three potential preferential trade agreements examined in this paper —the T-TIP, the TPP and the CETA—there appears to be no consensus among those negotiating that the social policy objective embedded in a labeling system based on *consumer's right to know* would outweigh the perceived trade benefits of the current WTO system. Neither the US nor Canada, for example, are at all willing to eschew the WTO system for labeling for one based on the *consumer's right to know*.

Hence, at this point, trade in the products of agricultural biotechnology will continue to be inhibited by some countries' inability to live up to their WTO commitments. The divergence in labeling policies contributes to this reduction in the benefits of trade. The current spate of preferential trade agreements do not appear able to bridge the gap among differing approaches to domestic regulation of GMOs, including labeling policies.

## References

Bessel, K.M., J.E. Hobbs, and W.A. Kerr. 2006. Food safety and private international law: Liability, traceability and transboundary marketing. *Journal of International Food and Agribusiness Marketing* 18 (1&2): 29–48.

Clark, L.F., C.D. Ryan, and W.A. Kerr. 2014. Direct democracy, state governments, and the re-energized GMO debate: Implications of California's Proposition 37. *AgBioForum* 16 (3): 177–186.

Coase, R.H. 1960. *The problem of social cost*. London: Palgrave Macmillan.

Gaisford, J.D., and W.A. Kerr. 2001. *Economic analysis for international trade negotiations*. Cheltenham: Edward Elgar.

Gaisford, J.D., J.E. Hobbs, W.A. Kerr, N. Perdikis, and M.D. Plunkett. 2001. *The economics of biotechnology*. Cheltenham: Edward Elgar.

Government of Canada. 2011. *Summary report on consultations with Canadians on regulatory cooperation between Canada and the United States*. Ottawa: Government of Canada. http://www.borderactionplan-plandactionfrontalier.gc.ca/psec-scep/assets/pdfs/RCC-CCR_Eng.pdf.

Hobbs, J.E. 1996. A transaction cost approach to supply chain management. *Supply Chain Management* 1 (2): 15–27.

Hobbs, J.E. 2007. Technical barriers to trade. In *Handbook on international trade policy*, ed. W.A. Kerr, and J.D. Gaisford, 394–403. Cheltenham: Edward Elgar.

Hobbs, J.E., W.A. Kerr, and S.J. Smyth. 2014. How low can you go? The consequences of zero tolerance. *AgBioForum* 16 (3): 207–221.

Isaac, G. E., M. Phillipson and W. A. Kerr. (2002). International Regulation of Trade in the Products of Biotechnology. Estey Centre Research Papers, No 2. Saskatoon: Estey Centre for Law and Economics in International Trade. http://law.usask.ca/documents/estey-journal/Isaac-Phillipson-Kerr%20-%20Biotechnology%20Regulation%20-%20Estey%20Study%202%20-%20Full%20Report.pdf.

Kerr, W.A. 2006. International harmonization and the gains from trade. *Journal of International Law and Trade Policy* 7 (2): 116–125.

Kerr, W.A. 2007. Introduction to trade policy. In *Handbook on international trade policy*, ed. W. A. Kerr, and J.D. Gaisford, 1–8. Cheltenham: Edward Elgar.

Kerr, W.A. 2010. *Conflict, chaos and confusion—the crisis in the international trading system.* Cheltenham: Edward Elgar.

Kerr, W.A. 2011. The preference for new preferential trade agreements: Does it lead to a good use of scarce resources? *Journal of International Law and Trade Policy* 12 (1): 1–11.

Kerr, W.A. 2015. Governance of international trade in genetically modified organisms: Is future global food security at risk? *Journal of International Law and Trade Policy* 16 (2): 60–77.

Kerr, W.A., and J.E. Hobbs. 2002. The North American-European Union dispute over beef produced using growth hormones: A major test for the new international trade regime. *The World Economy* 25 (2): 283–296.

Kerr, W.A., and J.E. Hobbs. 2015. A protectionist bargain?: Agriculture in the European Union-Canada trade agreement. *Journal of World Trade* 49 (3): 437–456.

Khorana, S., N. Perdikis, M.T. Yeung, and W.A. Kerr. 2010. *Bilateral trade agreements in the era of globalization.* Cheltenham: Edward Elgar.

Khorana, S., N. Perdikis, and W.A. Kerr. 2015. Global economies of scale in the EU-India trade agreement: Are they the key to a return to economic growth? *Asia Europe Journal* 13 (1): 41–55.

Maltsberger, R., and N. Kalaitzandonakes. 2000. Direct and hidden costs in identity preserved supply chains. *AgBioForum* 3 (4): 236–242.

Perdikis, N. 2007. Trade agreements: Depth of integration. In *Handbook on international trade policy*, ed. W.A. Kerr, and J.D. Gaisford, 106–119. Cheltenham: Edward Elgar.

Phillips, P.W.B. 2016. Market solutions to coexistence and regulatory asynchrony. In *The co-existence of genetically modified, organic and conventional foods: Government policies and market practices*, ed. N. Kalaitzandonakes, P.W.B. Phillips, J. Wesseler, and S.J. Smyth, 333–344. New York: Springer.

Read, R. 2007. Dispute settlement, compensation and retaliation under the WTO. In *Handbook on international trade policy*, ed. W.A. Kerr, and J.D. Gaisford, 497–508. Cheltenham: Edward Elgar.

Smyth, S.J., W.A. Kerr, and P.W.B. Phillips. 2011. Recent trends in the scientific basis of sanitary and phytosanitary trade rules and their potential impact on investment. *Journal of World Investment and Trade* 12 (1): 5–26.

Smyth, S.J., P.W.B. Phillips, and D. Castle (eds.). 2014. *Handbook on agriculture, biotechnology and development.* Cheltenham: Edward Elgar.

Smyth, S.J., P.W.B. Phillips, and W.A. Kerr. 2015. Food security and the evaluation of risk. *Global Food Security* 4: 16–23.

Sunstein, C. R. (2002–2003). The paralysing principle. *Regulation*, Winter: 32–37.

Van den Belt, H. 2003. Debating the precautionary principle: "Guilty Until Proven Innocent" or "Innocent Until Proven Guilty"? *Plant Physiology* 132 (July): 1122–1126.

Viju, C., and W.A. Kerr. 2011. Agriculture in the Canada-EU economic and trade agreement. *International Journal* 76 (3): 677–694.

Viju, C., M.T. Yeung, and W.A. Kerr. 2012. The trade implications of the post-moratorium European Union approval system for genetically modified organisms. *Journal of World Trade* 46 (5): 1207–1238.

Viju, C., M.T. Yeung, and W.A. Kerr. 2014. Zero tolerance for GM flax and the rules of trade. *World Economy* 37 (1): 137–150.

# Chapter 13
# Liability and Redress Regimes

**Abstract** This chapter examines the feasibility of establishing a liability fund that could be utilized to compensate exporting nations should exports be detected containing GM products and discusses the funding options. African commodity export data is analyzed to establish the range of values of trade that might be annually affected. Both the economic and political feasibility of this solution are examined to determine whether it would enable African producers and consumers to participate in the benefits of GM crops without the worry that their commercial interests will be harmed by an EU trade action.

## Introduction

Global trade in agricultural products has a notorious history of being subject to trade barriers and other distortionary policies. Nowhere has international trade been more distorted than in the trade of genetically modified (GM) products. In 2006, Argentina, Canada and the United States (US) were successful in a trade dispute with the European Union (EU) regarding the importation of GM products, with the World Trade Organization (WTO) finding that the EU had abrogated its obligations under the Sanitary and Phytosanitary (SPS) Agreement. In spite of this WTO ruling, little has changed, with the horizon looking equally bleak. The EU processes for approving GM crops are in disarray and the future production of any GM crops that might receive approval was dealt a severe blow in 2011 when the European Court of Justice determined that pollen is a food ingredient. Since the pollen from GM plants has not been approved for food use, and it is impossible to keep GM pollen out of honey, it is very difficult to fathom how GM crops can ever reach commercial production levels in the EU given the constraints of the existing policy on coexistence.

Given the apparently unbridgeable divide that has developed between the 25-odd countries that grow and trade GM crops and evolving EU regulatory hurdles, it may be time to consider alternatives to realizing a global market for agricultural products. Africa is one area of the world where the battle over GM agriculture is being



S.J. Smyth et al., *Biotechnology Regulation and Trade*, Natural Resource Management and Policy 51, DOI 10.1007/978-3-319-53295-0_13

played out, yet it is the continent where genetic modification could have the greatest positive impact. Numerous African nations, given their long-standing trade connections to European nations, are fearful of allowing the commercialization of GM crops due to the potential for comingling of GM and conventional products and, hence, the loss of export opportunities to the EU. These are legitimate concerns.

The present polarization of uptake and use of agricultural biotechnology between the EU and North America presents two ends of a spectrum; that is, European consumers and governments appear to remain predominantly opposed to the innovation, whereas Canadian and American consumers generally support the technology. However, such has not always been the case.

Much of the early genomic research regarding plant biotechnology was undertaken by firms in the EU. Belgium, France and Germany were all leading biotechnology countries in the late 1980s and early 1990s. The optimism of this period was, however, short-lived. The moratorium on products of agricultural biotechnology that prevented the production and/or import of food or feed products between 1999 and 2000, was a blow to the EU biotechnology industry, but not a death blow. Optimism still existed through the mid-2000s, but as regulatory decisions took longer and longer to be received for feed use and minimal production, this optimism waned. EU consumers became increasingly indifferent to the products, environmental opponents were relentless in their attacks and governments were no longer willing to support this innovation. The defining moment occurred early in 2012, when BASF announced that it was moving all of its research and development (R&D) capacity from Europe to the United States (BASF 2012). This represented a loss of over 200 research positions in Europe and signalled to other agricultural firms engaged in agricultural R&D that Europe is no longer a place to do business.

Finding ways of increasing food production appears no longer to be a global concern. Most of the developed world population eats too much—hunger is a distant memory. Resistance to biotechnology among the well-fed in the EU contrasts with significant sustaining concern in Africa. Civil society groups and governments in the EU express clear preferences for non-GM food products. Yet there is growing evidence that GM crop varieties will be needed to meet the food needs in Africa. In 2012 the Gates Foundation's Annual Letter, for example, asserted food security could only or at least best be realized through the use of all new plant breeding technologies (Gates Foundation 2012). The challenge would seem to be to commercialize GM crops in Africa without jeopardizing the value of African food exports to the EU. Many African nations have reservations about the commercial production of GM crops because they fear that their financially lucrative and economically vital crop export markets in the EU will be harmed if GM crops are produced (Paarlberg 2008). Finding a means of increasing the adoption of GM crops in Africa, while at the same time providing assurances that any potential export market externalities can be corrected through compensation, appears to be one pathway to overcoming the current stalemate.

## Compensation Funds as Ex Ante Commitments

With the initial GM crops being approved for production in 1996, 2016 represents the twentieth year of GM crop production. International political discussions, dialogues, consultations and negotiations have taken place in a variety of international institutions over the past two decades, all with the same result; adoption of agricultural biotechnology by developing African countries continues to lag that of many other developing nations and regions. Clearly, international political discourse is dominated by organizations or governments that are more opposed to increased adoption of GM crops than are concerned about food security (Aerni and Ermen 2011). The time may have come for an assessment of other options.

Paarlberg (2008) asserts many developing African nations are hesitant to adopt GM crop varieties due to the potential for the GM varieties to comingle with conventional and non-GM varieties, thereby jeopardizing export markets for these commodities in the EU. The European Union has a 0.9% threshold of coexistence for GM trait events that have been approved in the EU, a 0.1% threshold for GM trait events that have been approved in other jurisdictions but not in the EU and a zero percent threshold for GM traits that have not been approved in any jurisdiction. With the low level presence examples of rice and flax reported in previous chapters, the probability of human error when exporting agricultural commodities exists. The crux of this issue is to assure the governments and regulators of developing African nations that when a human error occurs and is detected, a mechanism exists that will secure the value of the African commodity export. Without assurance, it will be difficult to overcome the pressures to not commercialize GM crop varieties.

Not only is this a political reality, but also a market reality. Hobbs et al. (2013) identify that the cost to Bayer of LL601 rice being comingled in exports from the US to the EU was nearly US$800 million, while rice exports remain depressed by 70% compared to pre-detection export levels in 2005. Ryan and Smyth (2012) estimated that the international cost from the detection of Canadian GM flax in the EU market exceeded C$80 million. By some estimates, the zero tolerance policy on unapproved GM traits in the EU has cost international commodity traders in excess of US$1 billion in recent years. Given these costly market examples, it is only natural that African exporters would be cautious about adopting GM crops and facing potential disruption of their food markets. Currently, this would only apply to exporters of those crops that have GM varieties under production; corn and cotton are the only GM crops approved for production in any African country. Genetically modified bananas could also be included as GM bananas are expected to be approved for production in the near future (Kikulwe 2010).

Any agricultural commodity export from Africa to the EU that might be rejected due to the low level presence of GM comingling would have a secondary market, albeit at a lower price and possibly higher transaction or transportation costs. Hence, the entire value of an African commodity export would not be lost, but rather only a portion of the full export value. One option to address this would be to establish a compensation fund to ensure that any African nation that experiences a financial loss due to the unintentional comingling between GM and non-GM commodities could apply for remuneration of the lost exports. This would allow for developing African nations to approve and produce GM crop varieties for domestic use without inciting undue concern that GM crops will have detrimental impacts on the country's export markets.

Compensation funds have a long history of use as a means of providing a resource pool that can be accessed in times of need. There are two common forms of compensation fund—long-term, self-sustaining funds and short-term, specific-incident funds. Long-term self-sustaining funds are typically financed by the premiums from the purchase of insurance contracts (i.e. automobile insurance or crop insurance). Short-term specific incident funds are usually created and funded by governments, such as for compensating for unusual natural disasters (i.e. unusual floods or other adverse conditions). Obviously, funds of the latter nature are short lived, lasting for a few months or until all of the claims are settled, whereas insurance compensation funds usually intend to operate perpetually.

An example of a government funded liability mechanism can be found in a Danish example. Denmark has launched a compensation fund for those adversely affected by the production of GM crops (Smyth et al. 2010). However, Denmark has yet to allow the commercial production of any GM crops. The biotechnology industry has stepped forward and offered their own compensation mechanism. The "Compact", launched in 2010 by the leading crop development firms involves an agreement to compensate any state that can prove biodiversity damage due to the release of a living modified organism (LMO) (CropLife International 2010). This private sector fund offers one innovative approach by industry members to demonstrate their confidence in the quality of their products and their commitment to the environment.

Any form of compensation fund that could be established should be funded by stakeholders from adopting, industrialized countries—currently the US, Canada and Australia—largely because the developers, early adopters and key producers in those countries both are the main beneficiaries of the technology and probably have the most to gain from more widespread adoption of the technology. We propose that in this instance the revenue required to establish such a fund should be equally shared among stakeholders and have a clearly defined time limit. Premiums could be based on acreage, with the costs spread equally between technology developers, producers, philanthropic organizations and national governments.

In order to determine the size and structure of the fund, it is necessary to determine the scale of the potential problem. The next section examines the scope of bi-lateral African/EU trade at risk from comingling.

## African Agricultural Commodity Exports

The three GM crop commodities of corn, cotton and bananas are either commercialized or very close to being commercialized in Africa. Figure 13.1 presents the data for total world exports of African produced corn, cotton and bananas. The value of African corn exports is volatile, reaching highs of US$644 million in 1994 and US$622 million in 2008. Weather had a dramatic impact on production at the turn of the millennium with exports averaging only US$100 million during 1999–2001. Banana exports have climbed steadily over the past two decades, rising from US$60 million in 1990 to US$200 million from 2006 onwards. Cottonseed exports have been very consistent during this time, ranging from a low of US$13 million in 1992 to a high of US$34 million in 2008. In spite of the lower value of cottonseed exports, exports have doubled from approximately US$15 million in the early to mid-1990s to over US$30 million following 2005.

The data in Fig. 13.1 represents the international export of these three commodities to all markets. It is necessary to remove the non-EU export destinations as well as inter-Africa trade to fully appreciate the level of fiscal compensation that might be required in any particular year. Figure 13.2 presents the export situation for corn and cottonseed. The value of banana exports are reported in Table 13.1. Corn exports (not including sweet corn) to the EU peaked in 1996 at US$19 million and trended steadily downward for the rest of the decade, with a recent recovery to about US$8 million. Cottonseed exports broke through the US$2 million barrier for the first time in 2010, with previous levels ranging from US$625,000 to US$1.8 million.

The trade in fresh bananas from Africa to the EU is dominated by exports from Cameroon and the Ivory Coast. In most years, exports from these two nations account for well over 90% of the total banana exports from African countries to the EU. Ghana's banana exports, while scant in the early part of the last decade, have increased substantially. Somalia and South Africa also have minor amounts of banana exports. Somalia has exported bananas in three of the past ten years, with an

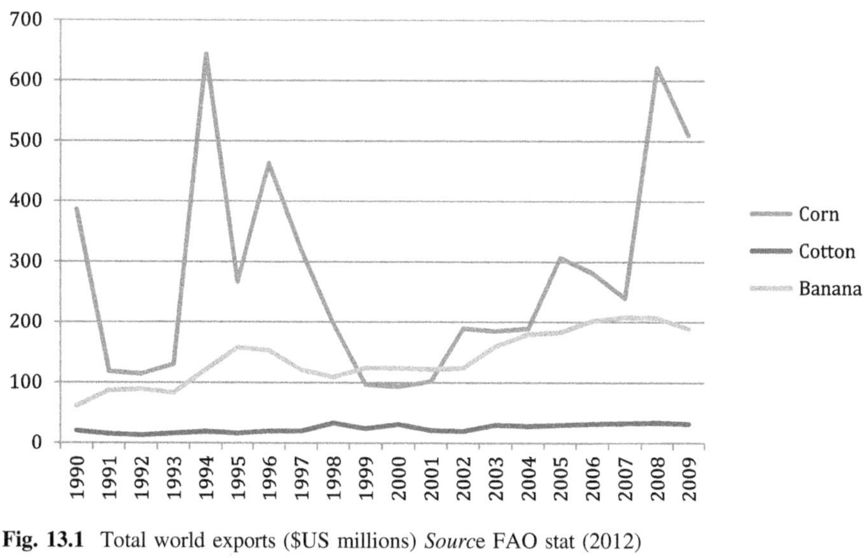


**Fig. 13.1** Total world exports ($US millions) *Source* FAO stat (2012)

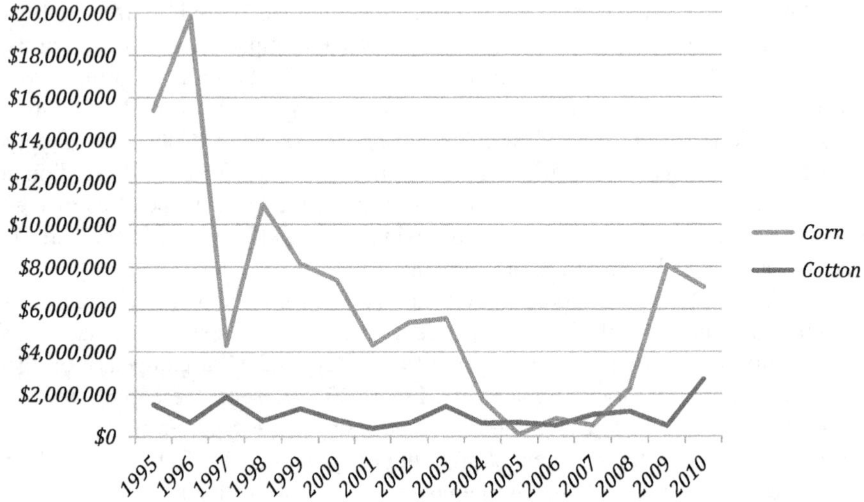


**Fig. 13.2** African corn and cotton exports to Europe ($US millions) *Source* UNCTAD stat (2012)

**Table 13.1** African banana exports to Europe ($US millions)

| Country | 2001 | 2002 | 2003 | 2004 | 2005 | 2006 | 2007 | 2008 | 2009 |
|---|---|---|---|---|---|---|---|---|---|
| Cameroon | 42 | 45 | 67 | 66 | 65 | 63 | 58 | 81 | 70 |
| Ghana | 4 | 4 | 0.6 | 2 | 3 | 15 | 42 | 13 | 10 |
| Ivory coast | 62 | 63 | 76 | 85 | 74 | 89 | 82 | 98 | 100 |
| Uganda | 0.06 | 0.02 | 0.04 | 0.08 | 0.16 | 0.3 | 0.6 | 0.9 | 0.3 |
| Total | 108 | 111 | 144 | 153 | 142 | 168 | 183 | 192 | 179 |

*Source* UNCOMTRADE (2012)

annual average of US$7700. South Africa exports bananas annually with the value ranging from a high of US$230,000 in 2002 to a low of US$5300 in 2009.

Total trade between Africa and the EU in those African commodities that have commercialized GM varieties (corn and cotton) equaled less than US$10 million in 2010 and there is little evidence of any major trend upwards. Trade in bananas is not affected at this point in time as GM banana varieties have not yet received regulatory approval. Indeed, the value of African exports of crops with commercialized GM varieties was highest at the time of the original release of GM varieties in Canada and the US (i.e. just above US$20 million). From 2005 to 2007, the value of annual sales at risk was less than US$2 million, with the 2005 value dropping to only US$750,000. While this value of trade is vital to the African economies, it is disturbing to think that the benefits of GM crop varieties are being denied in Africa over trade that often adds up to less than US$10 million annually.

The question then is what might be the scale of the funds needed to ensure that every single instance of the EU's import rejection of an African commodity would receive full and complete compensation?

One possibility is that ten percent of the full value of the original sale would be required to cover the price differences for alternative markets and transaction costs. In the current case of just GM corn and cotton, ten percent of the export value difference would amount to relatively small numbers. At no time since the first commercialization of GM crops has the export value of corn and cottonseed from Africa to the EU exceeded US$30 million and in 2010 it was just shy of $10 million. The worst case scenario would be the case where all $10 million of the 2010 trade was rejected by the EU, thus triggering compensation claims for nearly $1 million. With an investment in coexistence education and training of African producers, supply chain stakeholders and exporters (which could be supported by the fund), the expected level of unacceptable comingling detected in the EU should be drastically lower. Once GM bananas are commercialized, the potential costs of compensation could rise sharply. For example, assume that the secondary market value resulted in a loss of ten percent for total African banana exports in any given year; based on the data in Table 13.1 this would be nearly US$20 million. Again, this is the worst case scenario where 100% of African banana exports are rejected by the EU and redirected to secondary markets. The market reality for unacceptable imports is expected to be considerably below 100%.

So the resources needed would range from a *de minimus* token annual amount to as much as US$21 million per annum, in a worst case scenario, with the most likely draw in the US$10–12 million range.

## Dimensions and Structure of a Credible Liability Fund

Structuring a credible compensation fund is problematic. While some might be tempted to suggest that biotechnology companies alone, or perhaps the biotechnology companies in partnership with adopting farmers, should pay the levy, we believe there is a valid argument that a mix of developers, adopting producers, philanthropic foundations and governments of advanced agri-food exporting countries should all participate.

Clearly developers would derive benefits from greater and more widespread adoption of GM varieties, so their participation is obvious. Producers currently producing using GM varieties similarly derive value from more widespread adoption, as that would both broaden the resource base for new research efforts and would help to facilitate trade between developed, GM producing nations and fast-growing developing markets. Further matching could be generated from the various philanthropic organizations that currently seek to facilitate funding into food and health issues in developing countries and governments. It is envisioned that funding of this nature would fit with the increased food security objectives that are held by these organizations. Similarly, the federal governments of Australia,

Canada and the US currently have numerous programs to support both domestic agricultural research and agriculture and development in developing nations which might fit with such a venture.

So if those are the actors, the challenge is to develop a credible, ex ante commitment that would influence African nations. While a pay-as-you-go system, either formalized by contract, managed through tort law or based on best-efforts could work in theory, each of those systems would involve enough uncertainty or transactions costs to undercut their credibility. The likely best option might be some kind of funded compensation scheme. While the costs of such a program might be amortized over the expected life of the scheme, it would undoubtedly be more credible if there was some form of up-front payment that demonstrated the confidence of the supply chain.

We believe the optimal strategy would be a liability fund, secured in a three-year period to generate enough to fund any foreseeable risk. Given that the annual costs would most likely be in the US $10–12 million range, a fund would need approximately US$600 million capital to generate US$12 million annually at the prevailing two percent rate of return. While a worst case scenario could trigger payments in excess of US$20 million annually, it would be prudent to start smaller. While US$600 sounds large, it could be reached with only modest contributions annually from each of the proposed partners.

If one assumed each of the four partners would contribute equal shares, triggered by adoption and use, one could design the program using production data. James (2015) estimates that there were about 180 million hectares, 444 million acres) of GM crops produced in 28 different countries in 2015. James identifies that 19 of these countries are developing countries. Of the 10 industrial countries that produce GM crops, only Australia, Canada and the US can be considered as having sufficient production to merit participation in a compensation fund. Within these three countries, only four staple GM crops (canola, corn, cotton and soybeans) are considered as the platform for the funding scheme.

Early and aggressive adoption and production of GM crops has made the US the global leader. In 2015, the US cultivated 175 million acres of GM crops. Given that over 90% of the US canola is produced in North Dakota, where producers use the same varieties grown in Canada, it is expected that adoption is generally consistent with that in Canada. Soybean production is predominantly undertaken utilizing GM varieties while corn and cotton production have GM adoption rates that range from two-thirds to three-quarters.

Canada was one of two countries that first adopted GM crops in 1995 (parallel to developments in the US). Adoption of GM varieties of canola and corn advanced rapidly, with full adoption for corn and near full adoption for canola in recent years. Approximately two-thirds of soybean production in Canada is undertaken with GM varieties. In total, just over 27 million acres of GM crops are produced annually in Canada. Due to climate, GM canola is predominantly produced in Western Canada with GM corn and soybean production in Central Canada, although new GM corn and soybean varieties are beginning to enter Western Canada.

Australia has a long history of GM crop adoption staring with cotton. The first GM cotton varieties were approved in 1996 and by 2010 virtually all Australian cotton production used GM varieties. Following this, Australia revised the regulation of GM crops and in 2000 implemented the Gene Technology Act, creating the Office of the Gene Technology Regulator (OGTR), which centralized the regulation of GM crops. When the OGTR began in 2001, consumer attitudes were considerably different from 1996 and within a few years every state within Australia had enacted a moratorium against the commercial production of new GM food crops (the production of GM cotton was not affected by these moratoriums). The state moratoriums started in 2003, with the first removals in 2007. The state of South Australia still has a moratorium in place. Genetically modified canola began to be produced in Australia in 2008 following the first removal of state moratoriums. While adoption has been delayed, uptake and use is expected to increase over the next few years. In 2015, Australia had about 1.7 million acres of GM crop production.

If the fund was targeted at US$600 million, a $1/acre levy on about 200 million eligible acres annually in those three countries, shared equally between the four partners, would be enough to fully subscribe the fund within three years. At the end of the three-year period, the levy could be removed and the fund self-sustaining.

To be credible, funds would need to be invested in secure investments, generating a return that would be expected to cover the annual compensation costs. However, if the returns were insufficient, the capital could be drawn down to cover the compensation.

Basing the funding on a per acre basis allows the cost to be spread equitably between the partners. Farmers are accustomed to purchasing new seed on an annual basis and their share of the levy (25 cents per acre) could be included in the price of the seed (which costs more than $10/acre per crop). Participation by all of the technology development firms would raise the price of GM seed equally for all varieties, thereby not disadvantaging any one firm's varieties. The certified seed industry already remits revenues based on seed sales to private firms with which they have contracts to produce seed. No significant additional administrative costs would be created at this point as the certified seed growers could include the 25 cent levy with the existing levies they would annually owe to the technology development firms.

If the technology development firms received the funds from the certified seed industry, they would then know the total revenue generated by the 25 cent levy and could simply match that amount. Structured this way, every seeded acre of GM crops would generate 50 cents of revenue.

Further matching of this could be generated from both the philanthropic organizations and from governments. It is envisioned that funding of this nature would fit with the increased food security objectives that are held by these organizations. Philanthropic organizations could be invited to match the 25 cent levy that the technology developers provide to the compensation fund. Similarly, the federal governments of Australia, Canada and the US currently have numerous programs to support agriculture in developing countries. While fiscal cutbacks are currently common, significant resources are still flowing—Canada provides C$4.5–5 billion in foreign aid annually (CBC 2012). Figures are similar for Australia, although it has reported that Australia was expected to expand its commitments (ibid.). In the US, the Agency for International Development (USAID) actively manages and invests in development work, reporting budgetary resources in 2011 in excess of US$20 billion. Garnering fiscal support from the three federal governments will require a coordinated and organized approach to seek matching contributions.

When it comes to the administration of the fund, it is suggested that the most logical body to oversee such a fund would be the World Trade Organization (WTO). Given that the WTO evolved out of the General Agreement on Tariffs and Trade (GATT) and is the international body responsible for resolving disputes regarding the Agreements on Technical Barriers to Trade and Sanitary and Phytosanitary Measures, it is well suited to administer compensation regarding the trade of GM agricultural products between Africa and Europe.

Application for compensation and receipt of funds could be designed to be an efficient and speedy process. Through the use of trade contracts, presenting evidence of lost trade value could be quite a simple process. If an African nation has a trade contract with an EU-based buyer, the value of this contract would be evidence of the value of the trade in that commodity. Proof of comingling would be required —the African nation would be able to use evidence created by the EU as it tests products at point of entry. Testing by the rapid alert system for food and feed (RASFF) in the EU would identify levels of comingling that exceeded regulated levels, resulting in the rejection of any shipment as a non-GM shipment. It would be expected that an alternate market would be sought for the shipment and the contract for this sale would serve as the final piece of documentation, showing the difference in value between the final contract and the original contract. Additional testing could be sought to ensure that the test was not a false positive. Based on this, the African exporter could receive reimbursement for the difference in the value of the two contracts. The mechanism would have to be carefully structured to prevent problems associated with moral hazard and rent seeking.

In this manner, a fund would provide compensation for not only the GM crops presently on the market, but for new GM crop varieties, such as bananas, when they receive regulatory approval. We assume that setting up the fund could take about two years, with the result that if started in 2016–7, it could be fully operable in 2019–20. Using the annual return on investment for the first few years and then having to draw down on the principal would provide a life expectancy of such a compensation fund of over 25 years and possibly as long as 30 years. This would mean that developing African countries could commercialize GM crops without fear of lost export markets as far out as 2045 and possibly 2050. It would be hoped that with the growing evidence of the economic and environmental benefits of GM crops, that EU consumer attitudes would change enough over the coming generations to render the need for compensation obsolete prior to the exhaustion of the compensation funds.

A further suggestion to improve African food security would be that in any year where claims for compensation did not exceed the fund returns (conservatively estimated at US$12 million) the surplus funds could either be reinvested back into the fund or be invested in plant variety research and development programs for Africa, or some combination thereof. These surplus funds could additionally be used to fund regulatory training for government officials in the various African nations or provide coexistence training to supply-chain stakeholders of commodities that will soon experience GM varieties.

## Conclusions

The solution to improving food security globally is to ensure that all plant breeding techniques that currently exist are used to their full potential. This includes the technique of genetic modification, as well as new breeding techniques that are on the horizon. Consumers in the EU are currently holding up adoption, especially in Africa, which is a key supplier of EU consumers. We accept that EU consumers are entitled to choose the food products that they wish. But satisfying these choices should not impose excessive externalities on African nations struggling for food security. Every nation on the continent of Africa has the right to decide for themselves what crop production techniques can, and should, be utilized to feed their populations. To be able to make this decision fairly, they need to be secure in the knowledge that if they grant regulatory approval to GM crops within their country, that financial harm will not be inflicted on their commodity export industries.

To ensure that this option exists for developing African nations, we have proposed the establishment of a compensation fund to be equitably funded by developed world adopters of GM crop technologies. Of course, considerable work would be required to get the commitment of the multitude of partners and stakeholders needed to make such an endeavour successful. However, these obstacles are not insurmountable should the desire to make such a fund a reality exist.

Suggesting changing global trade structures to address concerns about GM agriculture is not something that we have entered into lightly. We do this mostly to open a dialogue about solutions to the stalemate. We encourage readers to engage in the debate, to highlight errors in our thinking, to suggest improvements in our approach, to suggest their own ideas and ultimately to get engaged regarding forward solutions to the regulatory and trade stalemate.

## References

Aerni, P. and D. Ermen. 2011. A global stakeholder survey on the potential of biotechnology to address the challenges of global climate change.In *Presentation at the 15th Annual Conference of the International Consortium on Applied Bioeconomy Research (ICABR)*, Ravello, Italy, June 26–29.

BASF. 2012. News release: BASF to concentrate plant biotechnology activities on main market in North and South America. http://www.basf.com/group/pressrelease/P-12-109.

Canadian Broadcasting Corporation. 2012. Does cutting foreign aid threaten Canada's reputation in the world? http://www.cbc.ca/m/touch/news/story/2012/04/01/f-international-aid-federal-budget.html#chart2.

CropLife International. 2010. The Compact. http://www.croplife.org/Files/Upload/Docs/Compact-%20Final%20-%20Overview%20-%20PUBLIC%20-%2025-05-10.pdf.

Food and Agriculture Organization. 2012. FAOSTAT database. http://faostat.fao.org/site/535/default.aspx#ancor.

Gates Foundation. 2012. Annual letter from Bill Gates. Available http://www.gatesfoundation.org/annual-letter/2012/Pages/home-en.aspx.

Hobbs, J.E., W.A. Kerr, and S.J. Smyth. 2013. How low can you go? *Consequences Zero Tolerance AgBioForum* 16 (3): 207–221.

James, C. 2015. 20th anniversary (1996 to 2015) of the global commercialization of biotech crops and biotech crop highlights in 2015. *ISAAA Brief* No. 51. ISAAA: Ithaca, NY.

Kikulwe, E.M. 2010. *On the introduction of genetically modified bananas in Uganda: Social benefits, costs, and consumer preferences*. Wageningen, NL: Wageningen University Press.

Paarlberg, R. 2008. *Starved for science: How biotechnology is being kept out of Africa*. Harvard, MA: Harvard University Press.

Ryan, C.D., and S.J. Smyth. 2012. Economic implications of low level Presence in a zero tolerance European import market: The case of Canadian Triffid flax. *AgBioForum* 15 (1): 21–30.

Smyth, S.J., A.B. Endres, T. Redick, and D. Kershen. 2010. *Innovation and liability in biotechnology: Transnational and comparative perspectives*. Cheltenham, UK: Edward Elgar Publishing Ltd.

United Nations Commodity Trade Statistics Database (UNCOMTRADE). 2012. http://comtrade.un.org/db/ce/ceSearch.aspx?it=banana&rg=1&r=97&p=384&y=2011&px=HS.

United Nations Conference on Trade and Development (UNCTDSTAT). 2012. http://unctadstat.unctad.org/TableViewer/tableView.aspx?ReportId=110.

# Chapter 14
# Alternate Futures for Food Security

**Abstract** The future for food security is more uncertain now than in any recent decade. After more than 70 years of increasing liberalization and globalization of markets and significant success in reducing food insecurity, there seems to be a pause that threatens to both destabilize the global world order and knock back the effort to mobilize science to create a more food secure world. Governments, industry and citizens are pulling back from using competitive markets to promote their interests, raising fears that we are moving into what some have coin a 'neo-medieval' world of special interests.

## Introduction

The 'bicycle theory' of trade policy, popularized in the 1970s by C. Fred Bergsten, the founder and director of the Peterson Institute for International Economics, a prominent pro-trade think tank in Washington, asserts that if negotiating momentum fails, the system could wobble and fall. Former United States Trade Representative Robert Zoellick in 2001 avowed that "if the trade liberalization process does not move forward, it will, like a bicycle, be pulled down by the political gravity of special interests" (Zoellick 2001) While this idea started in the policy community, it has been taken up and affirmed by academics, including Jagdish Bhagwati, who discusses it in his book *Protectionism*.[1]

In many ways the bicycle seems to be wobbling, with stalemate in both the marketplace and national and international regulatory systems. While a variety of strategies have been proposed, they face headwinds as governments, industry, citizens and consumers around the world are adjusting their expectations and approaches.

This chapter reviews where we are, assesses the prospects for resolution of the conflicting interests and outlines two alternate futures—one of muddling through to a more sustainable regulatory and commercial space and one of retrenchment and narrowing of the scope for science in agriculture.

[1] http://www.salon.com/2007/01/11/bicycle_theory/.

 
S.J. Smyth et al., *Biotechnology Regulation and Trade*, Natural Resource Management and Policy 51, DOI 10.1007/978-3-319-53295-0_14

## The State of Play in the World of GM Crops

This book explicitly focused on the challenge of facilitating expansion in food trade in order to generate a more food secure world. As cited by Fader et al. (2013), the task is large and growing. Currently more than 15% of the global population relies on international trade for their basic nutritional requirements and the forecast is that by 2030 about half of the world will depend on international trade to supply their daily nutrition. The task was to highlight the regulatory and trade challenges that could limit the achievement of global food security and offer a range of strategic responses to overcome these challenges. Scientific advances in the breeding of new crops is proceeding at a rapid pace, however the benefits of these technologies are only partly reaching food insecure markets due to existing regulatory and trade barriers and related commercial responses.

The global food system is only part way along a major transformation. Since the Second World War, the food industry, aided and partnered with governments around the world, has expanded supply enough to reduce the incidence of malnutrition and food insecurity significantly, in spite of a rapid rise in the global population. But despite good progress in expanding supply and meeting the increased demand for foods, the combination of weakening agri-food productivity, widespread soil depletion, increasing water shortages and climate change all threaten the achievements to date. As if this is not enough, demand is inexorably rising, both due to increasing global population and higher incomes that translate into increased demand for animal protein, which accelerates demand for feed and competition for food crops. In June 2011 US Agriculture Secretary Tom Vilsack concluded the Paris meeting of the G20 Agriculture ministers stating:

> The consensus reached today … marks an historic union of resolve in combating the pressing challenges of hunger and food prices confronting our world with greater regularity. Together, the G-20 nations are committed to increasing agricultural production through use of improved practices and technologies and a commitment to new and expanded research and development.

Technology and trade are inextricably linked in that one cannot advance and contribute to food security without the other. Technological advancement has clearly done its part, initially through development of modern improved varieties and hybrids that triggered the Green Revolution in the 1960s and more recently through the application of advanced breeding technologies and biotechnology. 2015 marked 20 years of commercial production using genetically modified (GM) crops and, while numerous non-governmental organizations continue to trumpet that ‘there are no benefits from biotechnology or GM crops’, the evidence suggests GM crops contribute significantly to productivity. We demonstrated through a review of the literature that the technology is generating benefits wherever it has been adapted and used (Chaps. 1 and 2). A wide range of positive economic, environmental and health benefits have been realized from GM crop adoptions, distributed among both adopting and non-adopting farmers, consumers and countries around the world. By some estimates this wave of technological innovations has generated more than $100 billion so far and continues to deliver large benefits annually.

The main challenge is that pushback from vested interests is rewriting the rules for commercial diffusion and use. Until the advent of GM crops, most countries were content to let the domestic regulators in the country of origin do the heavy lifting of assessing safety and efficacy of new cultivars. This system was fast, efficient and for the most part delivered safe and efficacious varieties for global use. Concerns over the effects on human and animal health from consumption of GM food crops, as well as about an array of potential impacts on the wider environment, have divided public opinion over the merits of using the technology. One consequence is that governments have, for the most part, been very cautious about accepting GM crops for cultivation and consumption. Instead of relying on the global system to ensure safety and value, most nations have renationalized the regulatory process, so that developers now need to secure approvals for each intended use in each national market. A concomitant effect of the expansion and duplication of regulatory oversight is that the costs involved (and time spent) in achieving regulatory approval for the commercial release of a GM crop have risen. This work has reviewed the evidence on the absolute and relative costs of gaining regulatory compliance within countries and assessed their aggregate effects. A range of studies show that there are significant costs of gaining regulatory compliance—product attributes, market context and technical aspects of measurement all have significant impact on the estimated costs of regulatory approval.

In a perfect world, one might imagine that all of the regulatory systems would work like clockwork, delivering consistent decisions simultaneously. Unfortunately, that is far from the case. Asynchronous regulatory approval and commercialization of new GM traits in a range of extensively-traded food crops is causing significant trade disruptions (Phillipson and Smyth 2016). This mismatch of expectations has generated quite a few instances of adventitious presence of unapproved or incompletely approved traits in canola, corn, rice and flax markets, causing disruption. Market losses from diversion to second-best markets, increasing marketing costs for securing quality-assured supplies, added regulatory compliance time and costs as well as costly and often destructive litigation have been linked to such events. Fears of similar events in wheat, soybeans and other staple crops have generated significant concern.

When countries and markets have had difficulties in the past, they have had the opportunity to elevate their concerns to the global level to resolve disputes and find long-term solutions. On paper, the current international trade architecture looks promising, but in practice it is far from effective.

Governments spent most of the post-Second World War period building an open and liberal trading regime—led by the US, the undisputed hegemonic power in the global market economy. Beyond the pressure from the US, there was a consensus among policy makers, academics, the media and most interest groups that it was the right thing to do. The creation of the World Trade Organization (WTO) in 1995 was the apex of achievement, realizing the full vision of the Bretton Woods authors with an authoritative, almost universal rules-based system with a capacity to deliver binding dispute resolution. Most importantly for our story, agriculture and food policy and trade was fully accommodated within the rules—it was no longer a *special case*. But almost before the ink was dry on the new agreement, governments began to backslide, especially in the context of trade in food. Governments renationalized review of both old and new foods and brokered no interference from abroad. When aggrieved countries launched disputes at the WTO, many of the decisions were rejected by litigants as illegitimate, raising concerns that consumers (and others) are disenfranchised. Trade in some key products has been in limbo for years. The result has been a loss of credibility for and at times demonization of the WTO.

In response to the challenge of governing transformative technologies such as GMOs, various international organizations have attempted to organize processes to generate fixes for the problem. The WTO, IPPC, *Codex Alimentarius Commission*, WIPO, OECD and Cartagena Protocol on Biosafety, among others, have attempted to develop the architecture to more effectively regulate the production, trade and marketing of biologically-derived crops, bioproducts and foods. The results of these global governance efforts have been disappointing. For every one apparent step forward, there have been two steps backwards. In short, the international trade rules based on science are not functioning efficiently. Considerable effort was expended in the latter half of the 20th century to remove politics from the frameworks that governed international trade. Some degree of success was achieved as numerous institutions were founded or their roles expanded. These institutions were established on the premise that science-based frameworks were essential to the efficient functioning of international commerce. The first decade and a half of the 21st century would seem to suggest that these institutions are foundering and that acceptance of science as foundation for international trade rules is on the decline. This disruption of international trade has had a profound effects on investment, as both public and commercial research have both concentrated on a narrow range of crops and traits that can meet their economic targets in such a disrupted market, to the exclusion of other efforts that might have a more direct effect on nutrition in food insecure parts of the world.

The advent of biotechnology products has triggered a spirited debate about how we should assess risks, who should undertake the task of defining what is acceptable, what rules we should draw upon and where we should vest the authority to decide. In the recent past there was a strong move to normalize and institutionalize a 'science-informed' system in national regulatory systems, international science and international treaties. Recently there has been a substantive pushback against the privileged role scientists and science institutions play in decision-making, especially in the introduction and use of GM crops. The underlying information, valuation and rules for deciding what get used or discarded are the focus of two discrete groups. The modernist, positivist community is advocating *science-informed* decision-making and rules while the post-modernist critics want to incorporate socio-economic considerations (SECs) to deepen the contextual base for decisions. While neither approach is beyond regulatory capture, the socio-economic methods that would be involved in exploring SECs are less well developed and codified than the scientific fields. This leaves them more open to capture and manipulation by vested interests, which would just compound the already difficult challenge of securing adequate nutrition for the burgeoning world population.

One result of this conflicted space is that we have developed a mix of venues with varying rules and overlapping and interlocking memberships. All are well intentioned but they are so different that they cannot by themselves reconcile their inconsistencies. In other times in other areas where there have been conflicting and contradictory rules and procedures, we have relied on lawyers to square the circle and create order from chaos. The 1969 Vienna Convention on the Law of Treaties was designed to provide direction and guidance regarding the complicated legality of primacy between international agreements. The difficulty is that three of the top five producers of GM crops (Argentina, Canada and the US) have not ratified the Cartagena Protocol on Biosafety (CPB) —the main venue for the key food importers and the locus of concern about the socio-economic impacts of new food technologies—and so relying on the Vienna Convention to sort out this mess may lead to more confusion than clarity.

In some ways, we face the irresistible force paradox, where the *unstoppable force* of GM technologies has met the *immovable object* of built regulatory and trade rules. GM foods are so integrated into the global food system that there is probably no going back. But by the same token, the international regulatory and trade system, anchored on the WTO, is so embedded in our global economic system that it is incapable of responding in any way that will convincingly resolve the conflict. So it is necessary to look to other venues to break the gridlock. Five venues or approaches have been assessed.

First, the concept of the risk analysis framework, which underpins the entire international regulatory and trade structure, is still quite new and may be open to further evolution or fine-tuning. The risk analysis framework (RAF) emerged in other areas for other purposes, but now drives design and decision making around the world. Our review of the changing methods of risk assessment for agricultural products suggests a deterioration in the likelihood that risk assessment exercises can or will be completed successfully. The changing nature of risk assessments is found to be inhibiting international market access, reducing trade and, hence, making investments in productivity enhancing technologies in agriculture less interesting.

Second, the European Union (EU) has invested heavily in an effort to create conditions to sustain coexistence between GM and GM-free crop and food production. At one level, the issue of coexistence is a problem of market-making. But both economic theory and commercial practice suggest that there are challenges with leaving market-making to private initiative. Meanwhile, given that the detection of an unapproved GM event results in the closing of borders, the issue inevitably moves to the WTO, which so far has shown little or no capacity to generate resolutions that satisfy all parties in disputes about new agri-food technologies.

Third, for the better part of a decade there has been an effort to use the Convention on Biological Diversity to open a whole new venue for resolving disputes about new food technologies. The Cartagena Protocol on Biosafety has moved the posts slightly, in that both firms and countries find their clearing house for regulatory and scientific information valuable, but so far it has not made much progress in reconciling the concerns about the technology itself.

Fourth, we are inveterate optimists and believe we should not give up on the possibility that we might be able to negotiate something better through one or more of the international trade institutions. A first best option is to design and maintain a global trade regime that meets the needs of both consumers and producers. There are options but the prospects are not good. We also reviewed the prospects for narrower, preferential trade negotiations that might offer either sui generis solutions to the regulatory bottlenecks, or at least might generate conditions for partial-equilibrium solutions involving more engaged labeling systems. Neither option seems likely.

Finally, we decided to reach outside the current architecture and offer the idea of a purpose built liability and redress regime, which might then relieve some of the pressures on national regulators, the WTO and other trade institutions to do everything possible to avoid a type I error. The idea is that a liability fund could be utilized to compensate exporting nations should their shipments be found to contain GM products. We examined African commodity export data to establish the range of trade that might be annually affected. Based on our analysis, it might be both economically and politically feasibility to create a system that would enable African producers and consumers to participate in the benefits of GM crops without the worry that their commercial interests will be harmed by EU trade action. This would be one of many possible ways of unblocking the system. Undoubtedly others exist.

Putting aside our hopes and fears, our simple, dispassionate conclusion is that nothing is likely to give way or change by simply repeating what has been tried. New ideas and new approaches are going to be needed, and as we discuss below, time is of the essence. The bicycle is beginning to wobble in ways that suggests we are heading for a fall.

## A New International Economic (Dis)order?

We are living in the shadow of the Bretton Woods system. Following the mayhem of global war, the victors accepted the advice of their experts that peace and prosperity were more likely to be realized through economic and political integration than through isolation and competition. The architects—especially John Maynard Keynes, who was involved in both the Versailles peace conference after the Great War and the Bretton Woods gathering near the end of the Second War—were convinced that the Second World War was a direct result of the punitive reparations and the general isolationism of imperial powers that emerged from the Great War. The US, as the undisputed hegemon at the end of the Second War, accepted that view and used its power and authority to refashion the global power system, rather than return to its historical isolationist role in the world.

The post-war system involved a mix of efforts to link former combatants both politically and economically. In the immediate post-war period the United Nations was founded, with all of the then-independent states, and a suite of international economic organizations were proposed, including a new global reserve bank (which became the International Monetary Fund, or IMF), the International Bank for Reconstruction and Development (which morphed into the World Bank Group) and an international trade organization (what was proposed in 1948 as the International Trade Organization). While the US backed and supported most of this package, it balked at the ITO and instead the General Agreement on Tariffs and Trade (GATT) emerged as a best-efforts arrangement to facilitate trade.

The US also supported and encouraged others to find ways to bind themselves closer together. The US was instrumental in nudging the former combatants on the European continent to form the Economic Coal and Steel Community in 1951, on the assumption that it would be decidedly more difficult for either France or Germany to rearm if they jointly managed the strategic steel and coal industries. They supported continued unification through the European Atomic Energy Commission in 1957 and the European Economic Community in 1958. While the US at times expressed concerns about the policies of the EEC as it grew from the six original members to its current 28 members, it generally supported the integration.

Since then the world has added an alphabet soup of bilateral and multilateral economic and political arrangements to help govern our international relations. For the most part, the 'bicycle theory' has never been far from the discussion. The memory of the devastation of two world wars propelled more and greater integration.

The demise of the colonial system, beginning in the late 1950s and still going on, multiplied tenfold the number of nations engaged in international affairs. New institutions with larger and more diverse membership emerged, including the FAO, WHO, Codex and a range of political, economic and trade entities. These new nations brought new issues and concerns to the table. We can trace back much of the angst about the modern power system, with a globalized economy and multilateralized political system, to the 1970s and the calls from members of UNCTAD and the non-aligned group of countries for a New International Economic Order, where the power and benefits of that order would be more equitably distributed. Their demand was that states collectively agree to rebalance the system to ensure different outcomes. While that early effort failed to make much headway, the aspirations for change remained.

In the past few years there is increasing evidence that the post-war consensus supporting a globalized, harmonized, multilateralized economic and political system may be unravelling. While the calls in the 1970s were for a New International Economic Order, our view is that we are facing pressures that could lead to a New International Economic (Dis)order. The evidence is scattered, but one can see pressures for disintegration and isolation emerging among governments, in industry and among citizens around the world.

In the first instance, while the international trade agenda is packed with efforts to advance the integration agenda, little success is expected. The Doha Round of the WTO is now in its 15th year of negotiating what would truly be an expansive agenda and more than 50 bilateral and multilateral trade agreements are in varying stages of negotiation and ratification. But appearances are deceiving. Each of those efforts, and many of the past successes, are in risk of collapse. By all accounts the Doha Round is a failure. Bureaucrats and politicians continue to meet, but little or no progress has been seen in the past five years, and the prospects are poor that anything will change in the coming years. Meanwhile the UK electorate in 2016 narrowly voted to withdraw from the European Union (what is euphemistically called the Brexit option). If and how the exit will be managed remains a major uncertainty in European affairs and will have knock-on effects on the WTO and all the other multilateral arrangements the EU has negotiated on behalf of the UK and its other members. Meanwhile, the European Union seems unable or unwilling to ratify and operationalize a range of major trade deals, including the Canada-EU Trade Agreement, which the region of Wallonia held up until it got a side deal. If or how the EU regroups is uncertain, especially given key national elections in Germany and France in 2017, where there are risks that new leaders may take the opportunity to refashion either the Union itself or their role in the Union. Given the poor performance of pollsters and prognosticators to predict outcomes in national elections and referenda, no one is making book on things remaining unchanged.

The election in late 2016 of Donald Trump, the Republican Party nominee, as President of the United States, promises to destabilize the global trading regime for the foreseeable future. The prospective president campaigned on the promise that he would revoke NAFTA, build a wall to isolate Mexico from mainland US, reinstate the economic embargo on Cuba, declare China a currency manipulator and impose countervailing tariffs, withdraw from the Trans Pacific Partnership (TPP) and renegotiate the Transatlantic Trade and Investment Partnership (T-TIP) between the EU and US. While he and his transition team have suggested some or all of these prospects are opening salvos as part of an effort to renegotiate the international economic system, just the suggestion of so much upheaval is causing politicians and business people around the world to look around for safe havens.

In the industrial setting, the biggest change we already see is in the number of firms willing to invest in advanced technologies in the food space. In the halcyon days of the early development of biotechnology in the food industry, all of the large national and multinational seed, chemical, handling, processed food and retail enterprises made efforts to assess the potential value from the technology. Meanwhile, a diverse range of scientific and entrepreneurial start-ups contested the industry, handsomely financed by the global venture capital industry. As controversies erupted and markets segmented, many of the large firms pulled back and retrenched in the conventional food space, while the remaining actors consolidated, to the point now that there will shortly be only two main commercial agri-food biotechnology companies—DuPont/Pioneer/Dow and Bayer/Monsanto—and a few major state-sponsored companies (most notably the merged ChemChina/Syngenta). Few if any start-ups are testing the market for GM foods. Instead, the entrepreneurial focus has shifted to data management and precision farming.

Normally we would take a deep breath and hope that the *wisdom of crowds* would see us through. It is far from clear that collective intelligence is still functioning. The dominant role the Internet, Twitter and blogs played in the UK referendum on Brexit and the 2016 US presidential election has raised the prospect of a 'post-truth' future, where facts, evidence and thoughtful discourse are rejected or ignored. In such as world, special interests have much greater opportunity to spin messages that convert our world into what political scientist Robert Gilpin (2001) has called a neo-medieval system where we are bound by a complex web of inwardly-looking, narrowly-focused, vested interests, analogous to the guild system of the Middle Ages.

We can and must do better than that with all the resources and capacity at hand. The millennial goals of food security for all is within our grasp if we can organize ourselves effectively.

## References

Fader, M., D. Gerten, M. Krause, W. Lucht, and W. Cramer. 2013. Spatial decoupling of agricultural production and consumption: Quantifying dependences of countries on food imports due to domestic land and water constraints. *Environmental Research Letters* 8(1):392 1–15.

Gilpin, R. 2001. *Global political economy: Understanding the international economic order*. Princeton: Princeton University Press.

Phillipson, M., and S.J. Smyth. 2016. The legal and international trade implications of regulatory lags in GM crop approvals. *The Estey Journal of International Law and Trade Policy* 17 (2): 76–90.

Zoellick, R. (2001). *The WTO and new global trade negotiations: What's at stake*, October 30. Washington, DC: Transcript of Talk to the Council on Foreign Relations. http://www.cfr.org/world/wto-new-global-trade-negotiations-s-stake/, p 4149.

# Index





S.J. Smyth et al., *Biotechnology Regulation and Trade*, Natural Resource Management and Policy 51, DOI 10.1007/978-3-319-53295-0

CPSIA information can be obtained
at www.ICGtesting.com
Printed in the USA
LVHW081936120419
613990LV00011B/293/P

9 783319 532936